T0248781

Protein Kinases: Comprehensive Researches

Edited by **Michelle McGuire**

New York

Published by Callisto Reference,
106 Park Avenue, Suite 200,
New York, NY 10016, USA
www.callistoreference.com

Protein Kinases: Comprehensive Researches
Edited by Michelle McGuire

International Standard Book Number: 978-1-63239-522-1 (Hardback)

Protein Kinases: Comprehensive Researches

Edited by **Michelle McGuire**

New York

Published by Callisto Reference,
106 Park Avenue, Suite 200,
New York, NY 10016, USA
www.callistoreference.com

Protein Kinases: Comprehensive Researches
Edited by Michelle McGuire

International Standard Book Number: 978-1-63239-522-1 (Hardback)

Printed in the United States of America.

Contents

Preface

This book of advanced researches aims to serve as a resource guide in the of protein kinases. The aim of this book is to educate readers on protein kinases. As regul of protein function, protein kinases are concerned with the control of cellular fun through complicated signaling pathways, enabling fine tuning of physiological funct book is an integrated effort, with contributions from experts in modern science from the globe. Existing literature is reviewed in this book, and occasionally, new data on ction of protein kinases in different systems is also provided. The book discusses broad related to the role of c-Src tyrosine kinase in bone metabolism, protein kinases and atic islet function, kinases in spinal plasticity etc.

The information contained in this book is the result of intensive hard done by researchers in this field. All due efforts have been made to make this borve as a complete guiding source for students and researchers. The topics in this b ave been comprehensively explained to help readers understand the growing trends i field.

I would like to thank the entire group of writers who made sincere efforts in book and my family who supported me in my efforts of working on this book. I take the ortunity to thank all those who have been a guiding force throughout my life.

Editor

The Crucial Role of c-Src Tyrosine Kinase in Bone Metabolism

Barbara Peruzzi[1], Nadia Rucci[2] and Anna Teti[2]

[1]Regenerative Medicine Unit, Ospedale Pediatrico Bambino Gesù. Rome
[2]Department of Experimental Medicine, University of L'Aquila
Italy

1. Introduction

c-Src belongs to the SRC Family of non receptor tyrosine kinases (SFKs), which includes at least ten members (Lyn, Fyn, Lck, Hck, Fgr, Blk, Yrk, Gfr, Yes and c-Src) sharing high homology in their domain structure (Brown & Cooper, 1996).

Due to its proto-oncogene nature, c-Src is the SFK most frequently associated with malignancy (Yeatman, 2004). Over 100 years ago, Peyton Rous observed that injection of cell-free extracts from tumours grown in chickens caused the development of the same type of tumour in host animals. This observation prompted the hypothesis that a filterable agent was the cause of the tumour (Rous, 1911a, 1911b). In support of this notion, in 1955 Rubin showed that the Rous'filterable agent was a virus, called Rous Sarcoma Virus (RSV), which was found to play a direct role in inducing cell malignancy (Rubin, 1955). In the '60s and '70s, the tools of modern molecular biology provided the genetic definition of v-Src, a viral oncogene included within the RSV genome. v-Src was observed not to be required for virus replication but to be the causative agent of cancer (Martin, 1970; Duesberg & Vogt, 1970). Shortly thereafter, it was shown that v-Src had a counterpart in eukaryotic cells, named c-Src (Takeda and Hanafusa, 1983). c-Src is involved in many physiological functions of the cells. It carries a regulatory domain lacking in v-Src (Fig. 1), therefore, its activity is under tight molecular control. c-Src was the first of several proto-oncogenes discovered in the vertebrate genome and, in 1989, this discovery earned Bishop and Varmus the Nobel Prize in Physiology or Medicine, for the description of "the cellular origin of retroviral oncogenes".

In 1977, Brugge and Erikson immunoprecipitated a 60-kDa phosphoprotein from RSV-transformed fibroblasts. The protein was called pp60[v-src], but it is now usually referred as v-Src. Both v- and c-Src have tyrosine kinase activity (Collett et al, 1978; Oppermann et al, 1979). c-Src also autophosphorylates itself at tyrosine residues (Hunter & Sefton, 1980), and is the prototype of a large family of kinases that we now know are involved in the regulation of cell growth and differentiation.

v-Src lacks the C-terminal domain that in c-Src has a negative regulatory role on its tyrosine kinase activity. Consequently, v-Src shows constitutive activity and transforming ability (Jove & Hanafusa, 1987). In addition, v-Src contains point mutations throughout its coding region that probably contribute to the high intrinsic activity and transforming potential of the protein.

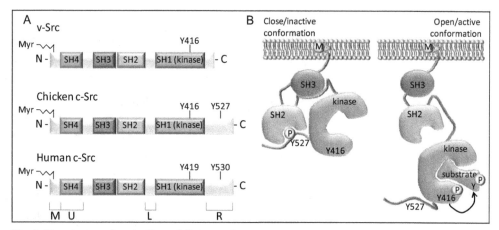

Fig. 1. Structure and activation of Src proteins
A) Comparison of protein structure of viral (v-)Src, chicken and human cellular (c-)Src, with indication of Src homology (SH) and membrane-binding (M), unique (U), linker (L) and regulatory (R) domains. (B) Representation of inactive (left) and active (right) conformation of chicken c-Src.

Aberrant activation of c-Src results in a wide variety of cellular phenotypic changes, including morphological transformation and acquisition of anchorage and growth-factor independence, that are implicated in the development, maintenance, progression, and metastatic spread of several human cancers, such as prostate, lung, breast, and colorectal carcinomas (Irby & Yeatman, 2000). Indeed, a number of human malignancies display increased c-Src expression and activation, confirming its involvement in oncogenesis (Alvarez et al, 2006).

c-Src is ubiquitous and physiologically expressed at high levels in a variety of cell types, including neurons and platelets (Brown & Cooper, 1996). Nonetheless, the very first mouse model of c-Src deficiency (Soriano, 1991) showed an unexpected prominent bone phenotype characterized by increased bone mass and lack of bone resorption, unveiling a previously unrecognized role of c-Src in bone cells (Boyce et al, 1992; Marzia et al, 2000).

In this review, we will describe the structure and function of c-Src and will highlight its crucial physiological and pathogenetic role in bone metabolism.

2. c-Src structure, activation and function

2.1 c-Src structure

c-Src shares with the other SFK members a conserved domain structure consisting of four consecutive Src Homology (SH) domains (Fig. 1A). The N-terminal segment includes the SH4 domain, as well as an "unique" domain of 50-70 residues that display the greatest divergence among the family members (Koegl et al., 1994 ; Resh, 1999). The SH3, SH2 and SH1 (catalytic) domains follow in order in the polypeptide chain. There is also a short, C-

terminal "tail" which includes a hallmark of Src kinases, that is an autoinhibitory phosphorylation site [Tyrosine (Tyr) 527 in chicken, Tyr 530 in human] (Cooper & King, 1986). This tail is not present in the v-Src isoform (Fig.1A).

The SH4 domain is a 15-amino acid sequence whose myristoylation allows binding of SFK members to the inner surface of the plasma membrane. The unique domain has been proposed to be important for mediating interactions with receptors or proteins that are specific for each family member. Serine and threonine phosphorylation sites have also been identified in the unique domains of c-Src and Lck (Winkler et al, 1993).

SH3 and SH2 are protein-binding domains widely present in other molecules, such as lipid kinases, protein and lipid phosphatases, cytoskeletal proteins, adaptor molecules and transcription factors (Mayer & Baltimore, 1993). The SH3 domain consists of small, β-barrel modules and is important for intra- as well as inter-molecular interactions, regulating c-Src catalytic activity, localization and recruitment of substrates. Proline-rich sequences in target molecules mediate the interactions with SH3 (Ren et al, 1993). The other domain regulating c-Src interaction with proteins is SH2, which preferentially binds to polypeptide segments containing a phosphotyrosine (Mayer et al, 1991; Pawson, 1995).

The catalytic domain (SH1) is the most conserved domain in all tyrosine kinases. It contains an ATP-binding pocket and the tyrosine-specific protein kinase activity. As it will be described in the next paragraph, the first step of c-Src activation is the autophosphorylation of Tyr416 (in chicken, Tyr419 in human), while phosphorylation of Tyr527 by c-Src kinase (CSK) and CSK homologous kinase (CHK) results in its inhibition (Kmiecik & Shalloway, 1987; Cartwright et al, 1987; Piwnica-Worms et al, 1987; Okada & Nakagawa, 1989).

2.2 c-Src activation

c-Src is normally maintained in an inactive or "closed" conformation, where the SH2 domain is engaged with the phosphorylated Tyr527, the SH3 domain binds the SH2-kinase linker sequence and the Tyr416 is dephosphorylated. Dephosphorylation of Tyr527 disrupts its intramolecular interaction with the SH2 domain and this open conformational state allows autophosphorylation of Tyr416, resulting in c-Src activation (Fig. 1B) (Yamaguchi & Hendrickson, 1996).

Phosphorylation of Tyr527 can be removed by several protein phosphatases that function as activators of c-Src, such as protein tyrosine phosphatase-α (PTPα) (Zheng et al, 1992), PTP1, SH2-containing phosphatase 1 (SHP1) and SHP2 (Jung & Kim, 2002). The most direct evidence for a role of c-Src activation in cancer among these phosphatases is for PTP1B, which is present at high levels in breast cancer cell lines (Jung & Kim, 2002). In addition, the direct binding of focal-adhesion kinase (FAK) (Schaller et al, 1994) or its molecular partner CRK-associated substrate (CAS, also known as p130CAS) to the SH2 and the SH3 domains of c-Src also results in the open, active configuration of c-Src, since the intramolecular interactions that maintain the closed configuration are displaced (Thomas et al, 1998).

2.3 c-Src functions

c-Src plays a key role in regulating the assembly and disassembly of cell-cell (adherens junctions) and cell-matrix (focal adhesions) adhesion (Yeatman, 2004) (Fig.2). Adherens junctions are maintained by homotypic interactions between E-cadherin molecules present on neighboring cells, and loss of E-cadherin is a key event in the epithelial-to-mesenchymal transition of cancer cells. It has been shown that increased c-Src signalling correlates with decreased E-cadherin expression and decreased cell-cell adhesion (Irby and Yeatman, 2002; Nam et al, 2002). Moreover, constitutively active c-Src can phosphorylate the cadherins, resulting in loss of the cadherin–catenin complex function, thereby promoting cell invasiveness (Irby & Yeatman, 2002; Behrens et al, 1989) (Fig.2).

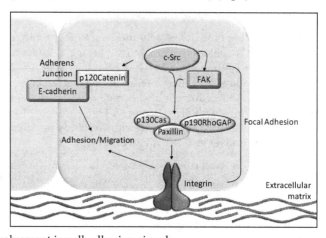

Fig. 2. c-Src involvement in cell adhesion signals.
Molecular interactions among c-Src and the components of both adherens junction and focal adhesion structure.

At the cell periphery, activated c-Src forms complexes with FAK, which in turn interacts with a multitude of substrates, including CAS, paxillin, and p190RhoGAP, that play critical roles in promoting actin remodelling and cell migration (Fig. 2) (Guarino, 2010; Playford & Schaller, 2004). In cancer, deregulated focal adhesion signalling has been implicated in increased invasion and metastasis, and decreased patient survival (McLean et al, 2005). c-Src can also be activated downstream of tyrosine kinase growth factor receptors, such as epidermal growth factor (EGF) (Tice et al, 1999), platelet-derived growth factor (PDGF) (De Mali et al, 1999; Bowman et al, 2001), insulin-like growth factor (IGF)-1 (Arbet-Engels et al, 1999), fibroblast growth factor (FGF) (Landgren et al, 1995), colony-stimulating factor (CSF)-1 (Courtneidge et al, 1993) and hepatocyte growth factor (HGF) receptors (Mao et al, 1997) (Fig. 3). Ligand binding to receptor tyrosine kinases leads to receptor dimerization, kinase activation, and autophosphorylation of tyrosine residues. These phosphorylated tyrosines then serve as docking sites for the SH2 domains of several signalling molecules, including c-Src (van der Geer et al, 1994). For instance, the EGF receptor can bind to c-Src and phosphorylate tyrosine sites on its C-terminal loop. Conversely, c-Src can directly bind to the EGF receptor and phosphorylate the Y845 residue, resulting in increased Ras/ERK/MAPK activity and enhanced cell mitogenesis and transformation (Biscardi et al, 2000).

c-Src has also been implicated in signalling activated by integrins and G-protein coupled receptors (GPCRs). Indeed, clustering of integrins can lead to downstream signalling pathways inducing activation of c-Src, FAK, Abl, and Syk (Miyamoto et al, 1995, Schlaepfer & Hunter, 1998). There is an increasing body of evidence for synergy between receptor tyrosine kinases and integrins, demonstrated by an increase in MAPK activation in response to various growth factors if integrins are preclustered (Miyamoto et al, 1996). The crosstalk between these pathways could be mediated by a common signalling molecule, including c-Src. A FAK-independent signalling pathway from integrins has also been described, in which caveolins act as adaptors, linking integrins and c-Src family kinases. Indeed, Wei et al. (1999) showed that caveolin is important for the association between β1 integrin and c-Src, and disruption of this interaction affected focal adhesions. On the other hand, c-Src can suppress the integrins attached to the extracellular matrix via phosphorylation of integrin subunits (Sakai et al, 2001; Datta et al, 2002). c-Src can also interrupt Rho-A function, which has an important role in actin filament assembly and stabilization of focal adhesions (Arthur & Burridge, 2001). c-Src activates FAK, Ras and phosphatidylinositol phosphate kinase, which indirectly affect integrin–actin cytoskeleton assembly (Brunton et al, 2004).

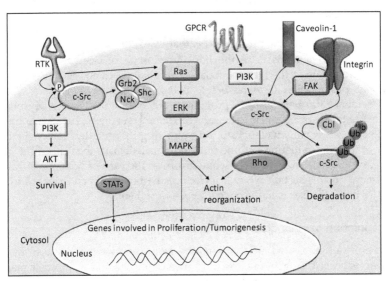

Fig. 3. c-Src-activating signals.
Extracellular signals involving Receptor Tyrosine Kinases (RTK), or G-Protein Coupled Receptors (GPCR) or integrins are shown to activate c-Src and its downstream pathways.

The localization of c-Src at the membrane–cytoskeletal interface in focal adhesions, lamellipodia and filopodia seems to be regulated by the small G-proteins RhoA, Rac1 and CdC42 (Timpson et al, 2001). The Cbl ubiquitin ligase has been shown to be important in suppressing v-Src transformation through ubiquitin-dependent protein degradation (Kim et al, 2004). Recent evidence indicates that the ubiquitin–proteasome pathway is deregulated in cancer cells, which might allow c-Src activation (Kamei et al, 2000).

Finally, there is also evidence that c-Src is activated through nitric-oxide signalling (Akhand et al, 1999) in mechanisms mainly implicated in cellular adhesion and motility (Gianni et al, 2010; Giannoni et al, 2005).

3. c-Src in the bone metabolism

Understanding the physiological role of SFKs has been aided by the advent of gene targeting and embryonic stem cell technology in the mouse. Targeted disruptions of all known mammalian SFK genes have been obtained in mice, with their phenotypes ranging from no overt defects to very distinct abnormalities in specific cell types and tissues. As stated above, the first SFK member to be disrupted was c-Src (Soriano et al, 1991). Although c-Src is ubiquitously expressed, mice lacking the *src* gene presented with overt alterations only in the bone tissue, suggesting a crucial role of this tyrosine kinase in the bone microenvironment that cannot be replaced by other SFK members, as probably occurs in the rest of the body.

The main phenotype associated with c-Src deletion is osteopetrosis, a bone remodelling disease in which excess of bone accumulates as a result of defective osteoclast bone resorption (Soriano et al, 1991). This mutation manifests itself by the failure of incisors to erupt, and the mutants have a much reduced survival rate after weaning. However, animals maintained on a soft food diet have been found to survive for at least a year and, on rare occasions, can breed, although some alterations in reproduction have been documented (Roby et al, 2005). In contrast with a general concept of c-Src involvement in cell proliferation, a detailed analysis of the bone phenotype of c-Src knock-out (KO) mice revealed the crucial role of the tyrosine kinase in regulating osteoclast activity, rather than formation and proliferation (Soriano et al, 1991). As discussed in more detail below, substantial evidence has been already provided in identifying c-Src as a key player in the correct cytoskeletal rearrangement necessary for bone resorption. Further studies pointed out the role of c-Src in bone metabolism, thus showing that the deletion of Src expression also enhances the differentiation and the function of osteoblasts, the cells of the bone tissue having osteogenic function, with a consequent further increase of bone mass (Marzia et al, 2000). Therefore, in this section we will introduce the bone cells and discuss in detail the multiple roles that c-Src exerts in the bone microenvironment.

3.1 c-Src regulation of osteoclast behaviour

Osteoclasts are multinucleated cells, originating from the myeloid tissue from which the mononuclear osteoclast progenitors arise and fuse into polykaria when their maturation is completed (Baron & Horne, 2005). They are terminally differentiated cells that resorb the mineralized matrix during physiological and pathological bone turnover by a peculiar extracellular mechanism involving specific domains of the plasma membrane. Indeed, during the bone resorbing process, the osteoclast is markedly polarized, in order to create three morphologically distinct areas of the plasma membrane: the basolateral membrane, which is not in contact with the bone; the tight sealing zone, which is closely apposed to the bone surface; and the ruffled border, a highly convoluted membrane that faces the resorbing surface (Baron & Horne, 2005).

The characterization of the phenotype of the c-Src KO mouse revealed that the most critical role of c-Src is related to osteoclast function rather than differentiation, since the number of osteoclasts in bones of c-Src KO mice is more than twice that in normal mice (Boyce et al, 1992). While osteoclasts also express other c-Src family kinases (Lowell et al, 1996), the deletion of any one of the genes encoding Fyn, Yes, Hck, and Fgr fails to produce osteopetrosis (Stein et al, 1994). Moreover, re-expression of c-Src in c-Src KO osteoclasts restores *in vitro* the bone-resorbing activity (Miyazaki et al, 2004), implying that c-Src performs some specific functions in osteoclasts that cannot be compensated by these other SFKs. A possible exception is Hck, since its expression is upregulated in c-Src KO osteoclasts and c-Src KO/Hck KO double-mutant mice are significantly more osteopetrotic than the c-Src KO animals (Lowell et al, 1996). At the cellular level, c-Src KO osteoclasts present with a critical feature, that is the absence of the ruffled border (Boyce et al, 1992), suggesting a c-Src contribution to the regulation of exocytic and/or endocytic vesicle trafficking, as well as to the attachment and motility mediated by the adhesion structures.

Osteoclastic bone resorption involves a series of regulatory phases: migration of osteoclasts to the resorption site, their attachment to the calcified tissue and development of the ruffled border and the clear zone, followed by the secretion of acids and lysosomal enzymes into the space beneath the ruffled border (reviewed in Peruzzi & Teti, 2011). The formation of the sealing zone is essential for the osteoclastic bone resorption, since it forms a diffusion barrier and permits the directional secretion of lysosomal enzymes into the space beneath the ruffled border. In the ruffled border membrane, the vacuolar-type proton ATPase mediates the transport of protons into the resorption lacunae. Lysosomal enzymes of osteoclasts, such as cathepsin K, and metalloproteinase-9 are also secreted through this membrane and degrade the organic matrix of bone. To organize these highly polarized cellular structures, osteoclasts must adhere to the bone surface as the initial and essential phase for their activity (Coxon & Taylor, 2008), which involves the interaction of integrins with the extracellular matrix proteins within the bone. Among several integrins, osteoclasts express very high levels of αVβ3 integrin, and lower levels of the collagen/laminin receptor α2β1 and the vitronectin/fibronectin receptor αVβ1 (Nakamura et al, 2007; Horton, 1997; Horton & Rodan, 1996).

Like all members of the αV integrin family, the αVβ3 receptor recognizes the RGD (Arg-Gly-Asp) adhesion motif present in several matrix proteins such as vitronectin, bone sialoprotein II and osteopontin (Rupp & Little, 2001; Wilder, 2002; Horton, 1997). This interaction induces an integrin conformational change leading to the so-called outside-in signalling, which in turn triggers a number of intracellular events, including changes in cytosolic calcium, protein tyrosine phosphorylation and cytoskeletal remodelling (Duong & Rodan, 2000; Teitelbaum, 2007; Faccio et al, 2003). The engagement of the matrix by the αVβ3 integrin in osteoclasts and osteoclast precursors activates the non-receptor tyrosine kinase Pyk2, a member of the FAK family, by a mechanism that involves an increase in cytosolic Ca^{2+} and the binding of Pyk2 to the cytoplasmic domain of the β subunit (Fig.4) (Faccio et al, 2003; Duong & Rodan, 2000).

Both the capacity of c-Src to bind the αVβ3 integrin and the subsequent activation of the kinase are mediated by Pyk2, which mobilizes c-Src to the integrin. αVβ3 integrin occupancy induces phosphorylation of Pyk2, which then binds the SH2 domain of c-Src. The proposed association between phosphorylated Pyk2 and c-Src would prevent c-Src-Y527 inactivating phosphorylation, thus relieving auto-inhibition of kinase function.

The signalling downstream of c-Src activation involves tyrosine phosphorylation of a distinct set of proteins, including Pyk2 itself, Cbl, PI3K, paxillin, cortactin, vinculin, talin, tensin, and p130Cas, which are present in the osteoclast adhesion structures, called podosomes (Thomas & Brugge, 1997; Linder & Aepfelbacher, 2003; Buccione et al, 2004).

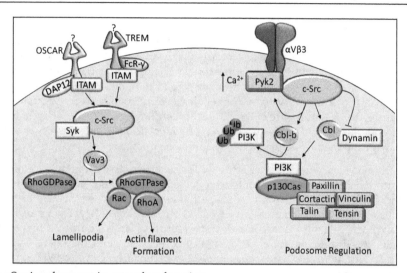

Fig. 4. c-Src involvement in osteoclast function.
Among several pathways regulating osteoclast activity, the cartoon shows c-Src activation and downstream effect depending on receptor signals.

Podosomes, which serve as attachment structures in osteoclasts and other highly motile cells, are more transient and dynamic than focal adhesion plaques (Destaing et al, 2003). As originally described (Marchisio et al, 1984; Marchisio et al, 1987), podosomes are small punctate structures with an F-actin-rich core surrounded by a ring of integrins and certain focal adhesion-associated proteins (e.g., paxillin, talin and vinculin). Cortactin, gelsolin, the actin-regulatory proteins Neuronal Wiskott–Aldrich Syndrome Protein (N-WASP), and Arp2/3 have also been identified in the podosome core (Linder & Aepfelbacher, 2003; Buccione et al, 2004). In osteoclasts, Src, Cbl, Pyk2, and various actin-associated proteins, including dynamin and filamin, are also associated with podosomes, whose rapid turnover (within minutes) is probably essential for the high mobility of the cells in which they occur (Fig. 4) (Destaing et al, 2003).

Among the substrates of c-Src activity involved in the mechanisms of bone resorption, Cbl plays a key role in promoting turnover or disassembly of podosomes (Sanjay et al, 2001). Indeed, c-Src, in association with Pyk2, recruits Cbl through its SH3 domain and promotes its activation by phosphorylation. Once activated, Cbl recruits PI3K and dynamin to the adhesion complex.

Since Cbl is an ubiquitin E3 ligase, it has been described to drive the negative feedback that has the potential to promote proteasomal degradation of the integrin-associated Pyk2/c-Src/Cbl complex (Fig. 4) (Yokouchi et al, 2001). Thus, Cbl is crucial in the integrin-mediated "inside-out" signalling, playing a key role in podosome detachment and subsequent disassembly. In this way, the c-Src/Pyk2/Cbl complex forms the basis for the cyclic attachment–detachment of single adhesion sites at the leading edge of lamellipodia in motile cells, and thereby participates in the assembly–disassembly of individual podosomes, ensuring cell adhesion while still allowing cell motility (Sanjay et al, 2001).

In the integrin-mediated outside-in signalling, also the non-receptor tyrosine kinase Syk plays a pivotal role in osteoclast activity, since it associates with the β3 integrin subunit domain in a region close to the c-Src binding site and is activated by c-Src itself, a key event in organizing the cytoskeleton (Zou W et al., 2007). Recently, it has been shown that c-Src and Syk are also involved in the signal downstream the immunoreceptor tyrosine-based activation motif (ITAM)-bearing co-receptors, DAP12 and FcRγ (Fig.4). DAP12 and FcRγ are associated with the immunoreceptors OSCAR/PIR-A and TREM2/SIRPβ1, respectively, recently identified on the osteoclast surface (Mócsai et al, 2004). The c-Src-mediated phosphorylation of Syk kinase leads to activation of a number of cytoskeleton-regulating proteins, including the Vav family of guanine nucleotide exchange factor (GEFs). These proteins convert Rho GTPases from their inactive GDP to their active GTP conformation. Among these proteins, Vav3 is expressed in osteoclasts (Faccio et al, 2005), where it is triggered upon matrix-induced Syk activation and regulates RhoGTPase-dependent effect on actin cytoskeleton (Zou et al, 2007). In this context, the small GTPases Rac and Rho exert distinctive effects on osteoclasts. Indeed, Rac stimulation in osteoclast precursors prompts the appearance of lamellipodia, thus forming the migratory front of the cell, while RhoA stimulates actin filament formation which, in osteoclasts, allow organization of the sealing zone (Fig.4) (Fukuda et al, 2005).

The integrin-activated c-Src signalling also functions in other processes necessary for normal osteoclast function, among which adenosine triphosphate (ATP)-dependent events, especially those involved in cell motility, proton secretion, and the maintenance of electrochemical homeostasis (Baron, 1993). Indeed, c-Src promotes the maintenance of energy stores in osteoclasts by phosphorylating cytochrome c oxidase within the mitochondria (Miyazaki et al, 2003), which are very abundant in osteoclasts, consistent with the energy requirements of their activity (Miyazaki et al, 2006).

3.2 c-Src regulation of osteoblast differentiation

Osteoblasts are mononucleated cells of mesenchymal origin that synthesize and mineralize the bone matrix during bone accrual and remodelling events. Bone formation involves osteoblast maturation that requires a spectrum of signalling proteins including morphogens, hormones, growth factors, cytokines, matrix proteins, transcription factors, and their co-regulatory proteins. They act coordinately to support the temporal expression (i.e., sequential activation, suppression, and modulation) of other genes that represent the phenotypic and functional properties of osteoblasts during the differentiation process from osteoblast precursors (Jiang et al, 2002). Pre-osteoblasts are also responsible of the production of cytokines regulating osteoclastogenesis, that is receptor activator of nuclear factor kappa B ligand (RANKL), osteoprotegerin (OPG) and CSF-1, thereby coupling osteoblast and osteoclast function. Given the osteoblast-mediated regulation of osteoclast differentiation and bone resorption (Rodan & Martin, 1981; Suda et al, 1997) and the bone phenotype resulted by c-Src distruption (Soriano et al, 1991), several studies aimed at investigating the involvement of osteoblasts in c-Src KO phenotype have been performed. In 1993, Lowe et al. demonstrated that osteoblasts derived from these mice successfully contributed to normal osteoclast differentiation and showed unremarkable morphological features relative to wild-type (WT) mice, suggesting that the inherited defect is independent of the bone marrow microenvironment (Lowe et al, 1993). The first evidence of an osteoblast involvement in c-Src KO mouse bone phenotype derived from Marzia and coworkers, who performed a detailed molecular analysis of the c-Src null osteoblasts (Marzia et al, 2000). This study clearly demonstrated that a decreased c-Src activity is responsible of enhanced osteoblast differentiation and *in vivo* bone formation, thereby highlighting the role of c-Src in maintaining osteoblasts in a poorly differentiated status.

Bone formation requires transcriptional mechanisms for sequential induction and repression of genes that support progressive osteoblast phenotype development. The Runx transcription factors and their co-regulators control cell differentiation and lineage commitment (Westendorf & Hiebert, 1999) by influencing the functional architecture of target gene promoters (Stein et al, 2000). Runx proteins are directed to subnuclear domains through the C-terminal nuclear matrix-targeting signal (NMTS) and interact with the DNA through the N-terminal runt homology domain (Zaidi et al, 2001). The Runx2 family member is essential for osteoblast maturation *in vivo* and its alteration is associated with the cleidocranial dysplasia (Komori et al, 1997). Runx2 is a target of several extracellular signals that regulate skeletal formation and homeostasis. The C-terminus of Runx2, which includes the NMTS, interacts with proteins involved in the transforming growth factor β/bone morphogenetic proteins (TGFβ/BMPs) (i.e., Smads), the transducin-like enhancer (TLE)/groucho and the c-Src/Yes tyrosine kinase (e.g., the Yes-associated protein, YAP) signalling pathways (Hanai et al, 1999; Yagi et al, 1999). Indeed, in response to c-Src/Yes signalling, YAP is phosphorylated and recruited by Runx2 to subnuclear sites of Runx2 target genes, resulting in their repression (Fig.5). Thus, c-Src controls osteoblast differentiation by regulation of Runx2–YAP interaction.

Another mechanism by which c-Src regulates osteoblast differentiation involves estrogens, which are known to control a variety of tissues, including the bone (Hall et al, 2001). Indeed, estrogen deficiency leads to accelerated bone loss which is the primary cause of postmenopausal osteoporosis (Manolagas et al, 2002). The estrogen receptors (ERs) belong to the nuclear receptor superfamily, acting as ligand-inducible transcription factors (Hall et al, 2001).

Indeed, ER expression is regulated by a c-Src/PKC-dependent mechanism involving osteoblast differentiation, with an increased responsiveness to estrogens in mature osteoblasts (Fig.5) (Longo et al., 2004). Estrogens are also responsible for an anti-apoptotic effect on osteoblasts, which is due to a rapid and sequential phosphorylation of the c-Src, Shc and ERK1/2 kinases. The c-Src/Shc/ERK signalling cascade rapidly phosphorylates the transcription factors Elk1, CREB and C/EBPβ with a mechanism that is retained when the receptor is localized outside the nucleus (Kousteni et al, 2003).

Beside the estrogen-mediated effect, other extracellular stimuli, such as mitogens and changes in the mechanical stress, are responsible of c-Src activation and of the downstream cascade involving the MAPK signalling. In this circumstance, the transcriptional target is the AP-1 complex, a heterodimer composed of members of the c-Fos, c-Jun, and activating transcription factor (ATF) families (Hess et al, 2004), which is an important regulator of bone development and homeostasis (Wagner & Eferl, 2005).

Of special interest in the regulation of c-Src activation and activity is the role of caveolae. They are small bulb-shaped invaginations located close to the cell surface representing specialized domains of the plasma membrane (Severs, 1988). Caveolin, a 21–24-kDa integral membrane protein, is a major structural and regulatory component of caveolae membranes (Rothberg et al., 1992). Several data suggest that caveolin may act as a scaffolding protein

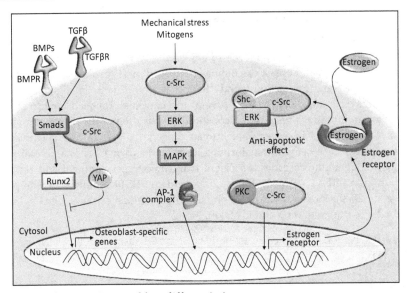

Fig. 5. c-Src involvement in osteoblast differentiation.
TGFβ/BMP signals, mechanical stress, mitogens and estrogen modulate c-Src activity and osteoblast differentiation.

within caveolae membranes, since both the N-terminal and C-terminal domains of caveolin face the cytoplasm (Dupree et al., 1993). Caveolin interacts with cytoplasmic signalling molecules including trimeric G proteins, Src family tyrosine kinases, and Ras-related GTPases. Thus, caveolin may serve as an oligomeric docking site for organizing and concentrating inactive signalling molecules within the caveolae membranes (Sargiacomo et al, 1995). Modification and/or inactivation of caveolin may be a common feature of the transformed phenotype. Caveolin can be phosphorylated by c-Src at Tyr14, an event that induces caveolar internalization by reorganizing the actin cytoskeleton (mediated by dynamin and PKC) (Mayor & Pagano, 2007).

On the other hand, caveolin has a role in regulating c-Src activation, since its interaction with c-Src, as well as with the other components of the SFKs, leads to the inhibition of auto-phosphorylation of these kinases, thus holding these molecules in the inactive conformation (Li et al, 1996).

Recently, we demonstrated that in osteoblasts c-Src regulates interleukin (IL)-6 and insulin-like growth factor binding protein (IGFBP)-5 expression (Peruzzi et al, 2012). More in details, c-Src controls IL-6 expression acting on STAT3, which is a downstream component of the IL-6 pathway and a transcription factor for IL-6 itself. At the same time, IL-6 stimulates the expression of IGFBP5 which, in turn, acts in an autocrine manner on osteoblasts inducing the c-Src activating-phosphorylation and inhibiting further osteoblast differentiation. On the other hand, in mature osteoblasts c-Src is barely expressed and therefore this loop is inactive, although IGFBP5 is still expressed under the control of Runx2. In this context, IGFBP5 has been observed to enhance osteoclast formation and bone resorption, thus unveiling its new role in the coupling between osteoblast and osteoclast activities (Peruzzi et al, 2012).

4. c-Src in cancer-related bone diseases

When deregulated, c-Src has the potential to participate in cancer pathogenesis and progression (Rucci et al, 2008). High levels of c-Src activity are found in numerous epithelial cancers, including colon and breast carcinomas (reviewed in Tsygankov & Shore, 2004), and a correlation between c-Src activity and the degree of tumour malignancy has been described (Aligayer et al, 2002). Among several mechanisms proposed for increased c-Src activity in cancer, mutations resulting in constitutive activation of the kinase domain of the protein have been found in a subset of human cancers (Irby et al, 1999). However, it is thought that rather than being a transforming agent on its own, c-Src is a pivotal regulator of a number of signalling cascades associated with tumour development and progression, including the JAK/STAT pathway (Silva, 2004) and the EGFR family member pathways (Biscardi et al, 2000). Indeed, increased c-Src activity upregulates these signals, leading to increased cell growth, migration and invasion.

Because of the central role of c-Src in both bone metabolism and tumourigenesis, c-Src signalling may be of particular importance in patients with cancer-related bone diseases. Therefore, in this section the role of c-Src in bone malignancies such as osteosarcoma and bone metastases will be elucidated.

Osteosarcoma is the most common and most often fatal primary malignant bone tumour, especially affecting children, adolescents and young adults (Raymond et al, 2002). It is a highly aggressive tumour that metastasizes primarily to the lung, with a consequent very poor prognosis (Shanley & Mulligan, 1991). Osteosarcoma arises from primitive transformed cells that exhibit osteoblastic features and produce malignant osteoid (Kansara & Thomas, 2007). Among the factors involved in the progression of this pathology, c-Src has been described as a mediator of tumour cell invasive features. Caveolin-1 downregulation has been proposed to function as a permissive mechanism by which c-Src signalling is activated, enabling osteosarcoma cells to become metastatic and invade the neighboring tissues (Cantiani et al, 2007).

Bone represents the principal site of relapse of many cancers, especially prostate and breast carcinomas. In the latter, skeletal metastases are osteolytic in nature, with a resulting bone destruction mediated by an exaggerated osteoclast activity. Tumour cells colonizing the bone produce factors that stimulate osteoclast formation, with a consequent increase of bone resorption. This leads to the release of growth factors normally stored in the bone matrix, which in turn stimulate tumour cell proliferation and survival. This mutual enhancement of tumour cell and osteoclast activities is termed "vicious cycle", which progressively increases both bone destruction and tumour burder (Weilbaecher et al, 2011). c-Src plays a role in the context of bone metastases. In fact, MDA-MB-231 breast cancer cells overexpressing a kinase dead dominant negative c-Src were found to form far less osteolytic metastases in immunocompromised mice than their wild-type counterpart (Myuoi et al, 2003; Rucci et al, 2006). Recent data demonstrate that c-Src is crucial for breast cancer cell growth in the bone marrow since it mediates AKT regulation and tumour cell survival in response to chemokine (C-X-C motif) ligand 12 CXCL12, thus confering resistance to the pro-apoptotic member of the tumour necrosis factor family, TRAIL (Zhang et al, 2009).

5. c-Src as pharmacological target in bone metastases

Several compounds have been tested in pre-clinical studies *in vitro* and *in vivo* to inhibit c-Src activity and reduce the incidence of metastases. In an model of experimentally induced breast cancer metastases, the pyrrolo-[2,3-d]pyrimidine c-Src inhibitor, CGP76030, proved effective at inhibiting the incidence of relapse in bone and visceral organs, thus improving mice survival. CGP76030 was observed to have potent anti-osteoclastic and anti-tumoral effects (Recchia et al, 2004; Rucci et al, 2006), explaining its *in vivo* effects on both fronts.

The efficacy of c-Src inhibitors in the treatment of breast cancer-induced bone metastases was strengthened by recent data showing that c-Src is a key signalling molecule for the growth of antiestrogen-resistant tumour cells. Indeed, it has been demonstrated that the expression of a constitutively active c-Src attenuated the sensitivity of MCF7 breast cancer cells to tamoxifen (Morgan et al, 2009). In the same cell line the synergistic interaction between ER, EGF receptor and c-Src enhanced tumour cell responsiveness to mitogenic stimuli, allowing their survival also in the presence of tamoxifen (Yue et al, 2007).

Another compound employed in preclinical studies is Dasatinib, a potent inhibitor of SFKs, which proved effective at inhibiting breast cancer cell growth, with a synergic effect when used in combination with chemotherapeutics (Nautiyal et al, 2009), while in mice inoculated with a triple-negative (for ER, progesterone receptor and Her2/neu expression) breast cancer cell line it prevented the formation of osteolytic metastases (Zhang et al, 2009).

Based on this body of evidence, clinical studies have recently been started using sarcatinib, which is a dual c-Src/Abl inhibitor. In a phase I study performed on patients with solid tumours, sarcatinib decreased the levels of bone resorption markers (Baselga et al, 2010). Sarcatinib is currently being tested in phase II clinical studies in patients with breast cancer bone metastases. Another dual c-Src/Abl inhibitor in clinical development with similar properties is Bosutinib.

6. Conclusions

c-Src is a non receptor tyrosine kinase ubiquitously expressed and involved in the regulation of many cellular functions, such as adhesion, growth, migration and survival. Indeed, its activity increases in response to several signals, especially downstream of growth factor and cytokine receptors, integrin receptors and G-coupled receptors. The proto-oncogene nature of c-Src is also well known, so that its aberrant activation is involved in the development and progression of many human cancers.

c-Src plays an unique role in bone metabolism, regulating the activity of both the bone-resorbing osteoclasts and the bone-forming osteoblasts. The interest on c-Src within the bone metabolism derived from the pioneer work of Soriano et al. (1991), who demonstrated that, despite the ubiquitous expression of c-Src, knock-out mice showed only a bone phenotype. This molecule is currently thought to be a suitable pharmacological target for bone metastases. Clinical studies are currently in progress to establish the efficacy of c-Src inhibitors as therapeutic agents in humans.

7. Acknowledgements

The original work was supported by a grant from the "Associazione Italiana per la Ricerca sul Cancro" (AIRC) to AT, and a grant from the European Calcified Tissue Society to BP. BP is recipient of the "Helga Baden" Fellowship provided by the "Fondazione Italiana per la Ricerca sul Cancro".

8. References

Akhand, AA., Pu, M., Senga, T., Kato, M., Suzuki, H., Miyata, T., Hamaguchi, M. & Nakashima, I. (1999). Nitric oxide controls src kinase activity through a sulfhydryl group modification-mediated Tyr-527- independent and Tyr-416-linked mechanism, *Journal of Biological Chemistry*, Vol.274, No.36, (September 1999), pp. 25821–25826, ISSN 0021-9258

Aligayer, H., Boyd, DD., Heiss, MM., Abdalla, EK., Curley, SA. & Gallick, GE. (2002). Activation of Src kinase in primary colorectal carcinoma: an indicator of poor clinical prognosis. *Cancer*, Vol.94, No.2, (January 2002), pp. 344–51, ISSN 0008-543X

Alvarez RH, Kantarjian HM, and Cortes JE. The role of Src in solid and hematologic malignancies: development of new-generation Src inhibitors. *Cancer* 107, 1918–1929. (2006) ISSN 0008-543X.

Arbet-Engels, C., Tartare-Deckert, S. & Eckhart, W. (1999). C-terminal Src kinase associates with ligand-stimulated insulin-like growth factor-I receptor. *Journal of Biological Chemistry*, Vol.274, No. 9, (February 1999), pp. 5422-5428, ISSN 0021-9258

Arthur, W T. & Burridge, K. (2001). RhoA inactivation by p190RhoGAP regulates cell spreading and migration by promoting membrane protrusion and polarity. *Molecular biology of the cell*, Vol.12, No. 9, (September 2001), pp. 2711–2720, ISSN 1059-1524

Baron, R. (1993). Cellular and molecular biology of the osteoclast, In: *Cellular and Molecular Biology of Bone*, Noda, M., Academic Press, ISBN 0-12-520225-3, San Diego

Baron, R. & Horne, WC. (2005). Regulation of osteoclast activity, In: *Topics in Bone Biology*, Vol. 2, Bronner, F., Farach-Carson, MC. & Rubin, J., Springer-Verlag, ISBN 9812562095, London

Baselga, J., Cervantes, A., Martinelli, E., Chirivella, I., Hoekman, K., Hurwitz, HI., Jodnell, DI., Hamberg, P., Casado, E., Elvin, P., Swazland, A., Iacone, R. & Tabernero, J. (2010). Phase Isafety, pharmacokinetics and inhibition of SRC activity study of sarcatinib in patients with solid tumors. *Clinical cancer research*, Vol.16, No.19, (October 2010), pp. 4876-4883, ISSN 1078-0432

Behrens, J., Mareel, MM., Van Roy, FM. & Birchmeier, W. (1989). Dissecting tumor cell invasion: epithelial cells acquire invasive properties after the loss of uvomorulin-mediated cell-cell adhesion. *Journal of Cell Biology*, Vol.108, No.6, (June 1989), pp. 2435–2447, ISSN 0021-9525

Biscardi, JS., Ishizawar, RC., Silva, CM. & Parsons, SJ. (2000). Tyrosine kinase signalling in breast cancer: epidermal growth factor receptor and c-Src interactions in breast cancer. *Breast Cancer Research*, Vol.2, No.3, (March 2000), pp. 203–210, ISSN 1465-5411

Bowman, T., Broome, MA., Sinibaldi, D., Wharton, W., Pledger, WJ., Sedivy, JM., Irby, R., Yeatman, T., Courtneidge, SA. & Jove, R. (2001). Stat3-mediated Myc expression is required for Src transformation and PDGF-induced mitogenesis. *Proceedings of the National Academy of Sciences of the United States of America*, Vol.98, No.13, (June 2001), pp. 7319-7324, ISSN 0027-8424

Boyce, BF., Yoneda, T., Lowe, C., Soriano, P. & Mundy, GR. (1992). Requirement of pp60c-src expression for osteoclasts to form ruffled borders and resorb bone in mice. *Journal of clinical investigation*, Vol.90, No.4, (October 1992), pp. 1622-1627, ISSN 0021-9738

Brown, MT. & Cooper, JA. (1996). Regulation, substrates and functions of src. *Biochimica et biophysica acta*, Vol.1287, No.2-3, (June 1996), pp. 121-149, ISSN 0006-3002

Brugge, JS. & Erikson, RL. (1977). Identification of a transformation-specific antigen induced by an avian sarcoma virus. *Nature*, Vol.269, No.5626, (September 1977), pp. 346-348, ISSN 0028-0836

Brunton, VG., MacPherson, IR. & Frame, MC. (2004). Cell adhesion receptors, tyrosine kinases and actin modulators: a complex three-way circuitry. *Biochimica et biophysica acta*, Vol. 1692, No.2-3, (July 2004), pp. 121-144, ISSN 0006-3002

Buccione, R., Orth, JD. & McNiven, MA. (2004). Foot and mouth: podosomes, invadopodia and circular dorsal ruffles. *Nature reviews. Molecular cell biology*, Vol.5, No.8, (August 2004), pp. 647-657, ISSN 1471-0072

Cantiani, L., Manara, MC., Zucchini, C., De Sanctis, P., Zuntini, M., Valvassori, L., Serra, M., Olivero, M., Di Renzo, MF., Colombo, MP., Picci, P. & Scotlandi, K. (2007). Caveolin-1 reduces osteosarcoma metastases by inhibiting c-Src activity and met signaling. *Cancer Research*, Vol.67, No.16, (August 2007), pp. 7675-7685, ISSN 0008-5472

Cartwright, CA., Eckhart, W., Simon, S. & Kaplan, PL. (1987). Cell transformation by pp60c-src mutated in the carboxyterminal regulatory domain. *Cell* Vol.49, No.1, (April 1987), pp. 83-91, ISSN 0092-8674

Collett, MS., Brugge, JS. & Erikson, RL. (1978). Characterization of a normal avian cell protein related to the avian sarcoma virus transforming gene product. *Cell*. Vol.15, No.4, (December 1978), pp. 1363-1369, ISSN 0092-8674

Cooper, JA. & King, CS. (1986). Dephosphorylation or antibody binding to the carboxy terminus stimulates pp60c-src. *Molecular and cellular biology*, Vol.6, No.12, (December 1986), pp. 4467-4477, ISSN 0270-7306

Courtneidge, SA., Dhand, R., Pilat, D., Twamley, GM., Waterfield, MD. & Roussel, MF. (1993). Activation of Src family kinases by colony stimulating factor-1, and their association with its receptor. *EMBO Journal*, Vol.12, No.3, (March 1993), pp. 943-950, ISSN 0261-4189

Coxon, FP. & Taylor, A. (2008). Vesicular trafficking in osteoclasts. *Seminars in cell & developmental biology*, Vol.19, No.5, (October 2008), pp. 424-433, ISSN 1084-9521

Datta, A., Huber, F. & Boettiger, D. (2002). Phosphorylation of beta3 integrin controls ligand binding strength. *The Journal of biological chemistry*. Vol.277, No.6, (February 2002), pp. 3943-3949, ISSN 0021-9258

De Mali, KA., Godwin, SL., Soltoff, SP. & Kazlauskas, A. (1999). Multiple roles for Src in a PDGF-stimulated cell. *Experimental cell research*, Vol.253, No.1, (November 1999), pp. 271-279, ISSN 0014-4827

Destaing, O., Saltel, F., Geminard, JC. & Jurdic, BF. (2003). Podosomes display actin turnover and dynamic self-organization in osteoclasts expressing actin-green fluorescent protein. *Molecular biology of the cell,* Vol.14, No.2, (February 2003), pp. 407–416, ISSN 1059-1524

Duesberg, PH. & Vogt, PK. (1970). Differences between the ribonucleic acids of transforming and nontransforming avian tumor viruses. *Proceedings of the National Academy of Sciences of the United States of America,* Vol.67, No.4, (December 1970), pp. 1673–1680, ISSN 0027-8424

Duong, LT. & Rodan, GA. (2000). PYK2 is an adhesion kinase in macrophages, localized in podosomes and activated by beta(2)-integrin ligation. *Cell motility and the cytoskeleton,* Vol.47, No.3, (November 2000), pp.174-88, ISSN 0886-1544

Dupree, P., Parton, RG., Raposo, G., Kurzchalia, TV. & Simons K. (1993). Caveolae and sorting in the trans-Golgi network of epithelial cells. *The EMBO journal,* Vol.12, No.4, (April 1993), pp. 1597-1605, ISSN 0261-4189

Faccio, R., Novack, DV., Zallone, A., Ross, FP. & Teitelbaum, SL. (2003). Dynamic changes in the osteoclast cytoskeleton in response to growth factors and cell attachment are controlled by beta3 integrin. *The Journal of cell biology,* Vol.162, No.3, (August 2003), pp. 499-509, ISSN 0021-9525

Faccio, R., Teitelbaum, SL., Fujikawa, K., Chappel, J., Zallone, A., Tybulewicz, VL., Ross, FP. & Swat, W. (2005). Vav3 regulates osteoclast function and bone mass. *Nature medicine,* Vol.11, No.3, (March 2005), pp. 284-290, ISSN 1078-8956

Fukuda, A., Hikita, A., Wakeyama, H., Akiyama, T., Oda, H., Nakamura, K. & Tanaka, S. (2005). Regulation of osteoclast apoptosis and motility by small GTPase binding protein Rac1. *Journal of bone and mineral research,* Vol.20, No.12, (December 2005), pp. 2245-2253, ISSN 0884-0431

Gianni, D., Taulet, N., DerMardirossian, C. & Bokoch, GM. (2010). c-Src-mediated phosphorylation of NoxA1 and Tks4 induces the reactive oxygen species (ROS)-dependent formation of functional invadopodia in human colon cancer cells. *Molecular biology of the cell,* Vol.21, No.23, (December 2010), pp. 4287-4298, ISSN 1059-1524

Giannoni, E., Buricchi, F., Raugei, G., Ramponi, G. & Chiarugi, P. (2005). Intracellular reactive oxygen species activate Src tyrosine kinase during cell adhesion and anchorage-dependent cell growth. *Molecular biology of the cell* Vol.25, No.15, (August 2005), pp. 6391-6403, ISSN 1059-1524

Guarino, M. (2010). Src signaling in cancer invasion. *Journal of cellular physiology,* Vol.223, No.1, (April 2010), pp. 14–26, ISSN 0021-9541

Hall, JM., Couse, JF. & Korach, KS. (2001). The multifaceted mechanisms of estradiol and estrogen receptor signaling. *Journal of Biological Chemistry,* Vol.276, No.40, (October 2001), pp. 36869–36872, ISSN 0021-9258

Hanai, J., Chen, LF., Kanno, T., Ohtani-Fujita, N., Kim, WY., Guo, WH., Imamura, T., Ishidou, Y., Fukuchi, M., Shi, MJ., Stavnezer, J., Kawabata, M., Miyazono, K. & Ito, Y. (1999). Interaction and functional cooperation of PEBP2/CBF with Smads. Synergistic induction of the immunoglobulin germline Calpha promoter. *Journal of Biological Chemistry,* Vol.274, No.44, (October 1999), pp. 31577–31582, ISSN 0021-9258

Hess, J., Angel, P. & Schorpp-Kistner, M. (2004). AP-1 subunits: quarrel and harmony among siblings. *Journal of cell science,* Vol.117, Pt.25, (December 2004), pp. 5965–5973, ISSN 0021-9533

Horton, MA. (1997). The αVβ3 integrins 'vitronectin receptor'. *The international journal of biochemistry & cell biology*, Vol.29, No.5, (May 1997), pp. 721–725, ISSN 1357-2725

Horton, MA. & Rodan, GA. (1996). Integrins as therapeutic targets in bone disease. In: *Adhesion Receptors as Therapeutic Targets*, Horton, MA., CRC Press Inc., ISBN 9780849376559, Boca Raton

Hunter, T. & Sefton, BM. (1980) Transforming gene product of Rous sarcoma virus phosphorylates tyrosine. *Proceedings of the National Academy of Sciences of the United States of America*, Vol.77, No.3, (March 1980), pp. 1311–1315, ISSN 0027-8424

Irby, RB. & Yeatman, TJ. (2002). Increased Src activity disrupts cadherin/catenin–mediated homotypic adhesion in human colon cancer and transformed rodent cells. *Cancer Research*, Vol.62, No.9, (May 2002), pp. 2669–2674, ISSN 0008-5472

Irby, RB., Mao, W., Coppola, D., Kang, J., Loubeau, JM., Trudeau, W., Karl, R., Fujita, DJ., Jove, R. & Yeatman, TJ. (1999). Activating SRC mutation in a subset of advanced human colon cancers. *Nature Genetics*, Vol.21, No.2, (February 1999), pp. 187–190, ISSN 1061-4036

Irby, RB. & Yeatman, TJ. (2000) Role of Src expression and activation in human cancer. *Oncogene*, Vol.19, No.49, (November 2000), pp. 5636–5642, ISSN 0950-9232

Jiang, Y., Jiang, Y., Jahagirdar, BN., Reinhardt, RL., Schwartz, RE., Keene, CD., Ortiz-Gonzalez, XR., Reyes, M., Lenvik, T., Lund, T., Blackstad, M., Du, J., Aldrich, S., Lisberg, A., Low, WC., Largaespada, DA. & Verfaillie, CM. (2002). Pluripotency of mesenchymal stem cells derived from adult marrow. *Nature*, Vol.418, No.6893, (July 2002), pp. 41-49, ISSN 0028-0836

Jove, R. & Hanafusa, H. (1987). Cell transformation by the viral *src* oncogene. *Annual review of cell biology*, Vol.3, (November 1987), pp. 31–56, ISSN 0743-4634

Jung, EJ. & Kim, CW. (2002). Interaction between chicken protein tyrosine phosphatase 1 (CPTP1)-like rat protein phosphatase 1 (PTP1) and p60(v-src) in v-src transformed Rat-1 fibroblasts. *Experimental & molecular medicine*, Vol.34, No.6, (December 2002), pp. 476–480, ISSN 1226-3613

Kamei, T. Machida K, Nimura Y, Senga T, Yamada I, Yoshii S, Matsuda S, Hamaguchi M. (2000). C-Cbl protein in human cancer tissues is frequently tyrosine phosphorylated in a tumor-specific manner. *International journal of oncology*, Vol.17, No.2, (August 2000), pp. 335–339, ISSN 1019-6439

Kansara, M. & Thomas, DM. (2007). Molecular pathogenesis of osteosarcoma. *DNA and cell biology*, Vol.26, No.1, (January 2007), pp. 1–18, ISSN 1044-5498

Kim, M., Tezuka, T., Tanaka, K. & Yamamoto, T. (2004). Cbl-c suppresses v-Src-induced transformation through ubiquitin-dependent protein degradation. *Oncogene*, Vol.23, No.9, (March 2004), pp. 1645–1655, ISSN 0950-9232

Kmiecik, TE. & Shalloway, D. (1987). Activation and suppression of pp60c-src transforming ability by mutation of its primary sites of tyrosine phosphorylation. *Cell*, Vol.49, No.1, (April 1987), pp. 65–73, ISSN 0092-8674

Koegl, M., Zlatkine, P., Ley, SC., Courtneidge, SA. & Magee, AI. (1994). Palmitoylation of multiple Src-family kinases at a homologous N-terminal motif. *The Biochemical journal*, Vol.303, Pt.3, (November 1994), pp. 749–753, ISSN 0264-6021

Komori, T., Yagi, H., Nomura, S., Yamaguchi, A., Sasaki, K., Deguchi, K., Shimizu, Y., Bronson, RT., Gao, YH., Inada, M., Sato, M., Okamoto, R., Kitamura, Y., Yoshiki, S. & Kishimoto, T. (1997). Targeted disruption of Cbfa1 results in a complete lack of bone formation owing to maturational arrest of osteoblasts. *Cell*, Vol.89, No.5, (May 1997), pp. 755–764, ISSN 0092-8674

Kousteni, S., Han, L., Chen, JR., Almeida, M., Plotkin, LI., Bellido, T. & Manolagas, SC. (2003). Kinase-mediated regulation of common transcription factors accounts for the bone-protective effects of sex steroids. *The Journal of clinical investigation*, Vol.111, No.11, (June 2003), pp. 1651-1664, ISSN 0021-9738

Landgren, E., Blume-Jensen, P., Courtneidge, SA. & Claesson-Welsh, L. (1995). Fibroblast growth factor receptor-1 regulation of Src family kinases. *Oncogene*, Vol.10, No.10, (May 1995), pp. 2027–2035, ISSN 0950-9232

Li, S., Couet, J. & Lisanti, MP. (1996). Src tyrosine kinases, Galpha subunits, and H-Ras share a common membrane-anchored scaffolding protein, caveolin. Caveolin binding negatively regulates the auto-activation of Src tyrosine kinases. *The Journal of biological chemistry*, Vol.271, No.46, (November 1996), pp. 29182-29190, ISSN 0021-9258

Linder, S. & Aepfelbacher, M. (2003). Podosomes: adhesion hot-spots of invasive cells. *Trends in cell biology*, Vol.13, No.7, (July 2003), pp. 376–385, ISSN 0962-8924

Longo, M., Brama, M., Marino, M., Bernardini, S., Korach, KS., Wetsel, WC., Scandurra, R., Faraggiana, T., Spera, G., Baron, R., Teti, A. & Migliaccio, S. (2004). Interaction of estrogen receptor alpha with protein kinase C alpha and c-Src in osteoblasts during differentiation. *Bone*, Vol.34, No.1, (January 2004), pp. 100-111, ISSN 8756-3282

Lowe, C., Yoneda, T., Boyce, BF., Chen, H., Mundy, GR. & Soriano, P. (1993). Osteopetrosis in Src deficient mice is due to an autonomous defect of osteoclasts. *Proceedings of the National Academy of Sciences of the United States of America*, Vol.90, No.10, (May 1993), pp. 4485–4489, ISSN 0027-8424

Lowell, CA., Niwa, M., Soriano, P. & Varmus, HE. (1996). Deficiency of the Hck and Src tyrosine kinases results in extreme levels of extramedullary hematopoiesis. *Blood*, Vol.87, No.5, (May 1996), pp. 1780–1792, ISSN 0006-4971

Manolagas, SC., Kousteni, S. & Jilka, RL. (2002). Sex steroids and bone. *Recent Progress in Hormone Research*, Vol.57, (2002), pp. 385–409, ISSN 0079-9963

Mao, W., Irby, R., Coppola, D., Fu, L., Wloch, M., Turner, J., Yu, H., Garcia, R., Jove, R. & Yeatman, TJ. (1997). Activation of c-Src by receptor tyrosine kinases in human colon cancer cells with high metastatic potential. *Oncogene*, Vol.15, No.25, (December 1997), pp. 3083–3090, ISSN 0950-9232

Marchisio, PC., Cirillo, D., Naldini, L., Primavera, MV., Teti, A. & Zambonin-Zallone, A. (1984). Cell–substratum interaction of cultured avian osteoclasts is mediated by specific adhesion structures. *The Journal of cell biology*, Vol.99, No.5, (November 1984), pp. 1696–1705, ISSN 0021-9525

Marchisio, PC., Cirillo, D., Teti, A., Zambonin-Zallone, A. & Tarone, G. (1987). Rous sarcoma virus transformed fibroblasts and cells of monocytic origin display a peculiar dot-like organization of cytoskeletal proteins involved in microfilament–membrane interactions. *Experimental cell research*, Vol.169, No.1, (March 1987), pp. 202–214, ISSN 0014-4827

Martin, GS. (1970). Rous sarcoma virus: a function required for the maintenance of the transformed state. *Nature*, Vol.227, No.5262, (September 1970), pp. 1021–1023, ISSN 0028-0836

Marzia, M., Sims, NA., Voit, S., Migliaccio, S., Taranta, A., Bernardini, S., Faraggiana, T., Yoneda, T., Mundy, GR., Boyce, BF., Baron, R. & Teti, A. (2000). Decreased c-Src expression enhances osteoblast differentiation and bone formation. *The Journal of cell biology*, Vol. 151, No.2, (October 2000), pp. 311-320, ISSN 0021-9525

Mayer, BJ., Jackson, PK. & Baltimore, D. (1991). The noncatalytic *src* homology region 2 segment of *abl* tyrosine kinase binds to tyrosine-phosphorylated cellular proteins with high affinity. *Proceedings of the National Academy of Sciences of the United States of America*, Vol.88, No.2, (January 1991), pp. 627–631, ISSN 0027-8424

Mayer, BJ. & Baltimore, D. (1993). Signalling through SH2 and SH3 domains. *Trends in cell biology*, Vol.3, No.1, (January 1993), pp. 8-13, ISSN 0962-8924

Mayor, S. & Pagano, RE. (2007). Pathways of clathrin independent endocytosis. *Nature reviews. Molecular cell biology*, Vol.8, No.8, (August 2007), pp. 603-612, ISSN 1471-0072

McLean, GW., Carragher, NO., Avizienyte, E., Evans, J., Brunton, VG. & Frame, MC. (2005). The role of focal-adhesion kinase in cancer—a new therapeutic opportunity. *Nature reviews. Cancer*, Vol.5, No.7, (July 2005), pp. 505–515, ISSN 1474-175X

Miyamoto, S., Akiyama, SK. & Yamada, KM. (1995). Synergistic roles for receptor occupancy and aggregation in integrin transmembrane function. *Science*, Vol.267, No.5199, (February 1995), pp. 883–885, ISSN 0036-8075

Miyamoto, S., Teramoto, H., Gutkind, JS. & Yamada, KM. (1996). Integrins can collaborate with growth factors for phosphorylation of receptor tyrosine kinases and MAP kinase activation: Roles of integrin aggregation and occupancy of receptors. *The Journal of cell biology*, Vol.135, No.6 Pt.1, (December 1996), pp. 1633–1642, ISSN 0021-9525

Miyazaki, T., Neff, L., Tanaka, S., Horne, WC. & Baron, R. (2003). Regulation of cytochrome c oxidase activity by c-Src in osteoclasts. *The Journal of cell biology*, Vol.160, No.5, (March 2003), pp. 709–718, ISSN 0021-9525

Miyazaki, T., Sanjay, A., Neff, L., Tanaka, S., Horne, WC. & Baron, R. (2004). Src kinase activity is essential for osteoclast function. *The Journal of biological chemistry*, Vol.279, No.17, (April 2004), pp. 17660–17666, ISSN 0021-9258

Miyazaki, T., Tanaka, S., Sanjay, A. & Baron, R. (2006). The role of c-Src kinase in the regulation of osteoclast function. *Modern rheumatology*, Vol.16, No.2, (2006), pp. 68-74, ISSN 1439-7595

Mocsai, A., Humphrey, MB., Van Ziffle, JA., Hu, Y., Burghardt, A., Spusta, SC., Majumdar, S., Lanier, LL., Lowell, CA., Nakamura, MC. (2004). The immunomodulatory adapter proteins DAP12 and Fc receptor gamma-chain (FcR gamma) regulate development of functional osteoclasts through the Syk tyrosine kinase. *Proceedings of the national academy of science USA*, Vol. 101, No.16, (April 2004) pp.6158-6163

Morgan, L., Gee, J., Pumford, S., Farrow, L., Finlay, P., Robertson, J., Ellis, I., Kawakatsu, H., Nicholson, R. & Hiscox, S. (2009). Elevated Src kinase activity attenuates Tamoxifen response in vitro and is associated with poor prognosis clinically. *Cancer biology & therapy*, Vol.8, No.16, (August 2009), pp. 1550-1558, ISSN 1538-4047

Myuoi, A., Nishimura, P., Williams, PJ., Tamura, D., Michigami, T., Mundy, GR. & Yoneda, T. (2003). c-Src tyrosine kinase activity is associated with tumor colonization in bone and lung in an animal model of human breast cancer bone metastasis. *Cancer research*, Vol.63, No.16, (August 2003), pp. 5028-5033, ISSN 0008-5472

Nakamura, I., Duong, le T., Rodan, SB. & Rodan, GA. (2007). Involvement of alpha(v)beta3 integrins in osteoclast function. *Journal of bone and mineral metabolism*, Vol.25, No.6, (October 2007), pp. 337-344, ISSN 0914-8779

Nam, JS., Ino, Y., Sakamoto, M. & Hirohashi, S. (2002). Src family kinase inhibitor PP2 restores the E-cadherin/catenin cell adhesion system in human cancer cells and reduces cancer metastasis. *Clinical cancer research*, Vol.8, No.7, (July 2002), pp. 2430–2436, ISSN 1078-0432

Nautiyal, J., Majumder, P., Patel, BB., Lee, FY. & Majumdar, AP. (2009). Src inhibitor dasatinib inhibits growth of breast cancer cells by modulating EGFR signaling. *Cancer letters*, Vol283, No.2, (October 2009), pp. 143-151, ISSN 0304-3835

Okada, M. & Nakagawa, H. (1989). A protein tyrosine kinase involved in regulation of pp60c-src function. *The Journal of biological chemistry*, Vol.264, No.35, (December 1989), pp. 20886–20893, ISSN 0021-9258

Oppermann, H., Levinson, AD., Varmus, HE., Levintow, L. & Bishop, JM. (1979). Uninfected vertebrate cells contain a protein that is closely related to the product of the avian sarcoma virus transforming gene (src). *Proceedings of the National Academy of Sciences of the United States of America*, Vol.76, No.4, (April 1979), pp. 1804–1808, ISSN 0027-8424

Pawson, T. (1995). Protein modules and signalling networks. *Nature*, Vol.373, No.6515, (February 1995), pp. 573-580, ISSN 0028-0836

Peruzzi B, Cappariello A, Del Fattore A, Rucci N, De Benedetti F, Teti A. (2012) c-Src and IL-6 inhibit osteoblast differentiation and integrate IGFBP5 signalling. *Nature Communications*, Vol. 3, (January 2012) article number 630. ISSN 2041-1723, doi: 10.1038/ncomms1651.

Peruzzi, B. & Teti, A. (2011). The Physiology and Pathophysiology of the Osteoclast, *Clinical Reviews in Bone and Mineral Metabolism*, ISSN 1534-8644, DOI: 10.1007/s12018-011-9086-6

Piwnica-Worms, H., Saunders, KB., Roberts, TM., Smith, AE. & Cheng, SH. (1987). Tyrosine phosphorylation regulates the biochemical and biological properties of pp60c-src. *Cell*, Vol.49, No.1, (April 1987), pp. 75–82, ISSN 0092-8674

Playford, MP. & Schaller, MD. (2004). The interplay between Src and integrins in normal and tumor biology. *Oncogene*, Vol.23, No.48, (October 2004), pp. 7928–7946, ISSN 0950-9232

Raymond, AK., Ayala, AG. & Knuutila, S. (2002). Conventional osteosarcoma. In: *World Health Organization classification of tumors*, Fletcher, CDM., Unni, KK. & Mertens, F., IARC Press, ISBN 92 832 2413 2, Lyon:, 2002. p. 264–70.

Recchia, I., Rucci, N., Funari, A., Migliaccio, S., Taranta, A., Longo, M., Kneissel, M., Susa, M., Fabbro, D. & Teti, A. (2004). Reduction of c-Src activity by substituted 5,7-diphenyl-pyrrolo[2,3-d]-pyrimidines induces osteoclast apoptosis in vivo and in vitro. Involvement of ERK1/2 pathway. *Bone*, Vol.34, No.1, (January 2004), pp. 65-79, ISSN 8756-3282

Ren, R., Mayer, BJ., Cicchetti, P. & Baltimore, D. (1993). Identification of a ten amino acid proline-rich SH3 binding site. *Science*, Vol.259, No.5098, (February 1993), pp. 1157-1161, ISSN 0036-8075

Resh, MD. (1999). Fatty acylation of proteins: new insights into membrane targeting of myristoylated and palmitoylated proteins. *Biochimica et biophysica acta*, Vol.1451, No.1, (August 1999), pp. 1–16, ISSN 0006-3002

Roby, KF., Son, DS., Taylor, CC., Montgomery-Rice, V., Kirchoff, J., Tang, S. & Terranova, PF. (2005). Alterations in reproductive function in SRC tyrosine kinase knockout mice. *Endocrine*, Vol. 26, No.2, (March 2005), pp. 169-176, ISSN 1355-008X

Rodan, GA. & Martin, TJ. (1981). Role of osteoblasts in hormonal control of bone resorption-- a hypothesis. *Calcified tissue International*, Vol.33, No.4, (1981), pp. 349-351, ISSN 0171-967X

Rothberg, KG., Heuser, JE., Donzell, WC., Ying, YS., Glenney, JR. & Anderson, RG. (1992). Caveolin, a protein component of caveolae membrane coats. *Cell*, Vol68, No. 4, (February 1992), pp. 673-682, ISSN 0092-8674

Rous, PA. (1911a). Transmission of a malignant new growth by means of a cell-free filtrate. *Journal of the American Medical Association*, Vol.56, (1911), pp. 98, ISSN 0098-7484

Rous, PA. (1911b). A sarcoma of the fowl transmissible by an agent separable from the tumor cells. *The Journal of experimental medicine*, Vol.13, No.4, (April 1911), pp. 397–411, ISSN 0022-1007

Rubin, H. (1995). Quantitative relations between causative virus and cell in the Rous No. 1 chicken sarcoma. *Virology*, Vol.1, No.5, (December 1955), pp. 445–473, ISSN 0042-6822

Rucci, N., Recchia, I., Angelucci, A., Alamanou, M., Del Fattore, A., Fortunati, D., Susa, M., Fabbro, D., Bologna, M. & Teti, A. (2006). Inhibition of protein kinase c-Src reduces th incidence of breast cancer metastases and increases survival in mice: implication for therapy. *The Journal of pharmacology and experimental therapeutics*, Vol. 318, No.1, (July 2006), pp. 161-172, ISSN 0022-3565

Rucci, N., Susa, M. & Teti, A. (2008). Inhibition of protein kinase c-Src as a therapeutic approach for cancer and bone metastases. *Anti-cancer agents in medicinal chemistry*, Vol.8, No.3, (April 2008), pp. 342–349, ISSN 1871-5206

Rupp, PA. & Little, CD. (2001). Integrins in vascular development. *Circulation research*, Vol.89, No.7, (September 2001), pp. 566–572, ISSN 0009-7330

Sakai, T., Jove, R., Fassler, R. & Mosher, DF. (2001). Role of the cytoplasmic tyrosines of beta 1A integrins in transformation by v-src. *Proceedings of the National Academy of Sciences of the United States of America*, Vol.98, No.7, (March 2001), pp. 3808–3813, ISSN 0027-8424

Sanjay, A., Houghton, A., Neff, L., DiDomenico, E., Bardelay, C., Antoine, E., Levy, J., Gailit, J., Bowtell, D., Horne, WC. & Baron, R. (2001). Cbl associates with Pyk2 and Src to regulate Src kinase activity, alpha(v)beta(3) integrin-mediated signaling, cell adhesion, and osteoclast motility. *The Journal of cell biology*, Vol.152, No. 1, (January 2001), pp. 181-195, ISSN 0021-9525

Sargiacomo, M., Scherer, PE., Tang, Z., Kübler, E., Song, KS., Sanders, MC. & Lisanti, MP. (1995). Oligomeric structure of caveolin: implications for caveolae membrane organization. *Proceedings of the National Academy of Sciences of the United States of America*, Vol.92, No.20, (September 1995), pp. 9407-9411, ISSN ISSN 0027-8424

Schaller, MD., Hildebrand, JD., Shannon, JD., Fox, JW., Vines, RR. & Parsons, JT. (1994). Autophosphorylation of the focal adhesion kinase, pp125FAK, directs SH2-

dependent binding of pp60src. *Molecular and cellular biology*, Vol.14, No.3, (March 1994), pp. 1680-1688, ISSN 0270-7306

Schlaepfer, DD. & Hunter, T. (1998). Integrin signaling and tyrosine phosphorylation: Just the FAKs? *Trends in cell biology*, Vol.8, No.4, (April 1998), pp. 151–157, ISSN 0962-8924

Severs, NJ. (1988). Caveolae: static inpocketings of the plasma membrane, dynamic vesicles or plain artifact? *Journal of cell science*, Vol.90, Pt.3, (July 1988), pp. 341-348, ISSN 0021-9533

Shanley, DJ. & Mulligan, ME. (1991). Osteosarcoma with isolated metastases to the pleura. *Pediatric radiology*, Vol.21, No.3, (1991), pp.226, ISSN 0301-0449

Silva, CM. (2004). Role of STATs as downstream signal transducers in Src family kinase-mediated tumorigenesis *Oncogene*, Vol.23, No.48, (October 2004), pp. 8017–8023, ISSN 0950-9232

Soriano, P., Montgomery C., Geske R. & Bradley A.. (1991). Targeted disruption of the c-src proto-oncogene leads to osteopetrosis in mice. *Cell*, Vol.64, No.4, (February 1991), pp. 693–702, ISSN 0092-8674

Stein, GS., van Wijnen, AJ., Stein, JL., Lian, JB., Montecino, M., Choi, JY., Zaidi, K. & Javed, A. (2000). Intranuclear trafficking of transcription factors: implications for biological control. *Journal of cell science*, Vol.113, Pt.14, (July 2000), pp. 2527–2533, ISSN 0021-9533

Stein, PL., Vogel, H. & Soriano, P. (1994). Combined deficiencies of Src, Fyn, and Yes tyrosine kinases in mutant mice. *Genes & development*, Vol.8, No.17, (September 1994), pp. 1999–2007, ISSN 0890-9369

Suda, T., Nakamura, I., Jimi, E. & Takahashi, N. (1997). Regulation of osteoclast function. *Journal of bone and mineral research*, Vol.12, No.6, (June 1997), pp. 869-879, ISSN 0884-0431

Takeya, T. & Hanafusa, H. (1983). Structure and sequence of the cellular gene homologous to the RSV *src* gene and the mechanism for generating the transforming virus. *Cell*, Vol.32, No.3, (March 1983), pp. 881–890, ISSN 0092-8674

Teitelbaum SL. (2007). Osteoclasts: what do they do and how do they do it? *The American journal of pathology*, Vol.170, No.2, (February 2007), pp. 427-435, ISSN 0002-9440

Thomas, SM. & Brugge, JS. (1997). Cellular functions regulated by Src family kinases. *Annual Review of Cell and Developmental Biology*, Vol.13, (November 1997), pp. 513–609, ISSN 1081-0706

Thomas, JW., Ellis, B., Boerner, RJ., Knight, WB., White, GC.II & Schaller, MD. (1998). SH2- and SH3-mediated interactions between focal adhesion kinase and Src. The *Journal of biological chemistry*, Vol.273, No.1, (January 1998), pp. 577–583, ISSN 0021-9258

Tice, DA., Biscardi, JS., Nickles, AL. & Parsons, SJ. (1999). Mechanism of biological synergy between cellular Src and epidermal growth factor receptor. *Proceedings of the National Academy of Sciences of the United States of America*, Vol.96, No.4, (February 1999), pp. 1415–1420, ISSN ISSN 0027-8424

Timpson, P., Jones, GE., Frame, MC. & Brunton, VG. (2001). Coordination of cell polarization and migration by the Rho family GTPases requires Src tyrosine kinase activity. *Current biology*, Vol.11, No.23, (November 2001), pp. 1836–1846, ISSN 0960-9822

Tsygankov, AY. & Shore, SK. Src: regulation, role in human carcinogenesis and pharmacological inhibitors. *Current pharmaceutical design*, Vol.10, No.15, (2004), pp. 1745–1756, ISSN 1381-6128

van der Geer, P., Hunter, T., & Lindberg, RA. (1994). Receptor protein-tyrosine kinases and their signal transduction pathways. *Annual review of cell biology*, Vol.10, (1994), pp. 251–337, ISSN 0743-4634

Wagner, EF. & Eferl R. (2005). Fos/AP-1 proteins in bone and the immune system. *Immunological reviews*, Vol.208, (December 2005), pp. 126–140, ISSN 0105-2896

Wei, Y., Yang, X., Liu, Q., Wilkins, JA., & Chapman, HA. (1999). A role for caveolin and the urokinase receptor in integrin-mediated adhesion and signaling. *The Journal of cell biology*, Vol.144, No.6, (March 1999), pp. 1285–1294, ISSN 0021-9525

Weilbaecher, KN., Guise, TA. & McCauley, LK. (2011). Cancer and bone: a fatal attraction. *Nature reviews. Cancer*, Vol.11, No.6, (June 2011), pp. 411-425, ISSN 1474-175X

Westendorf, JJ. & Hiebert, SW. (1999). Mammalian runt-domain proteins and their roles in hematopoiesis, osteogenesis, and leukemia. *Journal of cellular biochemistry*, Suppl. 32-33, (1999), pp. 51–58, ISSN 0730-2312

Wilder, RL. (2002). Integrin alpha V beta 3 as a target for treatment of rheumatoid arthritis and related rheumatic diseases. *Annals of the Rheumatic* Diseases, Vol.61, Suppl.2, (November 2002), pp. 96-99, ISSN 0003-4967

Winkler, DG., Park, I., Kim, T., Payne, NS., Walsh, CT., Strominger, JL. & Shin, J. (1993). Phosphorylation of Ser-42 and Ser-59 in the N-terminal region of the tyrosine kinase p56lck. *Proceedings of the National Academy of Sciences of the United States of America*, Vol.90, No.11, (June 1993), pp. 5176-5180, ISSN ISSN 0027-8424

Yagi, R., Chen, LF., Shigesada, K., Murakami, Y. & Ito, Y. (1999). A WW domain-containing yes-associated protein (YAP) is a novel transcriptional co-activator. *The EMBO Journal*, Vol.18, No.9, (May 1999), pp. 2551–2562, ISSN 0261-4189

Yamaguchi, H. & Hendrickson, WA. (1996). Structural basis for activation of human lymphocyte kinase Lck upon tyrosine phosphorylation. *Nature*, Vol.384, No.6608, (December 1996), pp. 484–489, ISSN 0028-0836

Yeatman, TJ. (2004). A renaissance for SRC. *Nature reviews. Cancer*, Vol.4, No.6, (June 2004), pp. 470–480, ISSN 1474-175X

Yokouchi, M., Kondo, T., Sanjay, A., Houghton, A., Yoshimura, A., Komiya, S., Zhang, H. & Baron, R. (2001). Src-catalyzed phosphorylation of c-Cbl leads to the interdependent ubiquitination of both proteins. *The Journal of biological chemistry*, Vol.276, No.37, (September 2001), pp. 35185-35193, ISSN 0021-9258

Yue, W., Fan, P., Wang, J., Li, Y. & Santen, RJ. (2007). Mechanisms of acquired resistance to endocrine therapy in hormone-dependent breast cancer cells. *The Journal of steroid biochemistry and molecular biology*, Vol.106, No.1-5, (August-September 2007), pp. 102-110,, ISSN 0960-0760

Zaidi, SK., Javed, A., Choi, JY., van Wijnen, AJ., Stein, JL., Lian, JB. & Stein, GS. (2001). A specific targeting signal directs Runx2/Cbfa1 to subnuclear domains and contributes to transactivation of the osteocalcin gene. *Journal of cell science*, Vol.114, Pt.17, (September 2001), pp. 3093–3102, ISSN 0021-9533

Zhang, XH., Wang, Q., Gerald, W., Hodis, CA., Norton, L., Smid, M., Foekens, JA. & Massagué, J. (2009). Latent bone metastasis in breast cancer tied to Src-dependent survival signals. *Cancer Cell*, Vol.16, No.1, (July 2009), pp. 67-78, ISSN 1535-6108

Zheng, XM., Wang, Y. & Pallen, CJ. (1992). Cell transformation and activation of pp60c-src by overexpression of a protein tyrosine phosphatase. Nature, Vol.359, No.6393, (September 1992), pp. 336–339, ISSN 0028-0836

Zou, W., Kitaura, H., Reeve, J., Long, F., Tybulewicz, VL., Shattil, SJ., Ginsberg, MH., Ross, FP. & Teitelbaum, SL. (2007). Syk, c-Src, the alphavbeta3 integrin, and ITAM immunoreceptors, in concert, regulate osteoclastic bone resorption. *The Journal of cell biology*, Vol.176, No.6, (March 2007), pp. 877-888, ISSN 0021-9525

Protein Kinases and Protein Phosphatases as Participants in Signal Transduction of Erythrocytes

Ana Maneva[1] and Lilia Maneva-Radicheva[2]
*[1]Medical University, Pharmaceutical Faculty,
Department of Chemistry and Biochemistry, Plovdiv,
[2]Medical University, Medical Faculty,
Department of Chemistry and Biochemistry, Sofia,
Bulgaria*

1. Introduction

Signal transduction is defined as the transfer of a signal starting from a primary messenger with a ligand that binds to specific receptors on the cell membrane. The signal then reaches the effector molecule(s) through a cascade involving various protein kinases and/or protein phosphatases as well as other molecules such as adaptor proteins, anchoring proteins, amplifier proteins, etc. Although erythrocytes are nucleus-free cells, they use elements of cell signal pathways that help them to maintain membrane integrity, ion transport, and metabolism. Aging, some "stress" conditions, and various diseases also may generate cell signals into erythrocytes (Minetti и Low, 1997; Antonelou et al. 2010; Berzosa et al. 2011; Pantaleo et al. 2010). Molecules from both erythrocyte membranes and from cytosol may take part in signal transduction. Membrane components involved in signal transduction include receptors, heterotrimeric G-proteins, adaptor and anchoring proteins, and some proteins of the Ras-superfamily. The erythrocyte intracellular signaling cascade is divided into protein kinases and phosphatases, second messengers and small GTPases, which could further be defined as cell surface and intracellular activities (Pasini et al. 2006). In general, the mechanisms of signal transduction in erythrocytes can be summarized in the following groups: 1) specific mechanisms triggering changes in the activity of erythrocyte enzymes, 2) specific mechanisms inducing changes in the activity of transport systems, 3) signal transduction related to membrane association/dissociation of cytoskeletal and integral proteins, 4) "stressful" conditions may induce activation of specific cellular signals into erythrocytes. Phosphorylation of the erythrocyte proteins by protein kinases and phosphatases is the key mechanism for controlling erythrocyte functions.

2. Binding of extracellular ligands to receptors on erythrocytes membrane initiates various processes of phosphorylation of erythrocyte proteins

2.1 Thyroid stimulating hormone

Thyroid stimulating hormone (TSH) binds to a specific TSH receptor (TSHR) which activates adenylate cyclase and increases cAMP levels in thyroid cells. Recent studies have reported that TSHR and Na/K-ATPase are localized on the membranes of both erythrocytes and erythrocyte ghosts. TSHR responds to TSH treatment by increasing intracellular cAMP levels from two to tenfold. The authors suggest a novel cell signalling pathway, potentially active in local circulatory control (Balzan et al. 2009).

2.2 Parathyroid hormone

Parathyroid hormone (PTH) operates with G-protein dependent receptors and has been shown to decrease erythrocyte deformability in a Ca^{2+}-dependent manner (Bogin et al. 1986). In pseudohypoparathyroidism, PTH resistance results from impairment caused by a deficiency of Gsα-signaling cascade due to a metylation defect of the GNAS gene and decrease in erythrocyte Gsα activity (Zazo et al. 2011).

2.3 Insulin

Insulin stimulates erythrocyte glycolysis, Na+/H+-antiport and Na+/K+-ATPase (Rizvi et al.1994), as well as membrane-associated NO-synthase (Bhattacharaya al. 2001). It has been shown that insulin uses signaling pathways involving both ζ-PKC and phospatidylinositol 3 kinase (PI3K), which leads to the activation of Na^+/H^+-antiports (Sauvage et al. 2000). Insulin is also used in the MAPK-signaling cascade of phosphorylation and activation of protein NHE1, responsible for Na^+/H^+ - antiport (Sartori et al. 1999).

Participation of C-peptide in insulin signaling pathways has also been discussed, including involvement of G-protein and Ca-dependent phosphatase. It has been reported that insulin restores the activity of Na+/K +-ATPase (De La Tour et al. 1998). The decrease in Na^+/K^+ adenosine triphosphatase (ATPase) in erythrocytes of type 1 diabetes is thought to play a role in the development of long-term complication. Infusion of insulin may restore this enzyme activity in red cells (Djemli-Shipkolye et al. 2000).

A reduced erythrocyte insulin receptor binding and tyrosine kinase activity was measured in hypertensive subjects with hyperinsulinemia (Corry et al. 2002). Erythrocytes from normal individuals showed increased pH and increased sodium influx (NHE1) after insulin stimulation. In contrast, insulin had no effect on NHE1 activity of eryhthrocyte from obese individuals (Kaloyianni et al. 2001). Insulin activation of insulin receptor kinase in erythrocytes is not altered in non-insulin-dependent diabetes and not influenced by hyperglycemia (Klein et al. 2000).

2.4 Insulin-like growth factor I

Erythrocytes possess receptors for insulin-like growth factor I (IGF-I). Binding to these receptors is dependent on cell age (Polychronakos et al. 1983). Acromegalic patients with higher plasma IGF-I and insulin levels presented lower IGF-I specific binding and affinity than normal adults. Growth hormone (GH)-deficient children showed higher IGF-I binding without significant affinity alterations than normal prepubertal children (el-Andere et al. 1995).

2.5 Leptin

The specific binding of leptin on erythrocytes is established by Scatchard analysis (Tsuda, 2006). NHE1 (Na^+/H^+-exchanger) activity increases in the presence of leptin but significantly less in the obese than in the control group. Since NHE1 activity is associated with insulin resistance and hypertension, the activation of this antiport by leptin may represent a link between adipose tissue hypertrophy and cardiovascular complication of obesity (Konstantinou-Tegou et al. 2001). It is possible to assume that leptin, similarly to insulin, might be able to activate NHE1 through MAPK activation (Sartori et al. 1999; Bianchini et al. 1991).

2.6 Adrenaline, Noradrenaline and DOPA

The β-2-adrenergic receptor coupled to the G-protein binds cateholamines and activates adenylate cyclase in human erythrocytes (Horga et al. 2000). The functional beta-receptor response depends to a large extent on Ca^{2+} concentrations (Horga et al. 2000). According to Muravyov et al. (2010) a crosstalk between adenylyl cyclase signaling pathway and Ca^{2+} regulatory mechanism exists. The potent beta-adrenergic agonist, isoproterenol (2 microM), epinephrine (10 microM) and norepinephrine (10 μM) stimulated the cAMP-dependent protein kinase in erythrocyte membranes, 38 +/- 7%, 31 +/- 6%, and 30 +/- 6%, respectively (Tsukamoto and Sonenberg, 1979). Micromolar concentration of noradrenaline (1 μM) increases the ^{32}P intake in band 2 with 70%, and with 40% in band 3 (Nelson et al. 1979).

Adrenaline and noradrenaline are both found to stimulate the erythrocyte Na^+/H^+-antiport (Perry et al. 1991; Paajaste and Nikinmar, 1991). Adrenaline stimulates Na^+/H^+-antiport through activation of NHE1 transport system with participation of PKC, since the effect has been reported to potentiate in the presence of phorbol ester (PKC activator), and being inhibited from calphostin (PKC inhibitor), respectively (Bourikas et al. 2003).

A specific erythrocyte DOPA transport protein was found who is also capable to transports choline. Its functions are regulated by insulin (Azoui et al. 1996).

Several catecholamines (phenylephrine, dobutamine and dopamine) inhibit the Cl-removal-activated Ca^{2+} entry into erythrocytes, thus preventing increase of cytosolic Ca^{2+} activity, subsequent cell shrinkage and activation of erythrocyte scramblase. The catecholamines thus counteract erythrocyte phosphatidylserine exposure and subsequent clearance of erythrocytes from circulating blood (Lang et al. 2005).

Exposure of RBCs to adrenaline resulted in a concentration-dependent increase in RBC filterability and authors supposed that adrenergic agonists may improve passage of erythrocytes through microvasculature (Muravyov et al. 2010). Rasmussen et al (1975) reported that the same low doses of adrenaline and isoproterenol induce a decrease of erythrocyte deformability.

2.7 Adenosine

Adenosine binds to the erythrocyte adenosine type1 receptors (A1AR) and adenosine type 2 receptors (A2AR), (Lu et al. 2004; Zhang et al. 2011). A1ARs are functionally coupled with pertussis toxin-sensitive G proteins and inhibit the activity of adenylate cyclase. A1ARs bind to erythrocyte membrane cytoskeletal protein 4.1G, which can inhibit A1-receptor action (Lu et al. 2004). A1ARs activation can also trigger the release intercellular Ca^{2+} (Lu et al. 2004). Increased adenosine levels promoted sickling, hemolysis and damage to multiple tissues in SCD transgenic mice and promoted sickling of human erythrocytes. (Zhang et al. 2011).

2.8 Prostaglandins (PG)

Binding of PGE2 to receptors coupled to the G-protein activates phospholipase C which in turn catalizes phospholipide turnover (Minetti et al. 1997) and/or stimulates a Ca^{2+}-dependent K^+ chanell in human erythrocytes and alters cell volume and filterability (Li et al. 1996). The PGE1 receptor coupled to the G-protein activates adenylate cyclase in human erythrocytes and increases erythrocyte deformality (Dutta-Roy et al. 1991). Prostacyclin (PGI) binding to human erythrocyte receptors stimulate cAMP synthesis and ATP release (Sprague et al. 2008).

2.9 Thyroid and steroid hormones

Lipophilic hormones take part in signal transduction also with "non-genomic" effects using signals starting from the plasma membrane (Falkenstein et al. 2000). Their role in erythrocyte signal transduction pathways and their activities have not been fully explored. Similar results have been reported for thyroid hormones (Angel et al.1989; Botta and Farias, 1985), and for estrogens (Gonçalves et al. 2001). Effect of thyroid hormones is probably related to calmodulin-dependent activation of erythrocyte membrane ATPase (Lawrence et al. 1993), that is inhibited in the presence of retinoic acid (Smith et al. 1989).

In vitro beta-estradiol 10^{-5} M decreased erythrocyte aggregation in blood samples of postmenopausal women undergoing hormone therapy, which could prevent high blood viscosity and, consequently, cardiovascular events (Gonçalves et al. 2001).

Results from X-ray diffraction studies revealed that cortisol and estradiol bind into the erythrocyte membrane bilayer, and exert opposite effect over Na^+/K^+-ATPase activity: cortisol diminishes its activity by 24%, but estradiol increases it by 18% (Golden et al. 1999). Other published *in vitro* studies showed that aldosterone stimulates Na^+/K^+-ATPase activity in human erythrocyte membranes (Hamlyn and Duffy, 1978). One possible explanation is that the incorporation leads to conformational changes and reorganization in the active center of the enzyme molecule of Na^+/K^+-ATPase. In addition to the delayed genomic steroid actions, increasing evidence for rapid, nongenomic steroid effects has been demonstrated for virtually all groups of steroids, and transmission by so far hypothetical specific membrane receptors is very likely. Nongenomic effects on cellular function involve conventional second messenger cascades (Falkenstein et al., 2000). Plasma selenium as well as plasma and erythrocyte gluthatione peroxidase activity increase with estrogen during the menstrual cycle. The mechanism is still unknown (Ha and Smith, 2003).

2.10 Cytokines

Biochemical evidence is provided for the presence of endothelin (ET) receptor subtype B in syckle and normal red cells. It was found that ET-1, PAF (Platelet Activating Factor), RANTES and IL-10 induce a significant increase in red cell density. These data suggest that activation of the Gardos channel is functionally coupled to receptors as C-X-C(PAF), C-C (RANTES) and ET receptors type B and the cell volume regulation or erythrocyte hydratation state might be altered by activation of the Gardos channel by cytokine in vivo (Rivera et al. 2002).

Human red cells bind specifically IL-8 (a neutrophil activating chemokine) with IL-8RA and IL-8RB receptors. Red cell absorbtion of IL-8 may function to limit stimulation of leucocytes by IL-8 released into blood (Darbonne et al. 1991). IL-8 released after acute myocardial infarction is mainly bound to erythrocytes (de Winter et al. 1997).

ICAM-4 belongs to the intercellular adhesion molecules and is an erythrocyte membrane component. ICAM-4 interacts specifically with platelet alphaIIbeta 3 integrin. RBCs are considered passively entrapped in fibrin polymers during trombosis through ICAM-4 (Hermand et al. 2003).

2.11 Thrombin

It is assumed that thrombin induces a signal that stimulates the formation of cAMP and PGE1 via activation of Gs-dependent adenylate cyclase and Ca^{2+}-independent PKC in eryhtrocyte progenitors. Signaling pathway is inhibited by amiloride and by PKC inhibitors such as GF-109203X, Go 6976 and staurosporine (Haslauer et al. 1998). However, whether or not these signals are also valid for mature erythrocytes has not been studied yet.

2.12 Lactoferrin

Lactoferrin (Lf) is a metal-binding glycoprotein with antioxidative (Cohen et al., 1992), anti-inflammatory, immunomodulatory (Legrand et al., 2004), anticancerogenic (Thotathil and Jameson, 2007) anti-bacterial (Weinberg, 2007), antiviral (Mistry et al., 2007), antiatherogenic (Kajikawa et al. 1994), and antithrombotic properties (Levy-Toledano et al., 1995).

Our previous studies showed that Lf binding with erythrocyte membrane receptors (Taleva et al., 1999) results in stimulation of glycolysis, antioxidative protection (Maneva et al., 2003) and activation of Na^+/K^+-ATPase activity (Maneva et al., 2007). Lf–receptor interaction might intervene in short–term effects of regulation, involving processes of changes in association, phosphorylation and oxidation of the membrane proteins.

Lf (10-50 nM) decreased the ATP content from 9 to 57% depending on the concentration used. There is a negative correlation found between the concentration of added Lf and formed ATP: y = 2.550 - 0.0076.x, r = - 0.993, p <0.001, n = 5 (Figure 1). The reason for the decrease of ATP in the presence of Lf could be the processes of phosphorylation and/or ion transport activated by Lf. Lf is an activator of protein phosphorylation (Maekawa et al. 2002; Curran et al. 2006) and an activator of ion transport (Sun et al. 1991). Such a decrease in ATP content was found by other authors (Assouline-Cohen and Beitner, 1999; Boadu & Sager, 2000) due to stimulation of vital processes of erythrocytes.

Lf could stimulate glycolysis by interfering with phosphorylation processes: 1) There is evidence that insulin activates the Na^+/H^+ - antiport in erythrocytes by PI3K-dependent signaling pathway (Sauvage et al. 2000) leading to activation of glycolysis (Madshus, 1988). Like insulin, Lf also activates Na^+/H^+ - antiport (Sun et al. 1991) and stimulates the lactate formation (Maneva et al. 2003). Therefore, it could be suggested that Lf uses the same PI3K - dependent mechanism for glycolysis activation (Boivin, 1988); 2) Phosphorylation of tyrosin residues in band 3 (Boivin, 1988) leads to dissociation of the complex with glycolytic enzymes, leading to their activation (Low et al. 1993). Lf could also stimulate the glycolytic enzymes by activating Src-kinase-dependent phosphorylation of band 3. There is data showing that Lf is an activator of tyrosine phosphorylation by Src-kinases (Takayama & Mizumachi, 2001).

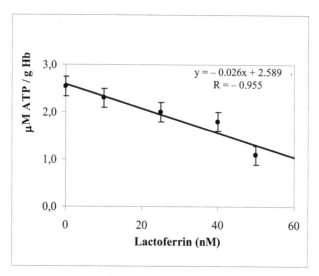

Fig. 1. Correlation and regression analysis of the relation between the concentration of Lf and content in erythrocyte ATP (unpublished data).

Sigma-Aldrich test kit has been used in this experiment. The method is based on using a reaction from oxidative phosphorylation (OP) in glycolysis where ATP is produced. OP is two–stage process involving the enzymes glyceraldehyde-3-phosphate dehydrogenase (GAPDH), and 3-phosphoglycerate kinase. The output products of OP are 3-phosphoglycerate, $NADH^+$ and ATP. The test uses GAPDH in reverse direction: a conversion of 1,3 bisphosphoglycerate into glyceraldehyde-3-phosphate, where the oxidation of $NADH^+$ to NAD^+ is directly proportional to the amounts of the ATP produced. The results are presented in μmols ATP/g.Hb.

Several possible mechanisms could be discussed in regard to how Lf binding to the receptors on the erythrocyte membrane might lead to Na^+/K^+-ATPase activation (Maneva al. 2007) via changes in levels of phosphorylation of membrane and cytosolic proteins: 1) Changes in the levels of phosphorylation of tyrosine residues in erythrocyte membrane modulates not only the activities of band 3 (Brunati et al. 2000), but also the activity of α-subunit of Na^+/K^+-ATPase (Done et al. 2002). α-subunit of Na^+/K^+-ATPase is phosphorylated by tyrosine kinases, as the target of phosphorylation is known to be the 537 tyrosne residue in the molecule. This phosphorylation is proved to be a prerequisite for the participation of Na^+/K-ATPase in signal transduction in experiments with renal cells (Done et al. 2002). Lf stimulates tyrosine phosphorylation of proteins (Takayama & Mizumachi, 2001). 2) It is known that Lf activates casein kinase 2 (CK2), and is phosphorylated by the latter (Maekawa et al. 2002). CK2 phosphorylates and activates a wide spectrum of erythrocyte membrane proteins associated in multiprotein complexes with Na^+/K^+-ATPase (Wei and Tao, 1993); 3) Non-receptor tyrosine kinase Src takes part in signal pathway that regulates Na-pump activity in other cell types (Haas et al. 2002). Lf is also able to activate Src (Takayama and Mizumashi, 2001). 4) Stimulating effect of Lf could be indirect as Lf activates Na^+/H^+ antiport (Sun et al. 1991). By this mechanism erythrocytes are loaded with Na^+ thus activating Na^+/K^+-ATPase which exports Na^+ out of the erythrocyte cell.

2.13 Polyamines

Membrane receptors for polyamines were proved to exist on erythrocyte membranes (Moulinoux et al. 1984). After binding with their receptors, polyamines enhance the protein–protein interactions among the cytoskeletal proteins and membrane lipid bilayer (Bratton, 1994; Farmer et al. 1985). Polyamine also inhibits transbilayer movement of plasma membrane phospholipids in the erythrocyte ghost (Bratton, 1994).

Polyamines are CK2 protein kinase activators (Leroy et al.1997). It was found that positively charged betaine and polyamines inhibit the activity of membrane Na^+/K^+-ATPase in erythrocytes. It has been demonstrated that the effect of polyamines is based on the competitive inhibition mechanism (Kanbak et al. 2001). A similar mechanism is most likely responsible for the poor ion transport in cancer patients whose tumor cells secrete polyamines (Villano et al. 2001). Raised erythrocyte polyamine levels are estimated in patients with diabetes type 1 and in non-insulin dependent diabetes mellitus with great vessel disease and albuminuria (Seghieri et al. 1997).

2.14 Caffeine

Caffeine binds to erythrocytes membrane proteins (Sato et al. 1990) and could induce changes in the erythrocyte protein complex's formation, thus modulating the activity of enzymes involved in signal transduction. Caffeine could interfere with the phosphorylation processes in erythrocytes as an inhibitor of erythrocyte CK2 (Lecomte et al. 1980), and PI3K (Buckley, 1977) as it was reported in other cell types. Methylxanthines are capable for inhibition of cyclin-dependent protein kinases found in both cytosol and erythrocyte membranes (Biovin, 1988).

Processes of phosphorylation-dephosphorylation of erythrocyte membrane and plasma proteins provide different levels of interaction and participate in maintaining the integrity of the erythrocyte membrane. They also exert control over important metabolic processes in erythrocytes. The literature data shows that signals in erythrocytes lead to changes in phosphorylation and association of integral membrane proteins and other intracellular proteins. Cytosolic protein kinases and protein phosphatases were found in erythrocyte membrane.

3. Protein kinases

3.1 Protein serine/threonine kinases

3.1.1 cAMP dependent protein kinases

Type 1 isoform of cAMP dependent protein kinase is localized in the membrane, while type 2 is located in the cytosol (Dreytuss et al. 1978). cAMP-protein kinases-dependent phosphorylation of isoforms of membrane Ca^{2+}-ATPase was found. Isoform 1 is shown to be better substrate in comparison to isoforms 2 and 4. Isoform 1 is susceptible to degradation by calpain (thiol-dependent protease) (Guerini et al. 2003).

In all cells, increasing in cAMP are regulated by the activity of phosphodiesterases (PDEs) (Sheppard and Tsien, 1975). In erythrocytes, activation of either beta adrenergic receptors (beta (2) AR) or the prostacyclin receptor (IPR) results in increases in cAMP and ATP release

(Sprague et al. 2001). Receptor-mediated increases in cAMP are tightly regulated by distinct PDEs associated with each signaling pathway, as shown by the finding that selective inhibitors of the PDEs localized to each pathway potentiate both increases in cAMP and ATP release (Adderley et al. 2010).

Adenylyl cyclase and cAMP are components of a signal-transduction pathway relating red blood cells (RBC) deformation to ATP release from human and rabbit RBCs (Sprague et al. 2008). Exposure of RBC to catecholamines (epinephrine, phenylephrine, an agonist of α1-adrenergic receptors, clonidine, an agonist of α2-adrenergic receptors and isoproterenol, an agonist of β-adrenergic receptors) led to change in the RBC microrheological properties. Forskolin (10 μM), an adenylate cyclase stimulator, increases the RBC deformability (RBCD). A somewhat more significant deformability rise appeares after RBC incubation with dibutyril-AMP. Red blood cell aggregation (RBCA) is significantly decreased under these conditions. All drugs having PDE inhibitory activity increase red cell deformability (Muravyov et al. 2009).

Ca^{2+} entry increase is accompanied by red cell aggregation rise, while adenylyl cyclase-cAMP system stimulation led to red cell deformability increase and its aggregation lowered. The crosstalk between two intracellular signaling systems is probably connected with phosphodiesterase activity. It was found that all four PDE inhibitors: IBMX, vinpocetine, rolipram, pentoxifylline decreased red cell aggregation significantly and, quite the contrary, they increased red cell deformability (Muravyov et al. 2010)

Erythrocytes are oxygen sensors and modulators of vascular tone (Ellsworth et al. 2009). It has become evident that erythrocytes participate in the regulation of vascular caliber in the microcirculation via release of the potent vasodilator, adenosine triphosphate (ATP). The regulated release of ATP from erythrocytes occurs via a defined signaling pathway and requires increases in cyclic 3',5'- adenosine monophosphate (cAMP) (Adderley et al. 2010) . Heterotrimeric G protein Gi is involved in a signal transduction pathway for ATP release from erythrocytes (Olearczyk et al. 2010). Insulin inhibits human erythrocyte cAMP accumulation and ATP release. The targets of insulin action are phosphodiesterase 3 and phosphoinositide 3-kinase (Hanson et al. 2010). TSH signalling pathway is cAMP-dependent and probably it is potentially active in local circulatory control (Balzan et al. 2009).

3.1.2 cGMP-pependent protein kinases

Petrov et al. (1994) supposed that human erythrocytes possess membrane and soluble guanylate-cyclase activity stimulated by atrial natriuretic peptide III (ANP-III) and that activation of Na$^+$/H$^+$ exchange by this peptide is mediated by cGMP (Petrov and Lijnen, 1996). There are no reported data considering participation of protein kinases and phosphatases in the signal pathway.

3.1.3 Casein kinase 1 and 2

Membrane proteins of human erythrocytes can be phosphorylated not only by membrane casein kinase but also by cytosolic casein kinases, resembling casein kinase 1 and 2 (CK1 and CK2), respectively. CK1 and CK2 phosphorylate serine and threonine residues in target proteins (Boivin, 1988). CK2 inactivation mediated by phosphatidylinositol-4, 5-bisphosphate, a substrate for phospholipase C which catalyzes formation of lipid mediators IP3 and DAG

(Chauhan et al. 1993) has been found. It was established that CK2 is inactivated by insulin (Sommerson et al. 1987).

An increased phosphorylation of the membrane proteins, promoted by the okadaic acid (strong inhibitor of P-Ser/Thr-protein phosphatase(s)), is accompanied by a release of CK from the membrane into the cytosol. Such an intracellular translocation might provide a feedback mechanism for the regulation of the CK catalyzed phosphorylation of membrane proteins in human erythrocytes (Bordin et al. 1994).

The membrane mechanical stability of erythrocytes is exclusively regulated by phosphorylation of β-spectrin by membrane bound CK1. Increased phosphorylation of β-spectrin decreased membrane mechanical stability while decreased phosphorylation increased membrane mechanical stability (Manno et al. 1995).

CK2 isolated from erythrocyte membrane and cytosolic fractions exhibited the same subunit composition ($αα1$) and the ability to utilize ATP and GTP as phosphate donors. Both kinases were found to catalize the phosphorylation of several erythrocyte membrane cytoskeletal proteins (spectrin, ankyrin, adductin, protein 4.1 and protein 4.9). Unlike CK1, CK2 did not phosphorylate band 3. Spermine, spermidine, and putrescine stimulated to varying degrees the activities of erythrocyte CK2, whereas heparin inhibited the kinase activities (Wei and Tao, 1993).

The phosphorylation sites of calmodulin are important for its ability to activate the human erythrocyte Ca^{2+}-ATPase. Phosphorylation of mammalian calmodulin on serine/threonine residues by casein kinase 2 decreased its affinity for Ca^{2+}-ATPase by two fold. In contrast, tyrosine phosphorylation of calmodulin by the insulin-receptor kinase did not significantly alter calmodulin-stimulated Ca^{2+}-ATPase activity (Sacks et al. 1996).

The COP9 signalosome (CSN) is a multimeric complex that is conserved from yeast to man (Bech-Otschir et al., 2002). Immunoprecipitation and far-western blots reveal that CK2 and PKD are associated with CSN. The COP9 signalosome (CSN) purified from human erythrocytes possesses kinase activity that phosphorylates proteins such as c-Jun and p53 with consequence for their ubiquitin (Ub)-dependent degradation. (Uhle et al., 2003)

3.1.4 MAPK (Mitogen Activated Protein Kinases)

It is well known that signaling pathways involving MAPKs are associated with control of cell growth and proliferation, but erythrocytes are mature highly differentiated cells. There are only a few known participants in the MARK-signaling pathways so far. Immunoblot with antiMAPK antibody revealed the two erythrocyte forms of MAPK–p44 (ERK1) and p42 (ERK2). Insulin and okadaic acid (inhibitor of serine/threonine protein phosphatases) stimulate MAPK activity. Insulin enhances the erythrocyte Na^+/H^+-exchanger through MAPK activation (Sartori et al. 1999; Bianchini et al. 1991). Membranes of human erythrocytes contain several proteins of the Ras superfamily (Ikeda et al. 1988; Damonte et al. 1990). One of them, RhoA, was detected in both cytosol and membrane fraction of the erythrocytes. Cytosolic Rho bound specifcally to the cytoplasmic surface of the erythrocyte membrane. The translocation of Rho to the membrane was absolutely GTP -dependent at low Mg^{2+} concentration (Boukharov et al. 1998).

3.2 Protein tyrosine kinases

Erythrocyte membrane associated tyrosine kinase activity has been established (Zylinska et al. 2002), together with surprisingly high levels of phosphorylated tyrosine in erythrocytes (Phan-Dinh-Tuy et al. 1983).

Two tyrosine kinases that are able to phosphorylate band 3 have been found. Involvement of the Src-kinases family in regulation of erythrocyte membrane transport has been suggested. That family represents an important class of non-receptor protein kinases participating in the regulation of cell communications, proliferation, differentiation, migration and survival (Minetti et al. 2004). Subsequent two-stage phosphorylation of band-3 from Syk- and Lyn-tyrosin kinase was found. (Brunati et al. 2000).

3.3 Ca^{2+} dependent phosphorylation

Since mature RBCs lack intracellular calcium stores, elevation in intracellular calcium must stem from calcium influx. At low intracellular Ca^{2+}, efflux of potassium and water predominates, leading to changes in erythrocyte rheology (Andrews et al. 2002). At higher Ca^{2+} content, activation of kinases and phosphatases was observed (Minetti et al.1996; Cohen and Gascard, 1992). Ca^{2+} ions are involved in regulation of phospholipase C, the enzyme generating inositol-1,4,5-triphosphate (InsP3), phosphatidylinositol-3-kinase (PI3K), the enzyme metabolizing InsP3 to InsP4 and in the regulation of protein kinase C (Carafoli, 1994; Carafoli, 2002).

3.3.1 Protein kinase C (PKC)

There are four isoforms of PKC established in the erythrocyte: α, ζ, ι and μ (PKD) (Govekar and Zingle, 2001). PKC translocates from cytosol to membrane – a process shown to be initiated by phorbol esters and diacylglycerols (Cohen and Foley, 1986). PKD and CK2 are associated with COP9 signalosome (CSN) (Uhle et al., 2003).

PKC phosphorylates serine residues in band 3, band 4.1 and band 4.8 (Govekar and Zingle, 2001). It has been reported that after stimulation with phorbol ester (PMA), PKC-α translocates towards the erythrocyte membrane (Govekar and Zingde, 2001), where it phosphorylates serine residues in band 2, and 4.1 and band 4.9 which in turn leads to thorough rearrangement of the cytoskeleton membrane network (Giraud et al. 1988; Ceolotto et al. 1998). PKC phosphorylates membrane Na$^+$/K$^+$-ATPase (Wright et al. 1993) and the carboxyl terminus of the plasma membrane Ca^{2+}-ATPase in human erythrocytes (Wang et al. 1991; Wright et al.1993; Smallwood et al.1988) and the effect of phosphorylation on the activity of the both enzymes depends on the isoenzyme form of protein kinase C.

The interactions between membrane, peripheral and cytoskeleton proteins are responsible for the maintenance of erythrocyte deformability and some of these interactions are modulated by PKC activity. Correlation was established between cytoskeleton proteins, PKC activity, band 3 phosphorylation degrees and erythrocyte deformability (de Oliveira et al. 2008). PKC participates in the processes of phosphorylation of band 3 that regulates its activity as a transport protein (Ceolotto et al. 1998). Protein kinase C mediates erythrocyte "programmed cell death" following glucose depletion (Klarl et al. 2006).

In erythrocytes, lead acetate stimulates the phosphorylation of membrane cytoskeletal proteins by a mechanism dependent on protein kinase C. Since levels of calcium or diacylglycerols did not increase, it appears that lead may activate the enzyme by a direct interaction (Belloni-Olivi et al. 1996).

Calpain is a Ca^{2+}-dependent thiol protease that translocates towards the erythrocytes membrane and activates with increase of intracellular calcium concentration (Glasser et al. 1994). Calmodulin binds proteins as calpain substrates (Wang et al. 1989). It is known that calpain causes transition from Ca^{2+}-dependent PKC form to Ca^{2+}-independent one, followed by decrease in its activity (Saido et al. 1994).

3.3.2 Calmodulin (CaM)

The increase in cellular Ca^{2+} results in reversible formation of Ca^{2+}/CaM complex that binds to a number of enzymes modulating their activity (Carafoli, 2002).

Calmodulin is a ubiquitous protein whose activity is regulated through phosphorylation and specific Ca^{2+}-binding (Benaim and Villalobo, 2002). Ca^{2+}-ATPase interacts with the carboxy-terminal half of calmodulin, which is the region that contains the majority of the phosphorylation sites in calmodulin (Bzdega and Kosk-Kosicka, 1992). CaM is phosphorylated by serine/threonine kinases such as CK2 and PKA (Benaim and Villalobo, 2002). Phosphorylation of a tyrosine residue in CaM (Tyr99) increases the affinity of its binding to the target proteins thus increasing their activity. A similar stimulating effect of CaM has been reported for Ca^{2+}-ATPase (Kosk-Kosicka et al. 1990), myosin light chain kinase (MLCK), Ca-CaM dependent kinase (CaM kinase II) and Ca-CaM dependent protein phosphatase 2B (calcineurin) (Corti et al. 1999). Elevated intracellular Ca^{2+}, in association with the Ca^{2+}-binding protein, calmodulin, stimulates erythrocytes phosphofructokinase (PFK) activity. This activation involves the detachment of the enzyme from erythrocyte membranes, which has been described as an important mechanism of glycolysis regulation on these cells (Zancan and Sola-Penna, 2005).

CaM is an inhibitor of PI3P-kinase (Villalonga et al. 2002), Na^+/H^+ antiport (Yingst et al. 1992) and of erythrocyte Na^+/K^+-ATPase (Yingst at al. 1992). The latter effect could be exerted via inducing of signal pathway involving CaM-dependent kinase that activates membrane phospholipase A2(PLA2), which in turn inhibits Na-pump (Okafor et al. 1997).

4. Protein phosphatases

Erythrocyte phosphatases that dephosphorylate phosphoserine, phosphothreonine and phosphotyrosine residues in different proteins were found. Cytoplasmic phosphatases, low-molecular-weight acid phosphatases, and neutral membrane-associated tyrosine phosphatases have been reported (Clari et al. 1987; Graham et al. 1974).

Protein phosphatases originated from two ancestor genes, one serving as the prototype for the phosphotyrosine phosphatases family (the PTP family), and the other for phosphoserine/phosphothreonine phosphatases (the PPP and PPM family). The PPP family groups the PP1, PP2, PP2B, PP4, PP5, PP6 and PP7 enzymes, whereas PP2C and bacterial enzymes like SpoII or PrpC belong to the PPM family (Batford, 1996; Kiener et al. 1987).

Band 3 is a target for tyrosine phosphatases SHP-1 and SHP-2 (Bordin et al. 2002) and PTP1B activities (Zipser and Kosower, 1996). Ca^{2+} promotes erythrocyte band 3 phosphorylation via dissociation of phosphotyrosine phosphatase from band 3 (Ziper et al. 2002)

SHP-2 is an essential soluble protein tyrosine phosphatase (PTP) containing two spectrin homologous domains - SH2. SHR-2 domain participates in multiple signaling pathways of growth factors (GF) and cytokines, and plays an important role in the release of signals from the cell surface to the nucleus. SH2 is the anchor (docking) place where the protein tyrosine kinases, e.g. Src family are attracted to and attached to, thus facilitating phosphorylation-dephosphorylation cycle. SHR-2 is located in the erythrocytes cytosol and it is translocated to the erythrocyte membrane when there is increased tyrosine phosphorylation of the transmembrane protein band 3, induced by PTP inhibitors such as pervanadate, and N-ethylmaleimide. Band 3 is both anchoring protein and substrate for the SHP-2 (Bordin et al. 2002). SHP-1 and SHP-2 ensure dephosphorylation of band 3 in different conditions: SHP-2 through interaction of its SH2 domain(s) to p-tyr protein is regulated by the band 3 tyr-phosphorylation level; SHP-1 dephosphorylates tyr-8, tyr-21 and tyr-904 and may be involved by simple membrane rearrangement (Bragadin et al. 2007).

Protein phosphatase-1 α (PP1alpha) is a selective substrate of peroxynitrite activated src family kinase frg and tyrosine phosphorylation of PP1alpha correspond to inhibition of its enzyme activity. The final effect of peroxynitrite is the ampliphication of tyrosine dependent signaling, a finding of general interest in nitrite oxide related pathophysiology (Mallozzi et al. 2005)

PTP1B is localized at the erythrocyte membrane associated with band 3. It is activated by Mg^{2+} and inhibited by Mn^{2+} and vanadate ions (VO_3^-), (Zipser and Kosower, 1996). PTP1B, unlike the other enzymes examined, was quantitatively conserved during erythrocyte aging (Minetti et al. 2004) and erythrocytes may undergo in vivo activation of the Ca^{2+}-dependent calpain system that proteolytically regulates PTP1B activity (Ciana et al. 2004).

The activities of Na^+/H^+ exchanger and Na-K-2Cl cotransporter in rat erythrocytes are regulated by protein phosphatases PP1 and PP2 and stimulated when protein dephosphorylation is inhibited (Ivanova et al. 2006).

5. Control over glycolytic enzymes

Glycolytic erythrocyte enzymes that form a complex with band 3 are under regulation by the processes of phosphorylation and oxidation (Campanella et al. 2005), and pyruvate kinase (PK) - by phosphorylation/dephosphorylation (Kiner and Westhead, 1980). Binding of phosphofructokinase (PPK), glyceraldehyde-3-phosphate dehydrogenase (GAPDH) and aldolase to band 3 lead to their reversible inhibition (Low et al. 1993), which is neutralized by tyrosine kinase phosphorylation of band 3 resulting in enzymes activation. (Harrison et al. 1991). p72[syk] and p56/53[lyn] tyrosine kinases are also involved in band 3 phosphorylation (Hubert et al. 2000), and phosphorylation/dephosphorylation cycle is maintained by protein phosphatases PTP1B and SHP-2 (Bordin et al. 2002; Zipser and Kosower, 1996; Minetti et al. 2004).

PK turns to an inactive form after phosphorylation by cAMP-dependent protein kinase, though it is unclear why since erythrocytes do not undergo gluconeogenesis. PK is activated by erythrocyte protein phosphatases (Kiener and Weathead, 1980).

CaM induces dimerisation of phosphofructokinase (PFK) and physiological levels of concentration of intracellular Ca^{2+} stimulates its catalytic activity (Marinho-Carvalho et al., 2006). However, results obtained from other authors showed that Ca^{2+}/calmodulin protein kinase (CaM-kinase) phosphorylation inhibits phosphofructokinase (PFK) in sheep heart (Mahrenholz et al. 1991)

Ferricyanide activates glycolysis in erythrocytes in two ways. Ferricyanide is the only known non-physiological extracellular agent who induces signal transduction leading to activation of cytoplasmic protein tyrosin kinase that phosphorylates enzyme-binding site in band 3, which in turn leads to release and activation of glycolytic enzymes GAPDH (glyceraldehyde-3-phosphate-dehydrogenase), PPK (phosphofructokinase) and aldolase (Low at al. 1990). It is proved that ferricyanide is a stimulator of glycolysis as end acceptor of electrons produced in glyceraldehyde-3-phosphate dehydrogenase reaction. Thus, ferricyanide is capable of recovering the level of oxidated NAD^+ which is important for maintaining the rate of glycolysis (Orringer and Roer, 1979). There are experiment-based evidence for the existence of erythrocyte e- transport transmembrane chain related to transferrin receptor (Orringer and Roer, 1979; Goldenberg et al. 1990), that is known to reduce $K_3Fe(CN)_6$ and removes protons from inside the cell through the cell membrane (Low et al. 1987). Our own experiments showed that $K_3Fe(CN)_6$ inhibits $^{59}FeLF$ binding to erythrocytes (Maneva et al. 2003) as most probably Lf and $K_3Fe(CN)_6$ compete for one and the same electron formed in oxidative phosphorylation in glycolysis; Lf is the physisological activator of the signal pathway used by ferricyanide that leads to tyrosine phosphorylation of band 3 and consequent glycolysis activation.

6. Modulation of components of erythrocyte membrane skeleton

In terms of proteins, the RBC membrane is a complex network of transporters, cytoskeletal molecules, and membrane-bound enzymes. The ability of transmembrane receptor proteins to change their cytoskeleton associations in response to ligand binding looks like a key mechanism for cell signaling through erythrocyte membrane. The erythrocyte membrane skeleton has 3 main components: spectrin molecules forming tetramers $\alpha_2\beta_2$, short actin oligomers containing 12-15 monomers, and band 4.1. A variety of membrane-associated enzymes, including several kinases (protein kinase A, protein kinase C, cdc kinase, casein kinase 1) are thought to regulate interactions within the network through induction of phosphorylation, methylation, myristoylation, palmytoylation, or farnesylation (Pasini et al. 2010).

Spectrin binding to actin is initiated from aductin and is significantly inhibited in the presence of Ca^{2+} and CaM. CaM is able to form weak bonds with spectrine and in the presence of Ca^{2+} could influence binding of protein 4.1 to actin, and probably spectrin phosphorylation as well (Manno et al. 1995).

Band 4.1 is mainly involved in membrane skeleton organization due to its ability to initiates spectrin/actin association. Band 4.1 binds to the membrane in at least two sites: one high affinity site localized on glycophorins, and another low-affinity site associated with band 3. Phosphorylation of Band 4.1 by PKC was found to modulate its binding to band 3. Most probably, this increases the flexibility of the membrane in response to mechanical stress as was reported for some other models where the elevated levels of membrane phosphorylation made cells more deformable and more resistant to mechanical stress (Danilov and Cohen, 1989). Association of band 4.1 with glycophorin is important for the maintenance of cell shape and is regulatated by polyphosphoinositide co-factor. Indeed, when a change in erythrocyte shape was found, a change in the phosphoinositide content was also found (Danilov et al. 1990). Band 4.1 phosphorylation regulates the assembling of spectrin, actin, and band 4.1 as phosphorylation diminish affinity of band 4.1 to spectrin. Band 4.1 and band 4.9 could be phosphorylated in different sites by PKC and by cAMP-dependent protein kinases (Horne et al.1990).

Cytoskeletal protein band 4.1G binds to the third intercellular loop of the A1 adenosine receptor and inhibits receptor action. (Lu et al., 2004)

7. Modulation of band 3

The two most important integral proteins in human erythrocyte memranes are glycophorin A and band 3. Band 3 is multifunctional protein whose N- and C-ends are localized on the cytoplasmic surface of the membrane and crosses the cell membrane 14 times. The C-terminal end is 52 kD domain (residues from 360 to 919) of band 3 and is responsible for anion transport through the membrane, while 40 kD from N-terminal end of the cytosolic domain plays a critical role in the binding of bilayer to spectrin-based skeletal net. Cytoplasmic domain of band 3 serves as a center of membrane organization, interacting with proteins, such as ankyrin, protein 4.1 and 4.4, hemoglobin, some glycolytic enzymes, tyrosine phosphatases, tyrosine kinase p72 (syk), and Na^+/K^+ - ATPase. There are about 1 million copies of band 3 per cell and they are presented as dimeric and tetrameric forms. Approximately 40-60% of band 3 is associated with spectrin-stabilizing cytoskeleton (van den Akker et al. 2010).

Band 3 (AE1) is a member of the family of anion-transportng proteins that maintain Cl-/HCO3 - exchange. It refers mainly to erythtocytes and plays a role in CO_2 transport among the tissues and lungs (Saldanha et al. 2007). Erythrocyte PKC phosphorylates serine residues in band 3 (Govekar and Zingle, 2001).

Band 3 modifications normally occur during physiological red blood cell (RBC) senescence and some pathological condition in humans (Santos-Silva et al. 1998). Band 3-tyrosine phosphorylation might be induce in "stress" conditions: "oxidative stress", inhibition of the phosphoprotein phosphatase activity by vanadate or by *thiol group blockers* (diamide, N-ethylmaleimide, etc.), protein tyrosine kinase (PTK) activation from hypertonic NaCl solution, or by intracellular increase of Ca^{2+} levels (Hecht & Zick, 1992). Erythrocyte thiol status is an intrinsic regulator of phospho tyrosine residues levels in band 3 by oxidation/reduction of band 3–associated phosphotyrosine protein phosphatases (PTP) (Zipser et al. 1997). It is considered that the increase in phosphorylation induced by Ca^{2+} ions, and including significant inhibition of PTP activity by dissociation of PTPase from band 3 may also play an important part in signal transduction pathways in some pathophysiological conditions accompanied with increased levels of intracellular Ca^{2+} (Minetti and Low, 1997).

Abnormal band 3 tyrosine phosphorylation has been observed in a number of red cell disorders (Terra et al. 1998). Hyper-phosphorylated band 3 showed a manifest tendency to cluster, indicating a change in its interactions with the cytoskeletal network. Irreversible band 3 tyrosine phosphorylation leads to membrane vesiculation in G6PD deficient red cells. Syk kinase inhibition largely prevents red cell membrane lysis and vesiculation, strongly suggesting a functional role of band 3 tyrosine phosphorylation in the red cell membrane destabilization (Pantaleo et al. 2011; Bordin et al. 2005).

Based on similarity in oligosaccharide components of band 3 and Lf (Ando et al. 1996), a competition for binding sites could be assumed (Beppu et al. 2000; Eda et al. 1996) together with restructuring of already established complexes. Our experiments revealed that Lf activates erythrocyte glycolysis (Maneva et al. 2003). This fact might be explained with some conformational changes leading to the release of glycolytic enzymes from their complex with band 3. Studies of the same investigators show that Lf, and erythrocyte band 3 occupy the same places on monocyte leukemic cell line THP-1 (Eda et al. 1996). Lf could also activate tyrosine phosphorylation of band 3 since Lf is already proved to be able to increase tyrosine phosphorylation of membrane proteins (Tanata et al. 1998).

8. Changes in phosphorylation in oxidative stress

The alteration of red cell thiol status affects the cell phosphotyrosine status and oxidative stress involves inhibition of PTP. Erythrocyte thiol alkylation by N-ethylmaleimide results in irreversible PTP inhibition and irreversible phosphorylation (Zipser et al. 1997).

Oxidation of erythrocyte membrane by diamide leads to formation of disulfide bonds and following conformational changes in band 3. Most probably those changes lead to the opening of cryptic sites that become accessible for the binding of anti-band 3 antibodies (Turrini et al. 1994). Macrophages recognize oxidatively damaged autologous erythrocytes, and cell surface fibronectin of macrophages enhances the recognition (Beppu et al. 1991). Ca^{2+} signaling including Ca^{2+} influx, calmodulin activation, and myosin light chain phosphorylation are involved in the fibronectin stimulation of the recognition of macrophages for oxidized erythrocytes (Beppu et al. 2000).

Pervanadate, N-ethylmaleimide and diamide strongly increase the two-stage phosphorylation of tyrosine residues of band 3 by Syk-kinase and Src-family (probably Lyn-kinase). An intriguing fact is that there are different mechanisms by which osmotic and oxidative stress activate PTK Syk: oxidative stress leads to autophosphorylation, but osmotic one-to SH-2 domain phosphorylation. It was demonstrated that the same agents strongly enhance interaction between SHP-2 and band 3, which take part simultaneously with the translocation of this phosphatase from cytosol to erythrocyte membrane. These events are most likely mediated by Src-phosphorylation, since both translocation of SHP-2 and phospatase interaction with the erythrocyte membrane are prevented by PP2, a specific Src inhibitor. SHP-2 binds to band 3 via its SH2 region(s). Authors suggest that the attraction of cytosolic SHP-2 to band 3 preceeds the next dephosphorylation of the transmembrane protein (Bordin et al. 2002).

Peroxynitrite (ONOO·) is a product of the reaction of nitric oxide and superoxide anion. It is able to nitrate protein tyrosine. If this modification occurs on phosphotyrosine kinase, substrates can down-regulate cell signaling. ONOO· at low concentrations is a stimulator of both band 3 tyrosine phosphorylation and erythrocyte glycolysis, but higher concentratons ONOO· induces cross-linking of membrane proteins, inhibition of band 3 phosphorylation, nitration of tyrosines in cytosolic domain of band 3, and irreversible inhibition of lactate production (Mallozi et al. 1997).

Deoxygenation and increase in intracellular Mg^{2+} content induce phosphorylation of tyrosine residues in human erythrocytes (Barbul et al. 1999).

Increasing in erythrocyte volume is combined with stimulation in the activities of the two PTK- p72 (syk) and p56 (lyn) that phosphorylate band 3 (Musch et al. 1999).

Oxidation of methionine residues in CaM prevents its activating effect on membrane Ca^{2+}-ATPase (Gaop et al. 2001).

The erythrocyte has a pool of flavonic compounds, which are considered a buffer, maintaining the antioxidative activity of the erythrocyte (Fiorani et al., 2003) Quercetin and resveratrol (piceatannol) take part in the regulation of the phosphorylation of band 3 in the presence of the physiological oxidant peroxinitrite. Quercetin decreases the Syk activity and particularly prevents mediated by the free radical peroxynitrite (ONOO·) phosphotyrosine phosphatase inhibition. Restraverol (whose anlogue is piceatannol) has another mechanism of action – it enables the mediated by peroxynitrite stimulation of tyrosine phosphorylation by another phosphotyrosine protein kinase – Lyn (Maccaglia et al., 2003).

Pleiotropic effects of resveratrol include antioxidant activity and inhibition of cyclooxygenase with decrease of PGE(2) formation. In erythrocytes, oxidation and PGE(2) activate Ca^{2+}-permeable cation channels. The Ca^{2+}-entry leads to activation of Ca^{2+}-sensitive K^+channels with subsequent cell shrinkage and cell membrane scrambling with phosphatidylserine (PS) exposure at the erythrocyte surface. Cell shrinkage and phosphatidylserine exposure are hallmarks of suicidal erythrocyte death or eryptosis. Eryptotic cells adhere to the vascular wall, thus compromising microcirculation, and are cleared from circulating blood, thus leading to anemia. Resveratrol is a potent inhibitor of suicidal erythrocyte death during energy depletion, oxidative stress, and isoosmotic cell shrinkage. The nutrient could thus counteract anemia and impairment of microcirculation under conditions with excessive eryptosis (Qadri al. 2009). However, in some cases, flavonoids have been suggested to work as prooxidants (Kitagawa et al. 2004)

9. Eryptosis (Apoptosis)

Two signaling pathways converge to trigger apoptosis: (1) formation of PGE2 leads to activation of Ca^{2+}-permeable cation channels; (2) the PLA2 (Phospholipase A 2) mediated release of PAF (platelet–activating factor) activates a sphingomielinase, leading to formation of ceramide. Increased cytosolic Ca^{2+} activity and enhanced cermide levels lead to membrane scrambling with subsequent PS exposure. Ca^{2+}-activated Ca^{2+}-sensitive K^+ channels leading to cellular KCl loss and cell shrinkage. Ca^{2+} stimulates the protease calpain, resulting in degradation of cytoskeleton (Foller et al. 2008; Lang et al. 2010).

Most triggers of eryptosis, except oxidative stress, are effective without activation of caspases. The involvement of Fas/caspase 8/caspase 3-dependent signaling in erythrocytes leads to PS externalization, a central feature of erythrophagocytosis and erythrocyte biology. The oxidatively stressed red cell recapitulated apoptotic events, including translocation of Fas into rafts, formation of a Fas-associated complex, and activation of caspases 8 and 3. The ROS (Radical Oxygen Species) scavenger N-acetylcysteine inhibits eryptosis (Mandal et al. 2005).

Protein kinase C mediates erythrocyte "programmed cell death" following glucose depletion (Klarl et al. 2006).

Eryptosis is stimulated in a wide variety of diseases including sepsis, haemolytic uremic syndrome, malaria, sickle-cell anemia, beta-thalassemia, glucose-6-phosphate dehydrogenase (G6PD)-deficiency, phosphate depletion, iron deficiency, and Wilson's disease. Excessive eryptosis is observed in erythrocytes lacking the cGMP-dependent protein kinase-1 (cGK1) or cAMP activated protein kinase (AMPK). Moreover, eryptosis is elicited by osmotic shock, oxidative stress, and energy depletion, as well as a wide variety of endogenous mediators and xenobiotics. Inhibitors of eryptosis include erythropoietin, nitric oxide NO, catecholamines, and high concentrations of urea (Lang et al. 2006).

10. Effect of modulators of phosphorylation on activity of erythrocyte glycolysis and sodium pump

Modulation of cellular signals from other agents is a leading pharmaceutical approach for developing a therapeutic strategy for many diseases. Of significant practical importance is to address the right questions and to find the appropriate answers in regard to whether certain agents would exert synergic or opposite effects over key cellular functions.

Erythrocyte membranes mediate the relation between ion transport and glycolysis. Na^+/K^+-ATPase is not only transport system but also participates in ouabain – initiated signal transduction (Haas et al. 2002; Xie and Cai, 2003; Mohammadi et al.2001), and inititated cell signals in cardiac myocytes that are independent from changes of intracellular concentrations of Na^+ and Ca^{2+} (Liu et al. 2000). It was demonstrated that glyceraldehyde-3-phosphate-dehydrogenase interacts with cytoplasmic surface of Na^+/K^+-ATPase, and that interaction is shown to be inhibited by ouabain (Fossel and Solomon, 1978). The α-subunit of Na^+/K^+-ATPase and band 3 form a heterodimeric structure (Martin and Sachs,. 1992). Similar structural integration allows shared control of both glycolysis and ion transport, and possible interference of modulators of ion transport in the control of glycolysis might be suggested (Fossel and Solomon, 1978). To verify our hypothesis we studied the effect of the same modulators of phosphorylation on both glycolysis (by measuring the amounts of its final product lactate) and Na^+/K^+-ATPase activity.

10.1 Methods

10.1.1 Isolation of the erythrocytes

Heparinized fresh drawn blood from healthy donors was centrifuged at 2,000 ×g for 5 min at 4°C and the pellet was resuspended in 4 volumes of phosphate buffered saline (PBS) pH 7.4. After three washes at 1,800, 1,500, 1,300 xg, the erythrocytes were isolated by density separation (Cohen et al., 1976). The erythrocyte fraction was resuspended in PBS pH 7.4 to obtain the same concentration as in the fresh blood. Cell concentration was counted in Burker's camera by a Standard KF2 microscope (Carl Zeiss, Jena, Germany). The suspension did not contain other cell species.

10.1.2 Preparation of the erythrocyte membranes

Five milliliters of packed RBCs were mixed with 15 ml cold PBS (0.144 g/l $KH_2PO_4.7H_2O$, 9.0 g/l NaCl, and 0.795 g/l $Na_2HPO_4.7H_2O$), pH 7.4, and centrifuged at 5,900 ×g for 10 min at 4°C. The supernatant was discarded, and cells were washed with 15 ml cold PBS, centrifuged as above, and resuspended in 5 ml of 5 mM Na_2HPO_4, pH 8.0, for hypotonic cell lysis. Lysed cells were centrifuged at 25,000 $\times g$ for 15 min at 4°C, after which the supernatant was gently aspirated and discarded. The RBC membrane pellet was repeatedly washed (4–5 times) with 5 mM Na_2HPO_4 until the pellet appeared white (indicating removal of Hb), and membranes were used for further experiments. The protein content of hemoglobin-free pellets was determined according to Bradford (1976), with human serum albumin as a calibrator. Samples were diluted to protein contents of 1.5 g/L.

10.1.3 Hemoglobin measurement

The hemoglobin (Hb) concentration of the erythrocyte suspension was determined according to Beutler (1975).

10.1.4 Cell treatment

The 50 μl erythrocyte suspension (2×10^7 cells/ml) was incubated 30 min at 25°C. Samples were performed in quadruplicates. In order to estimate the effect of cell signal modulators, samples were incubated either in the presence or in the absence of the agents.

10.1.5 Chemicals

All chemicals were purchased from Sigma-Aldrich Co, St Louis, MO USA. The mentioned below concentrations were chosen according the producer's prescription and data existing in the literature about their optimal effect: Go6976 (50 nM) and Go6983 (50 nM) – inhibitors of protein serine kinase PKC. Go6976 is an inhibitor of the classic isoforms of PKC and PKD, while Go6983 inhibits only the classic isoforms. Both inhibitors were used simultaneously to exclude the effect of PKD; Caffeine (20 mM) was used as a phosphodiesterase inhibitor; Okadaic acid (20 nM) and Calyculin A (50 nM) as inhibitors of serine/treonine protein phosphatases; N-(6-Aminohexyl)-5-chloro-1-naphtalene-sulfonamide (W-7) (30 μM) as a calmodulin antagonist.

10.1.6 Determination of lactate content in erythrocytes

After centrifugation for 10 min at 2,000 $\times g$ erythrocytes were resuspended in 0.4 ml 10% trichloroacetic acid (TCA). For obtaining of total precipitation the samples were cooled on ice 10 min and then centrifuged again at the same conditions. 0.1 ml from the supernatant was used further according the prescription of the test kit reagent obtained from "SIGMA Diagnostics" (St. Louis, USA). The method was based on the reaction of pyruvate oxidation in the presence of NADH and lactate dehydrogenase. To estimate lactate, the reaction was carried out with excess NAD^+ and the OD of NADH was measured at 340 nm.

50 μl of erythrocyte suspension (2×10^7 cells/ml) was incubated for 30 min at 25°C in the presence (samples) or absence (control) of the chosen agents. After 10 min of centrifugation at 2,000 ×g, erythorcytes were resuspended in 0.4 ml of 10% TCA. For total sedimentation, erythrocytes were cooled down for 10 min and the centrifugation step was repeated at the same conditions. 0.1 ml from the supernatant was added to a preliminary prepared solution of NAD^+ in 2.0 ml glycine buffer, 4.0 ml ddH_2O and 0.1 ml lactate dehydrogenase. Instead of

supernatant, 0.1 ml 10% TCA was added as a blank. After 30 min of incubation at room temperature, the extinction was read at OD=340 nm. Results were calculated using a standart curve and equation according to kit instruction. Results were presented in µg/ml.

10.1.7 Erythrocyte membrane ATPase activity

Here, a method based on enzyme kinetics was used for evaluation of ATPase activity based on the amounts of generatated NAD^+ in glycolytic reactions. The following reactions were explored: 1) conversion of phosphoenolpyruvate (PEP) to pyruvate and ATP by pyruvate kinase; 2) interconversion of pyruvate and lactate with concomitant interconversion of NADH and NAD+ catalyzed by lactate dehydrogenase (Vásárhelyi et al., 1997).

Samples (50 µl) with protein concentration 1.5 g/l of erythrocytes membranes were added to 450 µl of the following solution with final concentration per liter: 100 mmol of NaCl, 20 mmol of KCl, 2.5 mmol of $MgCl_2$, 0.5 mmol of EGTA, 50 mmol of Tris-HCl, pH 7.4, 1.0 mmol of ATP, 1.0 mmol of phosphoenolpyruvate, 0.16 mmol of NADH, 5 kU of pyruvate kinase, 12 kU of lactate dehydrogenase (all purchased from Sigma). Finally, the tested modulators were added in the samples in above mentioned concentrations. After 5 min, 5 µl of 10 mmol/l ouabain was added to inhibit the ouabain-sensitive ATPase activity. The change in absorbance was measured at OD=340 nm by a twin test (i.e., combination of two assays in one cuvette); Rate A (i.e., slope of total ATPase activity), 80–280 s; Rate B (i.e., slope of ouabain-resistant ATPase activity), 400–600 s. The difference between the two slopes was proportional to the Na^+/K^+-ATPase activity (Vásárhelyi *et al.* 1997). One unit of ATPase activity (1U) equals to one µmol oxidized NADH for 1 min. Calculations are based on the fact that NADH solution with concentration 1 mg/ml has 0.80 extinction at OD=340 nm (E_{340} = 0.80) (Vásárhelyi et al. 1997).

10.2 Results

PKC inhibitors (Go6876 and Go6983), protein phosphatase inhibitors (OA and calyculin A), caffeine, and W-7 increased reliable lactate formation. Calyculin A inhibits reliable Na^+/K^+-ATPase (Table 1 and 2). From all studied modulators only Go6976 and Go6983, which are PKCs inhibitors, stimulated reliably Na^+/K^+-ATPase, with 194% and 84%, respectively (Table 1 and 2; Figure 2 and 3).

Agents (n = 6)	Lactate µmol/g.Hb x ± SD	P
No agents (control)	5.15 ± 0.19	-
Go6976	8.23 ± 0.63	<0.001
Go6983	8.21 ± 3.24	<0.05
Caffeine	8.07 ± 0.65	<0.001
OA	7.84 ± 1.45	<0.001
Calyculin A	7.23 ± 0.37	<0.001
W-7	11.12 ± 0.30	<0.001

Table 1. Effect of modulators of phosphorylation on lactate formation

10.3 Discussion

10.3.1 Effect of PKC inhibitors

Four isoforms of PKC in erythrocytes have been found: α, ζ, ι and μ (PKD). Erythrocyte PKC phosphorylates serine residues in band 3, band 4.1 and band 4.8 (Govekar and Zingle, 2001). Go6976 and Go6983 are used to exclude the involvement of PKD in cell signals. Go6976 inhibits classical isoforms of PKC and PKD, but Go6983 inhibits classical isoforms only. The absence of differences in the effects of Go6976 and Go6983 on the lactate formation (Table 1) excludes the involvement of PKD in the control of glycolysis and may indicate involvement of classical PKCα as a negative regulator of glycolysis. It was demonstrated that αPKC translocated from the interior of erythrocytes to the membrane as a result of various stimuli (Govekar and Zingle, 2001).

Go6976 (inhibitor of classic isoforms and PKD), and also PKD "switch-off" inhibitor Go6983, reliably activate ATPase activity. This presents PKC as a negative regulator of Na$^+$/K$^+$-ATPase. The effect of stimulation with Go6976 is almost twice higher than stimulation with Go6983 (Table 2 and Fig.3). This probably suggests a specific function of PKD as an inhibitor of Na$^+$/K$^+$-ATPase. PKD has a catalytic domain, which shows more similarity to CaM-dependent kinases than to PKC. PKD efficiently phosphorylated synthetic substrates of Ca^{2+}/CaM -dependent kinase II, but does not catalyze phosphorylation of substrates

Agents ($n = 6$)	Na$^+$/K$^+$ ATPase [U/g. Protein]	p
No agents (control)	4.44 ± 0.98	-
Go6976	13.04 + 2.11	<0.001
Go6983	8.18 + 3.24	<0.05
Caffeine	5.68 + 1.57	>0.1
OA	5.18 ± 0.61	>0.1
Calyculin A	2.71 ± 0.98	<0.02
W-7	4.81 + 0.86	>0.1

Table 2. Effect of modulators of phosphorylation on Na$^+$/K$^+$-ATPase

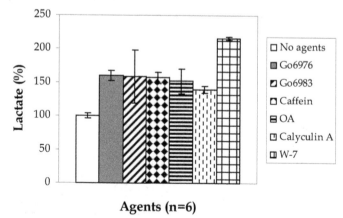

Fig. 2. Effect of modulators of phosphorylation on lactate formation (control=100%).

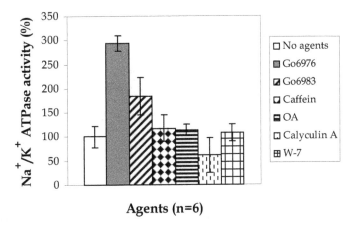

Agents (n=6)

Fig. 3. Effect of modulators of phosphorylation on Na+/K+-ATPase (control=100 %)

typical of PKC (Ron et al., 1999). CaM inhibits the sodium pump in erythrocytes (Okafor et al. 1997; Yingst et al. 1992), and because of similarities with the effects of CaM, PKD participation in erythrocyte phosphorylation could be assumed, leading to a decrease in activity of Na+/K+-ATPase. It is also proven an inhibitory effect of PKD on Na+/H+ - antiport in other cell types (Haworth and Sinnet-Smithh, 1999). It could be speculated that due to the antiport inhibition by PKD, the activity of Na+/K+-ATPase is being reduced.

10.3.2 Effect of caffeine

The mechanisms by which caffeine stimulates erythrocyte glycolysis (Table 1 and Fig.2) may include modulation of erythrocyte protein phosphorylation (Boivin, 1988) or changes in the conformation of the erythrocyte membrane proteins that facilitate the access of enzymes phosphorylating band 3 (Sato et al. 1990). It is known that methylxanthines are able to inhibit cyclic nucleotide-dependent protein kinases present in the cytosol and the erythrocyte membrane (Boivin, 1988). This inhibition may be due to competition for ATP (Boivin, 1988) and may be involved in the regulation of PK, whose active form is the dephosphorylated one (Kiener and Weathead, 1980; Garrillo et al., 2000; Nakashima et al., 1982). Caffeine can exert an indirect effect on lactate formation in erythrocytes through stimulation of Na+/K+-ATPase (Gusev et al. 2000), but our results showed that caffeine has no reliable effect on the activity of erythrocyte Na+/K+-ATPase (Table 2 and Fig 3). The identified differences may be due to the fact that in that study, erythrocytes from Rana temporaria were examined (Gusev et al, 2000).

10.3.3 Effect of protein phosphatases inhibitors

Okadaic acid (OA) is an inhibitor of serine/treonine phosphatases (Bordin et al. 1993). The incentive effect of OA on glycolysis (Table 1 and Fig.2) may be indirect, as a consequence of activation of MAPK-dependent way that increases Na+/H+-antiport (Sartori et al. 1999). It is known that one of the mechanisms for glycolysis stimulation is the activation of proton export through this antiport (Madshus, 1988). An interesting fact is that OA and various growth factors (GF), e.g. TGF and EGF, exert a stimulatory effect on Na^+/H^+-antiport (NHE-1) through phosphorylation of identical serine residues in its molecule (Sardet et al. 1991). Calyculin A and OA are likely to cause stimulation of glycolysis by using pathways involving protein phosphatase type II (PP2), Table 1 and Fig 2. Protein phosphatase type I (PP1) is highly sensitive to calyculin A (CalA), but not to OA. Protein phosphatase type II (PP2) is highly sensitive to both inhibitors, CalA and OA (Bize et al. 1999). OA has no effect on erythrocyte sodium pump (Table 2 and Fig.3), which probably means that phosphatase type II (PP2) is not involved in cellular signals engaged with the control of Na^+/K^+-ATPase, but Calyculin A inhibits reliably (with about 40%) the pump activity, which may be due to the involvement of PP1 in cell signals leading to activation of Na^+/K^+-ATPase (Table2 and Fig.3).

10.3.4 Effect of W-7

As calmodulin antagonist W-7 restores reduced physiological ability of erythrocytes to change their shape when loaded with Ca^{2+} (Murakami et al. 1986) and also exerts a vasodilator effect (Beresewiz, 1989). The beneficial effect of W-7 on erythrocyte rheology may be due to its stimulatory effect on glycolysis and improvement of erythrocyte bioenergetics (Table 1 and Fig.2). The stimulatory effect of W-7 on glycolysis may be due to blocking the inhibitory effect of calmodulin on Na^+/H^+-antiport (Yingst et al. 1992), or to decrease in the processes of phosphorylation/dephosphorylation, regulated by Ca^{2+}calmodulin (Benaim and Villalobo, 2002, Corti et al. 1999). W-7 increases by 8% the activity of Na^+/K^+-ATPase but the effect lacks statistical significance (Table 2 and Fig.3) in the absence of Ca^{2+}. There is evidence in the literature for an inhibitory effect of calmodulin on erythrocyte Na^+/K^+-ATPase, which occurs at 2 μM Ca^{2+} in the incubation medium (Yingst et al. 1992). The difference with our results could be explained probably with these specific experimental conditions. In our study, no extra ammounts of calcium were added into incubation medium. Erythrocytes have about 80 nM intracellular Ca^{2+} (Astarie et al. 1992). Perhaps the inhibitory effect of calmodulin (resp. stimulation with W-7) on the sodium pump acts as a regulatory mechanism when an increase in intracellular calcium content is presented, and primarily affects Ca^{2+}-binding capacity of calmodulin (Astarie et al. 1992).

10.4 Conclusions

Processes of phosphorylation-dephosphorylation of erythrocyte membrane and plasma proteins provide different levels of interaction, and also participate in maintaining the integrity of erythrocyte membrane and exerting control over important metabolic processes in erythrocytes. Summary of the existing data in literature shows they are more comprehensive in terms of the effect of primary messengers that bind to membrane receptors on erythrocytes and the consequent biological effects of the ligand-receptor interaction. Down-stream signaling pathways, though, where a major role in phosphorylation processes is played by protein kinases and protein phosphatases, remain less clear. Expansion of our knowledge and better understanding of signal transduction in erythrocytes will enable possible control to be exerted over their impact in maintenance of vascular tone, procoagulant activity and antioxidant status, erythrocyte aging, senescence, eryptosis (apoptosis), and erythrophagocytosis. Our results demonstrate that erythrocytes use different cell signals for regulation of glycolysis and activity of the sodium pump. They also reveal that different classes of PKC most likely taking part in regulation of glycolysis and Na+/K+-ATPase. Protein phosphatase inhibitors have opposite effects in the control of glycolysis and sodium pump, which indicates involvement of different protein phosphatases in cellular signal transduction that controls ion transport and cell energetics. Caffeine and W-7 have a reliable effect on the stimulation of glycolysis, but not on sodium pump.

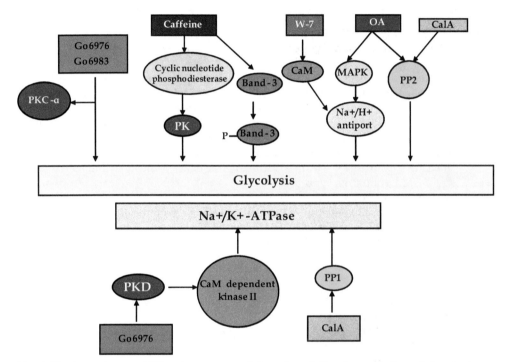

Fig. 4. Erythrocytes use different processes of phosphorylation in regulation of glycolysis and sodium pump.
(A). Glycolysis. Go6976 and Go6983 are inhibitors of PKC and stimulators of glycolysis. The target for this effect is PKC-α, which is probably a negative regulator of glycolysis. Caffeine increases the formation of lactate as: a) an activator of phosphodiesterase and inhibitor of c.AMP-dependent phosphorylation processes that keeps pyruvate kinase in dephosphorylated active form; b) induces conformational changes or activates phosphorylation of band 3. W-7 as a CaM antagonist blocks its inhibitory effect on Na+/H+-antiport, which activates glycolysis. Okadaic acid (OA) stimulates glycolysis as an inhibitor of protein phosphatase PP2, which is probably a negative regulator of glycolysis, and as a stimulator of MAPK-dependent phosphorylation of Na+/H+-antiport. Calyculin A (CalA) activates glycolysis as PP2 inhibitor that turns the proteins involved in glycolysis to their active dephosphorylated form. (B) Na+/K+-ATPase. PKD is likely a negative regulator of the sodium pump using CaM-dependent pathway since its specific inhibitor Go6976 significantly increases the activity of Na+/K+-ATPase. Cal A inhibits the activity of the pump reliably, which may be due to the involvement of PP1 in cell signals leading to activation of Na+/K+-ATPase.

11. Acknowledgments

The authors would like to thank Dr. Ilian Radichev (Staff Scientist at Sanford Research/USD), for his valuable comments, and to acknowledge the helpful assistance of Jacob Ellefson (Research Associate at Sanford Research/USD) on improving the language quality of this work.

12. References

Adderley, S.P.; Sprague, R.S.; Stephenson, A.H.; Hanson, M.S. Regulation of cAMP by phosphodiesterases in erythrocytes.(2010). *Pharmacological Reports*, vol. 62, № 3, pp 475-482

Ando, K.; Kikugawa, K.; Beppu, M. (1996). Binding of anti-band 3 autoantibody to sialylated poly-N-acetyllactosaminyl sugar chains of band 3 glycoprotein on polyvinylidene difluoride membrane and sepharose gel: further evidence for anti-band 3 autoantibody binding to the sugar chains of oxidized and senescent erythrocytes. *Journal of Biochemistry*, vol. 119, № 4, pp. 639-647

Andrews, D.A; Yang, L.; Low, P.S. (2002). Phorbol ester stimulates a protein kinase C-mediated agatoxin-TK-sensitive calcium permeability pathway in human red blood cells. *Blood*, vol.100, № 9, pp. 3392-3399

Angel, R.C.; Botta, J.A.; Farias, R.N. (1989). High affinity L-triiodothyronine binding to right-side-out and inside-out vesicles from rat and human erythrocyte membrane. *Journal of Biological Chemistry*, vol. 264, №32, pp.19143-19146

Antonelou, M.H; Kriebardis, A.G.; Papassideri, I.S. (2010). Aging and death signalling in mature red cells: from basic science to transfusion practice. *Blood Transfusion*, № 8, Suppl 3, 39-47

Assouline-Cohen, M.; Beitner, R. (1999). Effects of Ca2+ on erythrocyte membrane skeleton-bound phosphofructokinase, ATP levels, and hemolysis. *Molecular Genetics and Metabolism*, Vol. 66, № 1, pp. 56-61

Astarie, C.; Pernollet, M.G.; Del Pino, M.; Levenson, J.; Simon, A.; Devynck, M.A. (1992). Control of the erythrocyte free Ca2+ concentration in essential hypertension. *Hypertension*, vol.19, pp. 167-174

Azoui, R.; Cuche, J.L.; Renaud, J.F.; Safar, M.; Daher, G. (1996). A dopamine transporter in human erythrocytes: modulation by insulin. *Experimental Physiology* vol. 81, № 3, pp. 421- 434

Balzan, S.; Del Carratore, R., Nicolini, G.; Forini, F.; Lubrano, V.; Simili, M.; Benedetti, P.A.; Levrasi, G. (2009). TSH induces co-localization of TSH receptor and Na/K-ATPase in human erythrocytes. *Cell Biochemistry and Function*, vol. 27, № 5, pp. 259-263

Barbul, A.; Zipser, Y.; Nachles, A.; Korenstein, R. (1999). Deoxygenation and elevation of intracellular magnesium induce tyrosine phosphorylation of band 3 in human erythrocytes. *FEBS Letters*, vol. 455, № (1-2), pp. 87-91

Barford, D. (1996). Molecular mechanisms of the protein serine/threonine phosphatases. *Trends in Biochemical Sciences*, vol. 21, № 11, pp. 407–412

Bech-Otschir, D.; Seeger, M.; Dubiel, W. (2002). The COP9 signalosome: at the interface between signal transduction and ubiquitin-dependent proteolysis. *Journal of Cell Science*, vol. 115, (Pt. 3), pp. 467-473

Belloni-Olivi, L.; Annadata, M.; Goldstein, G.W.; Bressler, J. P. (1996). Phosphorylation of membrane proteins in erythrocytes treated with lead. *Biochemical Journal*, vol. 315, (Pt.2), pp. 401-406

Benaim, G.; Villalobo, A. (2002). Phosphorylation of calmodulin. Functional implications. *European Journal of Biochemistry*, vol. 269, № 15, pp. 3619-3631

Beppu, M.; Ando, K.; Saeki, M.; Yokoyama, N.; Kikugawa, K. (2000). Binding of oxidized Jurkat cells to THP-1 macrophages and antiband 3 IgG through sialylated poly-N-acetyllactosaminyl sugar chains. *Archivs of Biochemistry and Biophysics*, vol. 384, № 2, pp. 368-374

Beppu, M.; Masa H.; Hora, M.; Kikugawa, K (1991). Augmentation of macrophage recognition of oxidatively damaged erythrocytes by substratum-bound fibronectin and macrophage surface fibronectin. *FEBS Letters*, vol. 295, №(1-3), pp. 135-140

Beresewiz, A.. (1989), Anti-ischemic and membrane stabilizing activity of calmodulin inhibitors. *Basic Research in Cardiology*, vol. 84, № 6, pp. 631-645

Berzosa, C.; Gómez-Trullén, E.M.; Piedrafita, E.; Cebrián, I.; Martínez-Ballarín, E.; Miana-Mena, F.J.; Fuentes-Broto, L.; García, J.J. (2011). Erythrocyte membrane fluidity and indices of plasmatic oxidative damage after acute physical exercise in humans. *European Journal of Applied Physiology*, vol. 111, № 6, pp. 1127-1133

Beutler, E. (1975) Hemoglobin estimation. In: *Red Cell Metabolism. A Manual of Biochemical Methods. Basic techniques and equipment*. 2-nd Ed. (Grune and Stratton, ed.), New York, pp.112-114

Bhattacharaya, S.; Chakraborty, P.S.; Basu, R.S.; Kahn, N.N.; Sinha, A.K. (2001). Purification and properties of insulin-activated nitric oxide synhase from human erythrocyte membrane. *Archives of Physiology and Biochemistry*, vol. 109, № 5, pp.441-449

Bianchini, I.; Woodside, M.; Sardet, C.; Pouyssegur, J.; Takai, A.; Grinstein, S. (1991). Okadaic acid, a phosphatase inhibitor, induces activation and phosphorylation of the Na+/H+ antiport. *Journal of Biological Chemistry*, vol. 266, № 23, pp. 15406-15413

Bize, I.; Güvenç, B.; Robb, A.; Buchbinder, G.; Brugnara, C. (1999). Serine/ threonine protein phosphatases and regulation of K-Cl cotransport in human erythrocytes. *American Journal of Physiology*, vol. 277, (5 pt 1), pp. C926-C936

Boadu, E.; Sager, G. (2000). ATPase activity and transport by a cGMP transporter in human erythrocyte ghosts and proteoliposome-reconstituted membrane extracts. *Biochimica et Biophysica Acta*, vol. 1509, № (1-2), pp.467-474

Bogin, E.; Earon, Y.; Blum, M. (1986). Effect of parathyroid hormone and uremia on erythrocyte deformability. *Clinica Chimica Acta*, vol. 161, № 3, pp. 293-299

Boivin, P. (1988). Role of phosphorylation of red blood cell membrane proteins. *Biochemical Journal*, vol. 256, № 3, pp. 689-695

Bordin, L.; Brunati, A.M.; Donella-Deana A.; Baggio, B.; Toninello A.; Clari G. (2002). Band 3 is an anchor protein and a target for SHP-2 tyrosine phosphatase in human erythrocytes. *Blood*, vol. 100, № 1, pp. 276-282

Bordin, L.; Clari G.; Baggio, B.; Gambaro, G.; Moret, V. (1994). Relationship between membrane protein phosphorylation and intracellular translocation of casein kinase in human erythrocytes. *Biochemical and Biophysical Research Communication*, vol. 203, № 1 pp. 681-685

Bordin, L.; Clari, G.; Bellato, M.; Tessarin, C.; Moret, V. (1993). Effect of okadaic acid on membrane protein phosphorylation in human erythrocytes. *Biochemical and Biophysical Research Communication*, vol. 195, № 2, pp. 723-729

Bordin, L.; Zen, F.; Ion-Popa, F.; Barbetta, M.; Baggio, B.; Clari, G. (2005). Band 3 tyrphosphorylation in normal and glucose-6-phosphate dehydrogenase-deficient human erythrocytes. *Molecular Membrane Biology*, vol. 22, № 5, pp. 411- 420

Botta, J.A.; Farías, R. N. (1985). Solubilization of L-triiodothyronine binding site from human erythrocyte membrane. *Biochemical and Biophysical Reseach Communication*, vol. 133, № 2, pp. 442-448

Boukharov, A.A.; Cohen, C.M. (1998). Guanine nucleotide-dependent translocation of RhoA from cytosol to high affinity membrane binding sites in human erythrocytes. *Biochemical Journal*, vol. 330, (Pt 3), pp. 1391-1398

Bourikas, D.; Kaloyianni, M.; Bougolia, M.; Zolota, Z.; Koliakos, G. (2003). Modulation of the Na+-H+ antiport activity by adrenaline on erythrocytes from normal and obese patients. *Molecular Endocrinology*, vol. 205, № 1-2, pp. 141-150

Bradford, M.M. (1976). A rapid and sensitive method for quantitation of microgram quantities of protein-binding. *Analalitical Biochemistry*, vol. 72, pp. 248-254

Bragadin, M.; Ion-Popa, F.; Clari, G.; Bordin, L. (2007). SHP-1 tyrosine phosphatase in human erythrocytes. *Annual of the National Academy of Sciences of the United States of America*, vol. 1095, pp. 193-203

Bratton D.L. (1994). Polyamine inhibition of trans bilayer movement of plasma membrane phospholipids in the erythrocyte ghost. *Journal of Biological Chemistry*, vol. 269, № 36, pp. 22517-22523

Brunati, A.M.; Bordin, L.; Clari, G.; James, P.; Quadroni, M.; Baritono, E.; Pinna, L.A.; Donella-Deana, A. (2000). Sequential phophorylation of protein bind 3 by Syk and Lyn tyrosine kinases in intact human erythrocytes: identification of primary and secondary phosphorylation sites. *Blood*, vol. 96, № 4, pp. 1560-1567

Buckley J.T. (1977). Properties of human erythrocyte phosphatidylinositol kinase and inhibition by adenosine, ADP and related compounds. *Biochim Biophys Acta*. Vol. 498, № 1, pp. 1-9

Bzdega, T.; Kosk-Kosicka, D. (1992). Regulation of the erythrocyte Ca^{2+}-ATPase by mutant calmodulins with Glu-Ala substitutions in the Ca^{2+}-binding domains, *Journal of Biological Chemistry*, vol. 267, № 7, pp. 4394-4397

Campanella, M.E.; Chu, H.; Low, P.S. (2005). Assembly and regulation of a glycolytic enzyme complex on the human erythrocyte membrane. *Proceedings of the National Academy of Sciences of the United States of America*, vol. 102 , № 7, pp. 2402-2407

Carafoli, E. (1994). Plasma membrane calcium ATPase: 15 years of work on the purified enzyme, *Federation of American Societies for Experimental Biology (FASEB) Journal*, vol. 8, № 13, pp. 993-1002.

Carafoli, E. (2002) Calcium signaling: a tale for all seasons. *Proceedings of the National Academy of Sciences of the United States of America*, vol. 99, № 3, pp. 1115-1122

Ceolotto, G.; Corelin, P.; Clari, G.; Semplicini, A.; Canessa, M. (1998). Protein kinase C and insulin regulation of red blood cell Na+/H+ exchange and cytosolic free calcium in human. *American Journal of Hypertension*, vol. 11, (3 Pt 1), pp. 81-87

Chauhan, V.P.; Singh, S.S.; Chauhan, A.; Brockerhoff. H. (1993). Magnesium protects phosphatidylinositol-4,5- bisphosphate-mediated inactivation of casein kinase I in erythrocyte membrane. *Biochimica et Biophysica Acta*, vol.1177, № 3, pp.318-321

Ciana, A,; Minetti, G,; Balduini, C. (2004). Phosphotyrosine phosphatases acting on band 3 in human erythrocytes of different age: PTP1B processing during cell ageing. *Bioelectrochemistry*. Vol. 62, № 2, pp.169-173

Clari, G.; Brunati, A. M.; Moret, V. (1987). Membrane-bound phosphotyrosil-protein phosphatase activity in human erythrocytes. Dephosphorylation of band 3. *Biochemical and Biophysical Research . Communication*, vol. 142, № 2, pp. 587-594

Cohen NS, Ekholm JE, Luthra MG, Hanahan DJ. (1976). Biochemical characterisation of density-separated human erythrocytes. *Biochimica et Biophysica Acta*, vol. 419, pp. 229-237

Cohen, C.M.; Foley, S.F. (1986). Organization of the spectrin-actin-band 4.1 ternary complex and its regulation by band 4.1 phosphorylation. In: *Membrane and cytoskeletal-membrane association*, pp. 212-222, Alan R.Liss, Inc.N.Y

Cohen, C.M.; Gascard. P. (1992). Regulation and posttranslational modification of erythrocyte membrane- skeletal proteins. *Seminars in Hematology*, vol. 29, № 4, pp. 244-292

Cohen, M.S.; Mao, J.; Rasmussen, J.; Serodi, J.; C.; Brittigan, B.E. (1992). Interaction of lactoferrin and lipopolysacharide (LPS). Effect on the antioxidant property of lactoferrin and the ability of LPS to prime human neutrophils for enchanced superoxide formation. *Journal of Infection Disease*, vol. 166, № 6, pp. 1375- 1378

Gonçalves, I.; Saldanha, C.; Martins e Silva, J. (2001) Beta-estradiol effect on erythrocyte aggregation-a controlled in vitro study. *Clinical Hemorheology and Microcirculation*, vol. 25, № (3-4), pp. 127-134

Corry, D.H.; Joolhar, F.S.; Hori, M.T.; Tuck, M.L. (2002). Decreased erythrocyte insulin binding in hypertensive subjects with hyperinsulinemia. *American Journal of Hypertension*, vol. 15, (4Pt1), pp. 296-301

Corti, C.; Leclerc L'Hostis, E.; Quadroni, M.; Schmid, H.; Durussel, I.; Cox, J.; Dainese Hatt, P.; James, P.; Carafoli, E. (1999). Tyrosine phosphorylation modulates the interaction of calmodulin with its target proteins. *European Journal of Biochemistry*, vol. 262, № 3, pp.790-802

Curran C.S., Demick K.P. and Mansfield J.M. (2006). Lactoferrin activates macrophages via TLR4-dependent and -independent signaling pathways. *Cell Immunol.* vol.242, №1, pp. 23-30

Damonte, G.; Sdraffa, A.; Zocchi E.; Guida, L.; Polvani, C.; Tonetti, M.; Benatti, U.; Boquet, P.; De Flora, A. (1990). Multiple small molecular weight guanine nucleotide-binding proteins in human erythrocyte membranes. *Biochemical and Biophysical Research Communication*, Vol. 166, № 3 pp. 1398-13405

Danilov, Y.N.; Cohen, C.M. (1989). Wheat germ agglutinin but not concanavalin A modulates protein kinase C-mediated phosphorylation of red cell skeletal proteins. *FEBS Letters*, vol. 257, № 2, pp. 431-434

Danilov, Y.N.; Fennell, R.; Ling, E.; Cohen, C.M. (1990). Selective modulation of band 4.1 binding to erythrocyte membranes by protein kinase C. *Journal of Biological Chemistry*, vol. 265, № 5, pp.2556-2562

Darbonne, W.C.; Rice, G.C.; Mohler, M.A.; Apple, T.; Hebert, C.A.; Valente, A.J.; Baker, J.B. (1991). Red blood cells are a sink for interleukin 8, a leukocyte chemotaxin. *Journal of Clinical .Investigation*, vol. 88, № 4, pp. 1362-1369

De La Tour, D.D.; Raccah, D.; Jannot M..F.; Coste T.; Rougerie C.; Vague P. (1998). Erythrocyte Na+/K+ ATPase activity and disease: relationship with C-peptide level. *Diabetologia*, vol. 41, № 9, pp. 1080-1084

de Oliveira, S.; Silva-Herdade, A.S.; Saldanha, C. (2008). Modulation of erythrocyte deformability by PKC activity. *Clinical Hemorheology and Microcirculation*, vol. 39, № (1-4), pp.363-373

de Winter, R.J.; Manten, A.; de Jong, Y.P.; Adams, R.; van Deventer, S.J.; Lie, K.L. (1997). Interleukin-8 released after acute myocardial infarction is mainly bound to erythrocytes. *Heart*, vol. 78, № 6, pp.598-602

Djemli-Shipkolye, A.; Gallice, P.; Coste, T.; Jannot, M.F.; Tsimaratos, M.; Raccah, D.; Vague, P. (2000). The effects ex vivo and in vitro of insulin and C-peptide on Na/K adenosine triphosphatase activity in red blood cell membranes of type 1 diabetic patients. *Metabolism*, vol. 49, № 7, pp.868-872

Done´, S.C.; Leibiger, I.B.; Efendiev, R.; Katz, A.I.; Leibiger, B.; Berggren, P.O.; Pedemonte, C.H.; Bertorello A.M. (2002). Tyrosine 537 within the Na, K-ATPase-Subunit Is Essential for AP-2 Binding and Clathrin-dependent Endocytosis. *Journal of Biological Chemistry*, vol. 277, № 19, 17108–17111

Dreytuss, G.; Schwartz, K.J.; Blotet E.R. (1978). Compartmentalization of cyclic AMP-dependent protein kinases in human erythrocytes. *Proceedings of the National Academy of Sciences of the United States of America*, vol. 75, № 12, pp. 5926-5930

Dutta-Roy, A.K.; Kahn, N.N.; Sinha, A.K. (1991). Interaction of receptors for prostaglandin E1/prostacyclin and insulin in human erythrocytes and platelets. *Life Sciences*, vol. 49, № 16, pp.1129-1139.

Eda, S.; Kikugawa, K.; Beppu, M. (1996). Binding characteristics of human lactoferrin to the human monocytic leukemia cell line THP-1 differentiated into macrophages. *Biological & Pharmaceutical Bulletin*, vol. 19, № 2 pp.167-175

el-Andere, W.; Lerário, A.C.; Netto, D.G.; Wajchenberg, B.L. (1995). Erythrocyte insulin-like growth factor-I receptor evaluation in normal subjects, acromegalics, and growth hormone-deficient and insulin-dependent diabetic children. *Metabolism*, vol. 44, № 7, pp. 923-928

Ellsworth, M.L.; Ellis, C.G.; Goldman, D.; Stephenson, A.H.; Dietrich, H.H.; Sprague RS. (2009). Erythrocytes: oxygen sensors and modulators of vascular tone. *Physiology (Bethesda)*. Vol. 24, 107-116

Falkenstein,E.; Tillmann, H-C.; Christ, M.; Feuring, M.; Wehling, M. (2000). Multiple actions of steroid hormones - a focus on rapid, nongenomic effects. *Pharmacological Reviews*, vol. 52, № 4, pp. 513–555

Farmer, B.T. 2nd; Harmon, T.M.; Butterfield, D.A.(1985). ESR studies of the erythrocyte membrane skeletal protein network: influence of the state of aggregation of spectrin on the physical state of membrane protein, bilayer lipids, and cell surface carbohydrates. *Biochimica et Biophysica Acta*, vol. 821, № 3, pp. 420-430

Fiorani, M.; Accorsi, A.; Cantoni, O. (2003). Human red blood cells as a natural flavonoid reservoir. *Free Radical Research*, vol. 37, №12, pp.1331-1338

Föller, M.; Huber, S.M., Lang, F. (2008). Erythrocyte programmed cell death. *IUBMB Life.* vol. 60, № 10, pp. 661- 668

Fossel, E.T.; Solomon, A.K. (1978). Ouabain-sensitive interaction between human red cell membrane and glycolytic enzyme complex in cytosol. *Biochimica et Biophysica Acta,* vol. 510, № 1, pp. 99-111

Gaop J. ; Yao, Y.; Squier, T. C. (2001). Oxidatively modified calmodulin binds to the plasma membrane Ca-ATPase in a nonproductive and conformationally disordered complex. *Biophysical Journal,* vol. 80, № 4, pp. 1791-1801

Garrillo, J.J.; Ibares, B.; Esteban-Gamboa, A.; Felin, J.E. (2000). Involvement of both phosphatidylinositol 3-kinase and p44/p42 mitogen-activated protein kinase pathways in the short-term regulation of pyruvate kinase L by insulin. *Endocrynology,* vol. 142, № 3, pp. 1057-1064

Giraud, F.; Gascard P.; Sulpice J. C. (1988). Stimulation of polyphosphoinositide turnover upon activation of protein kinases in human erythrocytes. *Biochimica et Biophysyca Acta,* vol. 968, № 3, pp. 367-378.

Glasser, T.; Schwartz-Benmier, N.; Barnoy, S.; Barak, S.; Eshhar, Z.; Kosower, N.S.. (1994). Calpain (Ca2+-dependent thiol protease) in erythrocytes from young and old individuals. *Proceedings of the National Academy of Sciences of the United States of America,* vol. 91, № 17, pp. 7879-7893

Golden, G.A.; Mason, R.P.; Tulenko, T.N.; Zubenko, G.S.; Rubin, R.T. (1999). Rapid and opposite effects of cortisol and estradiol on human erythrocyte Na+/K+ -ATPase activity: intercalation into the cell membrane. *Life Sciences,* vol. 65, № 12, pp. 1247-1255

Goldenberg, H.; Dodel, B.; Seidl, D. (1990). Plasma membrane Fe2-transferrin reductase and iron uptake in K562 cells are not directly related. *European Journal of Biochemistry,* vol. 192, № 2, pp. 475-480

Gonçalves,, I.; Saldanha, C.; Martins e Silva, J. (2001). Beta-estradiol effect on erythrocyte aggregation - a controlled in vitro study. *Clinical Hemorheology and Microcirculation,* vol. 25, № (3-4), pp. 127-134

Govekar, R. B.; Zingde, S.M. (2001). Protein kinase C isoforms in human erythrocytes. *Annals of Hematology,* vol. 80, № 9, pp. 531-534

Graham, C.; Auruch, J.; Fairbanks, G. (1974). Phosphoprotein phosphatase of the human erythrocyte. *Biochemical and Biophysical Research Communication,* vol. 72, № 2, pp. 701-708

Guerini, D.; Pan, B.; Carafoli, E. (2003). Expression, purification, and characterization of isoform 1 of the plasma membrane Ca2+ pump: focus on calpain sensitivity. *Journal of Biological Chemistry,* vol. 278, № 40, pp. 38141-38148.

Gusev, G.P.; Agalakova, N.I. (2000). Na, K-Pump activation by isoproterenol, methylxanthines, and iodoacetate in erythrocytes of the frog Rana temporaria. *Zhurnal evoliutsionnoi biokhimii i fiziologii,* vol. 36, № 2 , pp.106-111

Ha, E.J.; Smith, A.M. (2003). Plasma selenium and plasma and erythrocyte gluthatione peroxidase activity increase with estrogen dunig the menstrual cycle. *Journal of the American College of Nutrition,* vol. 22, № 1, pp.43-51

Haas, M.; Wang, H.; Tian, J.; Xie, Z. (2002). Src-mediated inter-receptor cross-talk between the Na+/K+-ATPase and the epidermal growth factor receptor relays the signal from ouabain to mitogen-activated protein kinases. *Journal of Biological Chemistry*, vol. 277, № 21, pp. 18694-18702

Hamlyn, J.M.; Duffy, T. (1978). Direct stimulation of human erythrocyte membrane (Na+ +K+) ATPase activity in vitro by physiological concentration of d-aldosterone. *Biochemical and Biophysical .Research Communication.*, vol.84, №2, pp.458-464

Hanson, M.S.; Stephenson, A. H.; Bowles, E.A.; Sprague, R.S. (2010). Insulin inhibits human erythrocyte cAMP accumulation and ATP release: role of phosphodiesterase 3 and phosphoinositide 3-kinase. *Experimentsl Biology and Medicine (Maywood)*, vol. 235, № 2, pp. 256-262

Harrison, M.L.; Rathinavelu, P.; Arese, P.; Geahlen, R.L.; Low, P.S. (1991). Role of band 3 tyrosine phosphorylation in the regulation of erythrocyte glycolysis. *Journal of Biological Chemistry*, vol. 266, № 7, pp. 4106-4111

Haslauer, M.; Baltensperger, K.; Porzig, H. (1998). Thrombin and phorbol esters potentiate Gs-mediated c.AMP formation in intact human erythroid progenitors via two synergistic signaling pathways converging on adenylate cyclase type VII. *Molecular Pharmacology*, vol. 53, № 5, pp. 837-845

Haworth, RS.; Sinnet-Smithh, J. (1999). Protein kinase D inhibits plasma membrane Na+/H+ exchanger activity. *American Journal of Physiology*, vol. 277, № (6 pt 1), pp. C1202-C1209

Hecht, D.; Zick, Y. (1992). Selective inhibition of protein tyrosine phosphatase activities by H2O2 and vanadate in vitro. *Biochemical and Biophysical Research Communication*, vol. 188, № 2, pp.773-779

Hermand, P.; Gane, P.; Huet, M.; Jallu, V.; Kaplan, C.; Sonneborn, H.H.; Cartron, J.P.; Bailly, P. (2003) Red cell ICAM-4 is a novel ligand for platelet-activated alpha IIbbeta 3 integrin. *Journal of Biological Chemistry*, vol. 278, № 7, pp. 4892-4898

Horga, J.F,; Gisbert, J.; De Agustín, J.C.; Hernández, M.; Zapater, P. (2000). A beta-2-adrenergic receptor activates adenylate cyclase in human erythrocyte membranes at physiological calcium plasma concentrations. *Blood Cells, Molecules and Diseases*, vol. 26 № 3, pp. 223-228

Horne, W.C.; Prinz, W.C.; Tang, E.K. (1990). Identification of two cAMP-dependent phosphorylation sites on erythrocyte protein 4.1.*Biochimica et Biophysica Acta*, vol. 1055, № 1, pp. 87-92

Hubert, E.M.; Musch, M.W.; Goldstein, L. (2000). Inhibition of volume stimulated taurine efflux and tyrosine kinase activity in the skate red blood cells. *Pflugers Archiv*, vol. 440, № 1, pp.132-139

Ikeda, K.; Kikuchi, A.; Takai, Y. (1988). Small molecular weight GTP-binding proteins in human erythrocyte ghosts. *Biochem Biophys Res Commun.*, Vol. 156, № 2, pp. 889-897.

Ivanova, T.I.; Agalakova, N.I.; Gusev, G.P (2006) Activation of sodium transport in rat erythrocytes by inhibition of protein phosphatases 1 and 2A. *Comparative Biochemistry and Physiology: Part B - Biochemistry and Molecular Biology*, vol. 145, № 1 pp.60-67

Kajikawa, M.; Ohta, T.; Takase, M.; Kawase, K.; Shimamura, S.; Matsuda. I (1994). Lactoferrin inhibits cholesterol accumulation in macrophages mediated by acetylated or oxidized low-density lipoproteins. *Biochimica et Biophysica Acta*, vol. 1213, № 1, pp. 82-90.

Kaloyianni, M.; Bourikas, D.; Koliakos, G. (2001). The effect of insulin on Na+-H+ antiport activity of obese and normal subjects erythrocytes. *Cellular Physiology and Biochemistry*, vol. 11, № 5, pp.253-258

Kanbak, G.; Akynz, M.; Jnab, M. (2001). Preventive effect of betaine in ethanol-induced membrane lipid composition and membrane ATPases. *Archives of Toxicology*, vol. 75, № 1, pp. 59-61

Kiener, P.A.; Carroll, D.; Roth, B.J.; Westhead, E.W. (1987). Purification and characterization of a high molecular weight type I phosphoprotein posphatase from human erythrocytes. *Journal of Biological Chemistry*, vol. 262, № 5, pp. 2016-2024

Kiener P.A.; Westhead, E.W. (1980). Dephosphorylation and reactivation of phosphorylated pyruvate kinase by a cytosolic phosphoprotein phosphatases from human erythrocytes. *Biochemical and Biophysical Research Communication*, vol. 96, № 2, pp. 551-557

Kitagawa, S.; Sakamoto, H.; Tano, H. (2004). Inhibitory effects of flavonoids on free radical-induced hemolysis and their oxidative effects on hemoglobin. *Chemical & Pharmaceutical Bulletin (Tokyo)*, vol. 52, № 8, pp.999-1001.

Klarl, B.A.; Lang, P.A.; Kempe, D.S.; Niemoeller, O.M.; Akel, A.; Sobiesiak, M.; Eisele, K.; Podolski, M.; Huber, S.M.; Wieder, T.; Lang, F. (2006). Protein kinase C mediates erythrocyte "programmed cell death" following glucose depletion. *Americal Journal of Physiology- Cell Physiology*, vol. 290, № 1, pp. C244-C253.

Klein, H.H.; Muller, R.; Drenckhan, M.; Schutt, M.; Batge, B.; Fehm, H.L. (2000). Insulin activation of insulin receptor kinase in erythrocytes is not altered in non-insulin-dependent diabetes and not influenced by hyperglycemia. *Journal of Endocrinology*, vol. 166, № 2, pp. 275-281

Konstantinou-Tegou, A.; Kaloyianni, M.; Bourikas, D.; Koliakos, G. (2001). The effect of leptin on Na (+)-H(+) antiport (NHE1)activity of obese and normal subjects erythrocytes. *Molecular and Cellular Endocrinology*, vol. 25, № 183, pp. 11-18

Kosk-Kosicka, D.; Bzdega, T.; Johnson, J. D. (1990). Fluorescence studies on calmodulin binding to erythrocyte Ca2+'-ATPase in different oligomerization states, *Biochemistry*, vol. 29, № 7, pp. 1875 - 1879

Lang, F.; Gulbins, E.; Lang, P. A.; Zappulla, D.; Föller, M. (2010). Ceramide in suicidal death of erythrocytes. *Cellular Physiology and Biochemistry*, vol.26, № 1, pp. 21-28.

Lang, F.; Lang, K. S.; Lang, P. A.; Huber, S. M.; Wieder, T. (2006). Mechanisms and significance of eryptosis. *Antioxidants &Redox Signaling*, vol.8, № (7-8), pp.1183-1192

Lang, P.A.; Kempe, D.S.; Akel, A.; Klarl, B.A.; Eisele, K.; Podolski, M.; Hermle, T.; Niemoeller, O.M.; Attanasio, P.; Huber, S.M.; Wieder, T.; Lang, F.; Duranton, C. (2005). Inhibition of erythrocyte "apoptosis" by catecholamines. *Naunyn Schmiedeberg's Archives of Pharmacology*, vol. 372, № 3, pp. 228-235

Lawrence, W. D.; Deziel M.R.; Davis, P. J.; Schoenl, M.; Davis, F.B.; Blas, S. D. (1993). Thyroid hormone stimulates release of calmodulin-enhancing activity from human erythrocyte membranes in vitro. *Clinical Science*, (London). Vol. 84, № 2, pp. 217-223

Lecomte, M.C.; Galand, C.; Biovin, P. (1980). Inhibition of human erythrocyte casein kinase by methylxanthines. Study of inhibition mechanism. *FEBS Letters*, vol. 116, № 1, pp.45-47

Legrand, D.; Elass, E.; Pierce, A.; Mazurier, J. (2004). Lactoferrin and host defence: an overview of its immuno-modulating and anti-inflammatory properties. *Biometals*, Vol. 17, № 3 pp. 225-229

Leroy, D.; Filhol, O.; Delcros, J. G.; Pares, S.; Chambaz, E. M.; Cochet, C. (1997). Chemical features of the protein kinase CK2 polyamine binding site. *Biochemistry*, vol. 36, № 6, pp. 1242-1250

Lévy-Toledano, S.; Grelac, F.; Caen, J.P.; Maclouf, J. (1995). KRDS, a peptide derived from human lactotransferrin, inhibits thrombin-induced thromboxane synthesis by a cyclooxygenase-independent mechanism. *Journal of Thrombosis and Haemostasis*, Vol. 73, № 5, pp. 857-861

Li, Q.; Jungmann, V.; Kiyatkin, A.; Low, P.S (1996). Prostaglandin E2 stimulates a Ca2+-dependent K+ channel in human erythrocytes and alters cell volume and filterability. *Journal of Biological Chemistry*, vol. 271, № 31, pp. 18651-18656.

Liu, J.; Tian, J.; Haas, M.; Shapiro, J.I.; Askari, A.; Xie, Z. (2000). Ouabain interaction with cardiac Na+/K+-ATPase initiates signal cascades independent of changes in intracellular Na+ and Ca2+ concentrations. *J Biol Chem*. Vol. 275, № 36, pp. 27838-27844.

Low, P.S.; Ceahlen, R.L.; Mehler, E.; Harrison, M.L. (1990). Extracellular control of erythrocyte metabolism mediated by a cytoplasmic tyrosine kinase. *Biomedica Biochimica. Acta*, vol. 140, № (2-3), pp. 135-140

Low, H.; Grebing, C.; Lindgren, A.; Tally, M.; Sun, I. L.; Crane, F.L. (1987). Involvement of transferrin in the reduction of iron by the transplasma membrane electron transport system. *Journal of Bioenergetics and Biomembranes*, vol. 19, № 5, pp. 535-549

Low, P.S.; Rathinavelu, P.; Harrison, M. L. (1993). Regulation of glycolysis via reversible enzyme binding to the membrane protein, band 3. *Journal of Biological Chemistry*, vol. 268, № 20, pp.14627-14631

Lu, D.; Yan, H.; Othman, T.; Turner, C. P.; Woolf, T.; Rivkees, S.A. (2004). Cytoskeletal protein band 4.1G binds to the third intercellular loop of the a1 adenosine receptor and inhibits receptor action. *Biochemical Journal*, vol. 377, (Pt 1), pp. 51-59

Maccaglia, A.; Mallozzi, C.; Minetti, M. (2003). Differential effects of quercetin and resveratrol on Band 3 tyrosine phosphorylation signalling of red blood cells. *Biochemical and Biophyscal Research Communication*, vol. 305, № 3, pp. 541-547

Madshus, I.H. (1988). Regulation of intracellular pH in eukaryotic cells. *Biochemical Journal* , vol. 250, № 1, pp.1-8

Maekawa, T.; Fujihara, M.; Ohtsuki. K. (2002). Characterization of human lactoferricin as a potent protein kinase CK2 activator regulated by A-kinase in vitro. *Biological & Pharmaceutical Bulletin*, vol. 25, 118-121

Mahrenholz, A.M.; Lan, L.; Mansour, T. E. (1991). Phosphorylation of heart phosphofructokinase by Ca2+/calmodulin protein kinase. *Biochemical and Biophysical Research Commununication*, vol. 174, № 3 , pp. 1255-1259.

Mallozzi, C.; De Franceschi, L.; Brugnara, C.; Di Stasi, A. M. (2005). Protein phosphatase 1alpha is tyrosine-phosphorylated and inactivated by peroxynitrite in erythrocytes through the src family kinase fgr. *Free Radical Biology & Medicine*, vol. 38, № 12, pp. 1625-1636

Mallozi, C.; Di Stasi, A.M.; Minetti, M. (1997). Peroxynitrite modulates tyrosine-dependent signal transduction pathway of human erythrocyte band 3. *FASEB Journal*, vol. 11, № 14, 1281-1290

Mandal, D.; Mazumder, A.; Das, P.; Kundu, M.; Basu, J. (2005) Fas-, caspase 8-, and caspase 3-dependent signaling regulates the activity of the aminophospholipid translocase and phosphatidylserine externalization in human erythrocytes. *Journal of Biological Chemistry*, vol. 280, № 47, pp. 39460-39467

Maneva, A.; Angelova-Gateva, P.; Taleva, B.; Maneva-Radicheva, L. (2007). Lactoferrin stimulates erythrocyte Na+/K+ Adenosine Triphosphatase: Effect of some modulators of membrane phosphorylation. *Zeitschruft fur Naturforschung C*, vol. 62, № (11-12), pp. 897- 905

Maneva, A.; Taleva, B.; Maneva, L. (2003). Lactoferrin-Protector against oxidative stress and regulation of glycolysis in human erythrocytes. *Zeitschruft fur Naturforschung C*, vol. 58, № (3-4), pp. 256-262

Manno, S.; Takakuwa, Y.; Nagao, K.; Mohandas, N. (1995). Modulation of erythrocyte membrane mechanical function by beta-spectrin phosphorylation and dephosphorylation. *Journal of Biological Chemistry*, vol. 270, № 10, pp. 5659-5965

Marinho-Carvalho, M.M.; Zancan, P.; Sola-Penna, M. (2006). Modulation of 6-phosphofructo-1-kinase oligomeric equilibrium by calmodulin: formation of active dimers. *Molecular Genetics and Metabolism*, vol. 87, № 3, pp. 253-261

Martin, D.W.; Sachs, J.R. (1992). Cross-linking of the erythrocyte (Na+, K+) –ATPase. Chemical cross-linkers induce α-subunit-band 3 heterodimers and do not induce α-subunit homodimers. *Journal of Biological Chemistry*, vol. 267 № 33, pp. 23922-23929

Minetti, G.; Ciana, A.; Baldini, C. (2004). Differential sorting of tyrosine kinases and phosphotyrosine phosphatases acting on band 3 during vesiculation of human erythrocytes. *Biochemical Journal*, vol. 377, (Pt 2), pp. 487-497

Minetti, G.; Low, P.S. (1997). Erythrocyte signal transduction pathways and their possible functions. *Current Opinion in Hematology*, vol. 4, № 2, pp. 116-121

Minetti, G.; Piccinini, G.; Balduini C, Seppi C, Brovelli A. (1996). Tyrosine phosphorylation of band 3 in Ca2+/ A23187-treated human erythrocytes. *Biochemical Journal*, vol. 320, (Pt. 2), pp. 445-450

Mistry, N.; Drobni, P.; Näslund, J.; Sunkari, V.G.; Jenssen, H.; Evander, M. (2007). The anti-papillomavirus activity of human and bovine lactoferricin. *Antiviral Research*, vol. 75, № 3, pp. 258-265.

Mohammadi, K.; Kometiani P.; Xie, Z.; Askari, A. (2001). Role of protein kimase C in signal pathways that link Na+/K+-ATPase to ERK1/2. *Journal of Biological Chemistry*, vol. 276, № 45, pp. 42050-42056

Moulinoux, J.P.; Calve, M.; Quemener, V.; Quash, G. (1984). In vitro studies on the entry of polyamines into normal red blood cells. *Biochimie*, vol. 66, № 5, pp.385-393

Murakami, J.; Maeda, N.; Kon, K.; Shiga, T. (1986), A contribution of calmodulin to cellular deformability of calcium-loaded human erythrocytes. *Biochimica et Biophysica Acta*, vol. 863, No 1, pp. 23-32

Muravyov, A. V.; Tikhomirova, I. A.; Maimistova, A. A.; Bulaeva, S. V.; Zamishlayev, A. V.; Batalova, E. A. (2010). Crosstalk between adenylyl cyclase signaling pathway and Ca2+ regulatory mechanism under red blood cell microrheological changes. *Clinical Hemorheology and Microcirculation*, vol. 45, No (2-4), 337- 345

Muravyov, A.V.; Tikhomirova, I. A.; Maimistova, A. A.; Bulaeva, S.V. (2009). Extra- and intracellular signaling pathways under red blood cell aggregation and deformability changes. *Clinical Hemorheology and Microcirculation*, vol. 43, No 3, pp.223-232.

Musch, M.W.; Hubert, E. M.; Goldstein, L (1999). Volume expansion stimulates p72(syk) and p56(lyn) in skate erythrocytes. *Journal of Biological Chemistry*, vol. 274, No 12, pp. 7923-7928

Nakashima, K.; Fujii, S.; Kaku, K.; Kaneko, T. (1982). Calcium-calmodulin dependent phosphorylation of erythrocyte pyruvate kinase. *Biochimical and Biophysical Research Commununication*, Vol. 104, No 1, pp. 551-557

Nelson, M.J.; Ferrell J. E.; Huestis, W.H. (1979). Adrenergic stimulation of membrane protein phosphorylation in human erythrocytes. *Biochimica et Biophysica Acta*, vol. 558, No 1, pp. 136-140

Okafor, M.C.;.Schiebinger. R.J. Yingst, D.R. (1997). Evidence for a calmodulin-dependent phospholipase A2 that inhibits Na-K-ATPase. *American .Journal of Physiology*, vol. 272, (4 Pt 1), pp. C1365-C1372

Olearczyk, J.J.; Stephenson, A.H.; Lonigro, A.J.; Sprague, R.S. (2004). Heterotrimeric G protein Gi is involved in a signal transduction pathway for ATP release from erythrocytes. *American Journal of Physiology - Heart and Circulatory Physiology*, vol. 286, No 3, H940-H945

Orringer, E.P.; Roer, M.E. (1979). An ascorbate-mediated transmembrane-reducing system of the human erythrocyte. *Journal of Clinical Investigation*, vol. 63, No 1, pp. 53-58

Paajaste, M.; Nikinmar, M. (1991). Effect of noradrenaline on the methemoglobin concentration of rainbow trout red cells. *Journal of Experimental Zoology*, vol. 260, No 1, pp. 28-32

Pantaleo, A.; De Franceschi, L.; Ferru, E.; Vono, R.; Turrini, F. (2010). Current knowledge about the functional roles of phosphorylative changes of membrane proteins in normal and diseased red cells. *Journal of Proteomics*. Vol. 73, No 3, pp. 445-455

Pantaleo, A.; Ferru, E.; Carta F.; Mannu, F.; Simula, L. F.; Khadjavi, A.; Pippia, P.; Turrini, F. (2011). Irreversible AE1 Tyrosine Phosphorylation Leads to Mem brane Vesiculation in G6PD Deficient Red Cells . *PLoS ONE*, vol. 6, No 1, e15847, 1-8

Pasini, E.M.; Kirkegaard, M.; Mortensen, P.; Lutz, H.U.; Thomas, A.W.; Mann, M. (2006). In-depth analysis of the membrane and cytosolic proteome of red blood cells. *Blood*, vol. 108, No 3, pp. 791-801

Pasini, E.M.; Mann, M.; Thomas, A.W. (2010). Red blood cell proteomics. *Transfusion Clinique et Biologique*, vol. 17, No 3, pp. 151–164

Perry, S.F.; Wood, C.M.; Thomas, S.; Walsch, P.J. (1991). Adrenergic stimulation of carbon dioxide excretion by trout red blood cells in vitro is mediated by activation of Na+/H+ exchange. *Journal of Experimental Biology*, vol. 157, pp. 367-380

Petrov, V.; Amery, A.; Lijnen, P. (1994). Role of cyclic GMP in atrial-natriuretic-peptide stimulation of erythrocyte Na+/H+ exchange. *European Journal of Biochemistry*, vol. 221 No, pp. 195-1999

Petrov, V,; Lijnen P. (1996) Regulation of human erythrocyte Na+/H+ exchange by soluble and particulate guanylate cyclase. *American Journal of Physiology*; vol. 271, (5 Pt 1), pp. C1556-C1564

Phan-Dinh-Tuy, F.; Henry, J.; Rosenfeld, C.; Kahn, A.(1983), High tyrosine kinase activity in normal non proliferating cells. *Nature*, vol. 305, No 5933, pp. 435-440

Polychronakos, C.; Guyda, H. J.; Posner B. I. (1983). Receptors for the insulin-like growth factors on human erythrocytes. *Journal of Clinical Endocrinology and Metabolism*, vol. 57, No 2, pp. 436- 438.

Qadri, S.M; Föller, M.; Lang, F. (2009). Inhibition of suicidal erythrocyte death by resveratrol. *Life Sciences*, vol. 85, No (1-2), pp.33-38

Rasmussen, H.; Lake, W.; Allen, J.E. (1975). The effect of catecholamines and prostaglandins upon human and rat erythrocytes. *Biochimica et Biophys Acta*, vol. 411, No 1, pp. 63-73.

Rivera, A..; Jarolim, P.; Brugnara , C. (2002). Modulation of Gardos channel activity by cytokines in sickle erythrocytes. *Blood*, vol. 99, No 1, pp.357-603

Rizvi, S.I.; Incerpi, S.; Luly, P. (1994). Insulin modulation of Na/H antiport in rat red blood cells. *Indian Journal of Biochemistry and Biophysics*, vol. 31, No 2, pp. 127-130

Ron, D.; Kazanietz, M.G. (1999). New insight into the regulation of protein kinase C and novel phorbol ester receptors. *FASEB Journal*, vol. 13, No 13, pp. 1658-1676

Sacks, D.B.; Lopez, M.M.; Li, Z.; Kosk-Kosicka, D. (1996). Analysis of phosphorylation and mutation of tyrosine residues of calmodulin on its activation of the erythrocyte Ca (2+)-transporting ATPase. *European Journal of Biochemistry*, vol. 239, No 1, pp. 98-104.

Saido, T.C.; Sorimachi, H.; Suzuki, K. (1994). Calpain: new perspectives in molecular diversity and physiological/pathological involvement. *FASEB Journal*, vol. 8, No 11, pp. 814- 822

Saldanha, C.; Silva, A.S.; Gonçalves, S.; Martins-Silva, J. (2007). Modulation of erythrocyte hemorheological properties by band 3 phosphorylation and dephosphorylation. *Clinical Hemorheology and Microcirculation*, vol. 36, No 3, pp.183-194.

Santos-Silva, A.; Castro, E.M.; Teixeira, N.A.; Guerra, F.C; Quintanilha, A. (1998). Erythrocyte membrane band 3 profile imposed by cellular aging by activated neutrophils and by neutrophilic elastase. *Clinica Chimica Acta*, vol. 275, № 2, pp.185-196

Sardet, C.; Farounoux, P.; Poreysseger, J. (1991). Alpha-thrombin, epidermal growth factor, and okadaic acid activate the Na+/H+ exchanger, NHE-1, phosphorylating a set of common sites. *Journal of Biological Chemistry*, vol. 266, No 29, pp. 19166-19171

Sartori, M.; Ceolotto, G.; Semplieini, A. (1999). MAPKinase and regulation of the sodium-proton exchanger in human red cell. *Biochemica et Biophysica Acta*, vol. 1421, No 1, pp. 140-148

Sato, Y.; Miura, T. and Suzuki, Y. (1990). Interaction of pentoxifylline with human erythrocytes. I. Interaction of xanthine derivatives with human erythrocyte ghosts. *Chemical Pharmaceutical Bulletin (Tokyo)*, vol. 38, No 2, pp. 552-554

Sauvage, M.; Maziere, P.; Fathalah, H.; Girard, E.. (2000). Insulin stimulates NHE1 activity by sequential activation of phosphatidylinositol3-kinase and PKC zeta in human erythrocytes. *European Journal of Biochemistry*, vol. 267, No 4, pp. 955-962

Seghieri, G.; Anichini, R.; Ciuti, M.; Gironi, A.; Bennardini, F.; Franconi, F. (1997). Raised erythrocyte polyamine levels in non-insulin -dependent diabetes mellitus with great vessel disease and albuminuria. *Diabetes Research and Clinical Practice*, vol. 37, No 1, pp. 15-20

Sheppard, H.; Tsien, W.H. (1975). Alteration in the hydrolytic activity, inhibitor sensitivity and molecular size of the rat erythrocyte cyclic AMP-phosphodiesterase by calcium and hypotonic sodium chloride. *Journal of Cyclic Nucleotide Research,* vol. 1, No 4, pp. 237-242

Smallwood, J.I.; Gugi, B.; Rasmussen, H. (1988). Regulation of erythrocyte Ca2+ pump activity by protein kinase C. *Journal of Biological Chemistry,* vol. 263, No 5, pp. 2195-2202

Smith, T.J.; Davis, F.B.; Davis, P.J. (1989). Retinoic Acid Is a Modulator of Thyroid Hormone Activation of Ca2+- ATPase in the Human Erythrocyte Membrane. *Journal of Biological Chemistry,* vol. 264, No 2, pp. 687-689

Sommerson, J.; Mulligan, J.A.; Lozeman, F.J.; Krebs, E.G. (1987). Activation of casein kinase in response to insulin and to epidermal growth factor. *Proceedings of National Academy of Sciences of the United States of America,* vol. 84, No 24, pp. 8834-8838

Sprague, R.S.; Bowles, E.A.; Hanson, M.S.; DuFaux, E.A.; Sridharan, M.; Adderley, S.; Ellsworth, M.L.; Stephenson, A.H. (2008). Prostacyclin analogs stimulate receptor-mediated cAMP synthesis and ATP release from rabbit and human erythrocytes. *Microcirculation.* vol. 15, No 5, pp. 461-471

Sprague, R. S.; Ellsworth, M. L.; Stephenson, A H.; Lonigro, A. J. (2001). Participation of cAMP in a signal-transduction pathway relating erythrocyte deformation to ATP release *American Journal of Physiology- Cell Physiology,* vol. 281,No 4, C1158–C1164

Sun, I.L.; Crane, F.L.; Morré, D.J.; Löw, H.; Faulk, W.P. (1991). Lactoferrin activates plasma membrane oxidase and Na+/H+ antiport activity. *Biochemical and Biophysical Reseach Communication,* vol. 176, № 1, pp. 498-504

Takayama, Y.; Mizumachi, K. (2001). Effects of lactoferrin on collagen gel contractile activity and myosin light chain phosphorylation in human fibroblasts. *FEBS Letters,* vol. 508, No 1, pp.111-116.

Taleva, B.; Maneva, A.; Sirakov. L. (1999). Essential metal ions alter the lactoferrin binding to the erythrocyte plasma membrane receptors. *Biological Trace Element Research,* vol. 68, No 1, pp. 2-23

Tanaka, T.; Omata, Y.; Isamida, T.; Saito, A.., Shimazaki, K.; Yamauchi, K.; Suzuki, N. (1998). Growth inhibitory effect of bovine lactoferrin to Toxoplasma gonadii achyzoites in murine mactophages: tyrosine phosphorylation in murine macrophages induced by bovine lactoferrin. *Journal of Veterinary .Medicine,* vol. 60, № 3, pp. 369-371

Terra, H.T., Saad, M.J.; Carvalho, C.R.; Vicentin, D.L.; Costa, F.F.; Saad, S.T. (1998). Increased tyrosine phosphorylation of band 3 in hemoglobinopathies. *American Journal of Hematology,* vol. 58, No 3, pp. 224–230.

Thotathil, Z.; Jameson, M. B. (2007). Early experience with novel immunomodulators for cancer treatment. *Expert Opinion on Investigational Drugs,* Vol. 16, No 9, pp. 1391-1403.

Tsuda, K. (2006). Leptin receptor and membrane microviscosity of erythrocytes in essential hypertension. *American Joournal of Hypertension,* vol. 19, No 8, 874-875.

Tsukamoto, T.; Sonenberg, M. (1979). Catecholamine regulation of human erythrocyte membrane protein kinase. *Journal of Clinical Investation,* vol. 64, № 2, pp. 534-540.

Turrini, F.; Mannu, F.; Cappadero, M.; Ulliers, D.; Giribaldi, G.; Arese, P. (1994). Binding of naturally occurring antibodies to oxidatively modified erythrocyte band 3. *Bichimica et Biophysica Acta,* vol. 1190, No 2, pp. 287-303

Uhle, S.; Medalia, O.; Waldron, R.; Dumdey, R.; Henklein, P.; Bech-Otschir, D.; Huang, X.; Berse, M.; Sperling, J.; Schade R.; Dubiel, W. (2003). Protein kinase CK2 and protein kinase D are associated with the COP9 signalosome *The EMBO Journal*, vol. 22, No. 6, pp. 1302-1312

van den Akker, E.; Satchwell, T. J.; Williamson, R.C.; Toye, A. M. (2010). Band 3 multiprotein complexes in the red cell membrane; of mice and men. *Blood Cells Mol Dis*. Vol. 45, No 1, pp.1-8.

Vásárhelyi, B.; Szabó, T.; Vér, Á.; Tulassay, T. (1997). Measurement of Na^+/K^+-ATPase activity with an automated analyzer. *Clinical Chemistry*, vol. 43, No. 10, pp. 1986-1987

Villalonga, P.; Lopez-Alcala, C.; Chiloeches, A.; Gil, J.; Marais, R.., Bachs, O.; Agell N. (2002). Calmodulin prevents activation of Ras by PKC in 3T3 fibroblasts. *Journal of Biological Chemistry*, vol. 277, No 40, pp. 37929-37935

Villano, P.J.; Gallice, P.M.; Nicoara, A.E.; Honore, S.G.; Owczarczak, K.; Favre, R.G.; Briand, C.M. (2001). Polyamine secreted by cancer cells possibly account for the impairment of the human erythrocyte sodium pump. *Cell and Molecular Biology*, vol. 47, No 2, pp. 305-312

Wang, K.K.W.; Villalobo, A.; Roufogalis, B.D. (1989). Calmodulin binding proteins as calpain substrates. *Biochemical Journal*, vol. 262, No 3, pp. 693-706

Wang, K.K.; Wright, L.C.; Machan, C.L.; Allen, B.G.; Conigrave, A.D.; Roufogalis, B.D. (1991). Protein kinase C phosphorylates the carboxyl terminus of the plasma membrane Ca(2+)-ATPase from human erythrocytes. *Journal of Biological Chemistry*, vol. 266, No 14, pp. 9078-9085

Wei, T.; Tao, M. (1993). Human erythrocyte casein kinase II: characterization and phosphorylation of membrane cytoskeletal proteins. *Archives of Biochemistry and Biophysics*, Vol. 307, No 1, pp. 206-216

Weinberg, E.D. (2007). Antibiotic properties and applications of lactoferrin. *Current Pharmieutical Design*, vol. 13, No 8, pp. 801-811

Wright, L.C.; Chen, S.; Roufogalis. B.D. (1993). Regulation of the activity and phosphorylation of the plasma membrane Ca(2+)-ATPase by protein kinase C in intact human erythrocytes. *Archives of Biochemistry and Biophysics*, vol. 306, No 1, pp. 277-284.

Yingst, D.R.; Ye-Hu, J.; Chen, H.; Barrett, V. (1992). Calmodulin increases Ca^{2+}-dependent inhibition of Na, K-ATPase in human red blood cells. *Archives of Biochemistry and Biophysics*, vol. 295, pp. 49-54

Zancan, P.; Sola-Penna, M. (2005). Calcium influx: a possible role for insulin modulation of intracellular distribution and activity of 6-phosphofructo-1-kinase in human erythrocytes. *Molecular Genetics and Metabolism*, Vol. 86, No 3, pp. 392-400

Zazo, C.;Thiele, S.; Martín, C.; Fernandez-Rebollo, E.; Martinez-Indart, L.; Werner, R.; Garin, I.; Group, S.P.; Hiort, O.; Perez de Nanclares, G. (2011). Gsα activity is reduced in erythrocyte membranes of patients with psedohypoparathyroidism due to epigenetic alterations at the GNAS locus. *Journal of Bone and Mineral Research*, Vol. 26, No 8, pp.1864-1870.

Zhang, Y.; Dai, Y.; Wen, J.; Zhang, W.;Grenz, A.; Sun, H.; Tao, L.; Lu, G.; Alexander, D.C.; Milburn, M.V.; Carter-Dawson, L.; Lewis, D.E.; Zhang, W.; Eltzschig, H.K.;

Kellems, R.E.; Blackburn, M.R.;Juneja, H.S.; Xia, Y. (2011). Detrimental effects of adenosine signaling in sickle cell disease. *Nat Med.* Vol. 17, No 1, pp 79-86.

Zipser, Y.; Kosower. N.S. (1996). Phosphotyrosine phosphatase associated with band 3 protein in the human erythrocyte membrane. *Biochemical Journal*, vol. 314, (Pt.3), pp. 881-887

Zipser, Y.; Piade, A.; Barbul, A.; Korenstein, R.; Kosower, N.S. (2002). Ca2+ promotes erythrocyte band 3 tyrosine phosphorylation via dissociation of phosphotyrosine phosphatase from band 3. *Biochemical Journal*, vol. 368, (Pt 1), pp. 137- 144

Zipser, Y.; Plade, A.; Kosower, N.S. (1997). Erythrocyte thiol status regulates band phosphotyrosine level via oxidation/reduction of band 3 associated phosphotyrosine phosphatase. *FEBS Letters*, vol. 406, No (1-2), pp. 126-130

Zylinska, L.; Sobolewska, B.; Gulczynska, E.; Ochedalski, T.; Soszynski, M. (2002). Protein kinases activities in erythrocyte membranes of asphyxiated newborns. *Clinical Biochemistry*, vol. 35, No 2, pp. 93-98

Xie, Z.; Cai, T. (2003) Na+-K+- ATPase-mediated signal transduction: from protein interaction to cellular function. *Molecular Interventions*, vol. 3, No 3, pp.157-168

Role of Protein Kinase Network in Excitation-Contraction Coupling in Smooth Muscle Cell

Etienne Roux[1,2], Prisca Mbikou[3] and Ales Fajmut[4]

[1] *Univ. de Bordeaux, Adaptation Cardiovasculaire à l'ischémie, Pessac,*
[2]*INSERM, Adaptation Cardiovasculaire à l'ischémie, Pessac,*
[3]*Institute of Biomedical Technologies, Auckland University of Technology, Auckland,*
[4]*University of Maribor, Medical Faculty, Faculty of Natural Sciences and Mathematics
and Faculty of Health Sciences,*
[1,2]*France*
[3]*New Zealand*
[4]*Slovenia*

1. Introduction

The aim of this chapter is to present a review of the main protein kinases involved in the signalling pathways between the stimulation of smooth muscle cell and the resulting dynamic contraction.

As in striated muscle cells, contraction in smooth muscle cells (SMC) is primarily triggered by intracytosolic Ca^{2+} ($[Ca^{2+}]_i$) increase. However, by contrast with striated muscle, in SMCs $[Ca^{2+}]_i$ increase generates contraction by activation of the myosin light chain kinase (MLCK) via the formation of the Ca^{2+}-calmodulin-MLCK complex. Activated MLCK phosphorylates the 20kDa regulatory light chain (MLC_{20}) of the thick filament. Phosphorylated MLC_{20} allows myosin to bind to actin, and this phosphorylation is critical for SMC contraction, since its inhibition generally abolishes agonist-dependent contraction, whereas relaxation is induced by MLC_{20} dephosphorylation by myosin light chain phosphatase (MLCP). Hence, excitation-contraction coupling in SMCs critically depends on Ca^{2+}-dependent MLCK activation and the balance of MLCK/MLCP activity. Moreover, it has been shown that these two major enzymes can be modulated by several protein kinases such as protein kinase A (PKA), protein kinase C (PKC) and Rho kinase (RhoK). As a consequence, these protein kinases indirectly modulate the activity of the thick filament of myosin. Additionally, actin-myosin interaction can be modulated by proteins associated to the thin filament of actin such as caldesmon and calponin, which modulation depends on their phosphorylation by several protein kinases. It appears then that in SMCs, the excitation-contraction coupling is determined by interacting signalling pathways involving various protein kinases, so that the canonical MLCK/MLCP enzymatic balance is embedded in a complex network of protein kinases acting both on the thick and thin filaments of the contractile apparatus. The chapter will present the functional structure of the contractile apparatus of SMC, and detail its activation by the Ca^{2+}-calmodulin-MLCK complex and down regulation by MLCP, and the action of the main PK that have been shown to modulate the sensitivity of the contractile apparatus to Ca^{2+}. The resultant behaviour of the SMC stimulated by contractile agonists not only depends on the structure of the regulatory network that modulates the contractile apparatus but also on the dynamics of the reactions. Mathematical modelling of this signalling network is of great help to decipher how different protein kinases involved in this network participate to the time-dependent behaviour of the contractile system. Several theoretical models have been developed in this sense. The chapter will present the general principles of these models, their predictions and how they help in understanding the role of PK in the time course of the contractile response of SCMs.

In this chapter, examples are taken from airway smooth muscle cells, though this chapter is not limited to this tissue, and describes general mechanisms present in other SMC types.

2. Ca^{2+} signalling and the contractile apparatus

2.1 Ca^{2+} signalling in airway smooth muscle cells

In smooth muscle cells, as in heart and skeletal muscle, $[Ca^{2+}]_i$ is the primary intracellular messenger that generates contraction. Stimulation of smooth muscle cells by contractile agonists results in an increase in $[Ca^{2+}]_i$, which in turn activates the contractile apparatus. However, the mechanisms by which Ca^{2+} stimulates the formation of actin-myosin bridges critically differ in smooth muscle cells from striated muscle cells (see below). Though Ca^{2+} signalling is not primarily triggered via direct protein kinase activation by contractile agonists, several agents that contribute to the Ca^{2+} signal can be modulated by protein kinases and, on the other hand, activation of several protein kinases involved in the modulation of the contractile apparatus is Ca^{2+} sensitive. Hence, this section will present a general overview of the mechanisms of Ca^{2+} signalling in smooth muscle cells, including in airways.

$[Ca^{2+}]_i$ value in basal conditions is maintained around 100 nM, a low value compared to extracellular medium, around 1-2 mM, and intracellular organelles like the sarcoplasmic reticulum, by active mechanisms of Ca^{2+} efflux through the plasma membrane and Ca^{2+} pumping into intracellular Ca^{2+} stores. When cells are stimulated, $[Ca^{2+}]_i$ is increased via different mechanisms which relative importance depends on smooth muscle cell types and/or contractile agonists. Basically, the general mechanisms of $[Ca^{2+}]_i$ increase involve either intracellular Ca^{2+} influx or intracellular Ca^{2+} release from internal Ca^{2+} stores, or both (Sanders, 2001; Somlyo et al., 1994). Voltage-dependent Ca^{2+} occurs via L-type voltage-operated Ca^{2+} channel (VOC). Normal resting membrane potential in airway myocytes is around -60 mV (Roux et al., 2006), and is highly dependent on basal K^+ conductance. Some K^+ channels are active at rest, and contribute to the resting K^+ conductance and hence resting membrane potential. Closure of K^+ channels tends to depolarize the plasma membrane which in turn induces extracellular Ca^{2+} influx via VOC. By contrast, additional activation of K^+ channels tends to hyperpolarize the plasma membrane, to inhibit VOC-dependent Ca^{2+} influx and hence to induce relaxation. Voltage-dependent Ca^{2+} influx and subsequent contraction is called the electromechanical coupling. In parallel to voltage-operated Ca^{2+} entry, Ca^{2+} influx can be generated by voltage-independent membrane channels. Receptor-operated channels can be opened by direct binding of the agonist on the membrane receptor, like P2X purinergic receptors (Mounkaila et al., 2005). Another voltage-independent source of Ca^{2+} entry is Ca^{2+} influx through store-operated Ca^{2+} channels (SOC). These channels are activated by emptying of intracellular Ca^{2+} stores (Marthan, 2004).

The other origin of $[Ca^{2+}]_i$ increase is Ca^{2+} release from intracellular organelles (Sanders, 2001). The major one from which Ca^{2+} is released upon contractile stimulation is the sarcoplasmic reticulum (SR). Ca^{2+} release from the sarcoplasmic reticulum, with internal Ca^{2+} concentration in the millimolar range, occurs via two main types of sarcoplasmic receptors, the inositol-trisphosphate receptors (InsP$_3$Rs) and the ryanodine receptors (RyRs). InsP$_3$Rs are activated by InsP$_3$, which is produced from phosphatidylinositol phosphate by phospholipase C (PLC) upon stimulation by contractile agonists (Roux *et al.*, 1998). Activation of contraction via primary InsP$_3$ production and InsP$_3$R-operated Ca^{2+} release, a consequence of enzymatic activation of PLC, is called the pharmacomechanical coupling, in opposition with the electromechanical coupling described above. RyRs are physiologically activated by an increase in Ca^{2+} concentration on the cytosolic face of the RyR, or by direct mechanical coupling between L-type VOC, and it has been shown that cyclic ADP ribose is a co-agonist of RyR (Prakash *et al.*, 1998). RyR activation by Ca^{2+} self-amplifies $[Ca^{2+}]_i$ increase, whatever its initial mechanism. This Ca^{2+}-induced Ca^{2+} release (CICR), however, is not observed in all smooth muscle cells. In human bronchial smooth muscle, for example, RyR, though present and functional, do not participate in the acetylcholine-induced Ca^{2+} response (Hyvelin *et al.*, 2000a).

Basal maintenance of low $[Ca^{2+}]_i$ and Ca^{2+} removal form the cytosol upon and after stimulation is due to active mechanisms that either extrude Ca^{2+} in the extracellular medium or uptake Ca^{2+} into intracellular Ca^{2+} stores. Ca^{2+} extrusion is mainly due to the activity of the plasma membrane Ca^{2+} ATPase (PMCA), and the Na^+-Ca^{2+} exchanger (NCX) (Sanders, 2001).The main mechanisms of Ca^{2+} uptake from the cytosol are Ca^{2+} pumping back into the SR by sarcoendoplasmic Ca^{2+} ATPase (SERCA), Ca^{2+} uptake into the mitochondria. Also, several Ca^{2+}-binding proteins can buffer cytosolic Ca^{2+} and hence decrease $[Ca^{2+}]_i$ (Gunter *et al.*, 2000; Roux *et al.*, 2004).

The Ca^{2+} signal is not only the overall increase in the cellular Ca^{2+} pool and subsequent $[Ca^{2+}]_i$ but also, and most importantly, the pattern of the dynamic change in $[Ca^{2+}]_i$. Indeed, the time-dependent variation in the shape of $[Ca^{2+}]_i$ is the actual "Ca^{2+} signal", since Ca^{2+} binding to signalling protein, e. g., calmodulin, depends on cytosolic Ca^{2+} concentration. Overall $[Ca^{2+}]_i$ measurements generally show a transient increase in $[Ca^{2+}]_i$ (Ca^{2+} peak) followed by a progressive decay to a steady-state value (Ca^{2+} plateau) and/or by Ca^{2+} oscillations (Bergner *et al.*, 2002; Hyvelin *et al.*, 2000b; Kajita *et al.*, 1993; Liu *et al.*, 1996). Changes in $[Ca^{2+}]_i$ are not uniform within the cytosol, and studies have evidenced the role of local change in Ca^{2+} signalling (Prakash *et al.*, 2000). Ca^{2+} signalling should hence be defined as spatiotemporal changes in $[Ca^{2+}]_i$, from which depends the activity of the contractile apparatus, and other cell functions. The shape of this Ca^{2+} signal critically depends on the dynamics of Ca^{2+} fluxes between intra- and extracellular media and also between intracellular Ca^{2+} compartments. The dynamic relationship between the calcium signal and MLCK activity and its consequence on force development will be discussed in more detail in the section "Theoretical modelling of PK and ASMC contraction".

2.2 Smooth muscle contractile apparatus

2.2.1 Components and general organization

The contractile apparatus of smooth muscle is basically constituted of thick filaments of myosin II and thin filament of actin and associated proteins, the thin/thick filament ratio being about 20/1 to 30/1 (Kuo *et al.*, 2003; Somlyo *et al.*, 1983). These filaments are not organized in sarcomeres and do not form well individualized myofibrils. Thick filaments are anchored on dense bodies in the cell and dense area on the plasma membrane and actin filaments are positioned between thick filaments. Dense bodies and filaments are connected by non-contractile intermediate filaments that constitute an intracellular network.

Thick filaments are 1.5 μm long and 12-14 nm in diameter, and are composed of polymerized monomers of myosin. Each monomer of myosin is formed by the association of 2 identical heavy chains (MHC) complexed to 2 pairs of light chains (MLC). Each of the 2 heavy chains has a C-terminal extended α-helix twisted to form a single tail and an N-terminal head. 4 isoforms of MHC have been described in smooth muscle cells, corresponding to 4 alternatively-spliced variants derived from the single smooth muscle myosin heavy chain gene MYH11, differing in their amino-terminal and carboxy-terminal portions (Hong *et al.*, 2011). Polymerized and monomeric myosins are in equilibrium and the stimulation of the myocyte increases myosin polymerization. The distal portion of the myosin head has the ATPase enzymatic activity required for its motor function, and a binding site for actin. A pair of light chains, a 17 kDa one and a 20 kDa one, is complexed with each of the head of the myosin heavy chain. The role of MLC_{17} is unclear, and it is thought to contribute to the stability of the molecule of myosin. Phosphorylation of the so-called regulatory MLC_{20} is required for actin-myosin binding, and hence phosphorylation/dephosphorylation of MLC_{20} regulates actin-myosin cross bridge and contraction. 2 residues, located in the amino-terminal portion of the protein, threonine 18 and serine 19, can be phosphorylated by MLCK. The major site of phosphorylation by MLCK in intact myocyte is ser19, though thr18 can also be phosphorylated by MLCK. It has been shown that Rho kinase can also directly phosphorylate MLC_{20}, but, at least in airway smooth muscle, Rho kinase regulates phosphorylated/dephosphorylated MLC_{20} ratio by acting on MLCP (Mbikou *et al.*, 2011). It has been shown that PKC may also phosphorylate MLC_{20} but on different residues, serine 1, serine 2 and threonine 9 (Barany *et al.*, 1996). However, it is generally admitted that PKC contributes to smooth muscle contraction mainly via MLCP inhibition.

Smooth muscle thin filaments are formed by a double helix of F-actin complexed with tropomyosin, caldesmon and calponin. Globular smooth muscle α-actin molecules assemble into filamentous polymers of F-actin, and 2 parallel strands of F-actin rotate on each other to form a double helix structure. F-actin filaments are about 1 μm long and 5-7 nmn in diameter. As in striated muscle, tropomyosin binds to F-actin in the furrow formed by the double helix of actin, and contributes to actin filament stabilization (Shah *et al.*, 2001). In contrast with striated muscle, here is no troponin on the thin filament of actin in smooth muscle, and other proteins, caldesmon and calponin, are associated with actin. Caldesmon and calponin are regulatory proteins that can be phosphorylated by several kinases and their phosphorylation modulates the formation of the actin-myosin bridge. Regulation of contraction by caldesmon and calponin is detailed below.

2.2.2 Actin-mysosin bridge cycling

Basically, muscle contraction is based on cycling attachment and detachment between thick filaments of myosin and thin filaments of actin. Interaction between actin and myosin, which corresponds to the formation of the actin-myosin crossbridge, is triggered by the head of myosin. As seen above, in smooth muscle, phosphorylation of MLC_{20} is the key event that allows the formation of the actin-myosin bridge, a step required for contraction to occur (Harnett *et al.*, 2003; Wingard *et al.*, 2001). MLC_{20} phosphorylation induces a conformational change of the neck of myosin so that the head of myosin can bind to actin. This allows the rotation of the head of myosin responsible for the "sliding" of actin and myosin filaments. MLC_{20} phosphorylation also activates the ATPase activity of the myosin head, followed by actin and myosin detachment. When MLC_{20} remains phosphorylated all along the crossbridge cycle, crossbridge cycling is rapid, and associated with high ATP consumption. However, sustained contraction can occur even if $[Ca^{2+}]_i$ and subsequent MLC_{20} phosphorylation decrease (Mbikou *et al.*, 2006), and this is generally attributed to the fact that if dephosphorylation of MLC_{20} occurs after the attachment of myosin on actin, crossbridge cycle goes on but at a slower rate, in particular in the stage where dephosphorylated myosin detaches actin. These maintained dephosphorylated crossbridges that cycle at a slow rate are termed latch-bridges.

Hence, 2 types of crossbridges cycling can occur in smooth muscles, a fast, phosphorylated one (during which MLC_{20} remains phosphorylated) and a slow, partially dephosphorylated one (during which MLC_{20} is phosphorylated after actin-myosin attachment), and, accordingly, a 4-state model of the contractile apparatus has been proposed, initially by Hai and Murphy (Hai *et al.*, 1988a). This model is presented in Figure 1. The first state corresponds to unbound actin and unphosphorylated MLC_{20}, the second one to unbound actin and phosphorylated MLC_{20}, the third one to actin bond to phosphorylated MLC_{20}, and the fourth one to actin bond to dephosphorylated MLC_{20}. Transition between state 1 and 2, and 3 and 4, is reversible and depends on MLC_{20} phosphorylation/dephosphorylation. Transition between state 2 and 3 is reversible and corresponds to the phosphorylated crossbridge cycling. Rate constant from state 1 to 4 is very slow, and it can be considered that transition from state 4 to 1 is almost irreversible. State 1-2-3-4-1 cycling corresponds to the latch-bridge cycling.

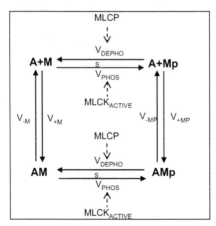

Fig. 1. **4 state model of smooth muscle contractile apparatus.** A, actin; M, unphosphorylated myosin; Mp, phosphorylated myosin; AM, unphosphorylated actin-myosin bridge; AMp, phosphorylated actin-myosin bridge; MLCK, myosin light chain kinase; MLCK, myosin light chain phosphatase; CaM, calmodulin. V, rate constants of the reactions.

The main enzyme by which MLC_{20} is phosphorylated is MLCK. Dephosphorylation of phosphorylated MLC_{20} is catalysed by MLCP. Though other enzymes can contribute to smooth muscle contraction, muscle contraction critically depends on the balance between MLCK and MLCP activity.

2.3 Modulation of MLC_{20} phosphorylation: The MLCK/MLCP balance

2.3.1 MLCK

MLCK has a ubiquitous distribution and is present in non-muscular cells as well as cardiac, skeletal and smooth muscle cells. Several genes encode for distinct isoforms of MLCK. In humans, MYLK2 and MYLK3 genes encode for the MLCK isoforms expressed in skeletal and cardiac myocytes, respectively, and the smooth muscle isoform is encoded by the MYLK1 gene. This gene has 2 initiation sites generating a short isoform corresponding to the smooth muscle MLCK and a long isoform corresponding to nonmuscle MLCK. The domain structure of smooth muscle mammalian MLCK is shown in Figure 2. MLCK has 2 CaM-binding domains, one at its aminoterminal portion, composed of 3 DFRxxL motifs, and the other at its carboxyterminal portion, before the IgT domain located at the N-terminus. The DFrxxL domain is also a binding site to actin, and Ca^{2+}-CaM binding to this domain results in weakened actin binding. Some results suggest that Ig1/Ig2 domains bind to actin. IgT domain is a binding site to myosin. The kinase domain is the catalytic portion of MLCK and includes a binding site for ATP and for MLC_{20}. The role of the other domains is unknown (Hong *et al.*, 2011). Hence, MLCK interacts with myosin on two sites. The catalytic core of MLCK interacts with the aminoterminal portion of MLC_{20}, which allows phosphorylation MLC_{20}. The IgT domain of MLCK may also interact with myosin at the head-neck junction. It is thought that IgT binding to myosin near the MLC_{20} increases the catalytic activity of MLCK. The concentration of MLCK is likely to vary from muscle to muscle, but typical values range from 1 to 8 μM, which is low compared with the typical concentration of myosin, classically ranging from 50 to 100 μM.

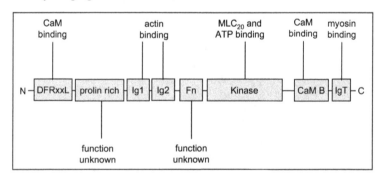

Fig. 2. **Structural scheme of myosin light chain kinase (MLCK) and its different domains.** In the absence of calmodulin (AcM) binding, the catalytic site (Kinase) is masked by the N-terminal CaM binding and IgT domains and hence inactive. Binding of Ca2+-CaM complex induces a conformational change and unmasks the kinase domain hence active.

The major substrate of MLCK in smooth muscle is the smooth muscle myosin, but recent results showed that non muscle myosin, which is also expressed in smooth muscle myosin, can be a substrate for MLCK (Yuen *et al.*, 2009). MLCK phosphorylates MLC_{20} on serine 19 and, with lower enzymatic activity, on threonine 18 (Deng *et al.*, 2001). MLCK activity is essential for airway smooth muscle contraction, as it has been evidenced by pharmacological inhibition of MLCK in rat trachea (Mbikou *et al.*, 2006) and knockout mice (Zhang *et al.*, 2010).

The most important regulator of MLCK activity is the Ca^{2+}-CaM complex. MLCK unbound to Ca^{2+}-CaM is in an auto-inhibitory state. The IgT domain and the adjacent CaM binding domain constitute an auto-inhibitory sequence that masks the catalytic core. Ca^{2+}-CaM binding to the auto-inhibitory site induces a conformational change. This displacement of the auto-inhibitory sequence unmasks the catalytic core and thereby reveals the enzymatic activity of MLCK. On the other hand, phosphorylation of MLCK on serine 512 by several kinases such as protein kinase A in the C-terminal CaM binding domain decreases MLCK affinity for the Ca^{2+}CaM complex, thereby inactivating MLCK even in the presence of Ca^{2+}, a mechanism responsible for MLCK desensitization (Stull *et al.*, 1993). CaM is a rather small protein (16700 Da) consisting of a single polypeptidic chain of 148 to 149 aminoacids. 3 distinct domains can be identified: an N-terminal domain, a central helix domain, and a C-terminal domain. Both amino- and carboxy-terminal domain possesses 2 binding sites for Ca^{2+}. The 4 sites have a high affinity for Ca^{2+}, though the affinity of the C-terminal sites is higher. These sites have an EF-hand structure. Ca^{2+} binding induces a conformational change of the F-helix of the EF-hand motif. In this conformation CaM binds to several proteins and modulates their activity. Amongst almost 40 different proteins that have been shown to be regulated by CaM, the most important in smooth muscle are MLCK and CaM-kinase II.

2.3.2 MLCP

MLCP is a holoenzyme composed of three subunits: the 38 kDa phosphatase catalytic subunit PP1cδ, the 110 kDa regulatory subunit MYPT1, and a 20 kDa noncatalytic subunit, M-20, which role remains unclear (Hartshorne *et al.*, 1999; Hartshorne *et al.*, 1998). PP1cδ is a serine/threonine phosphatase that dephosphorylates MLC_{20} on serine 19 and threonine 18. Disruption of the quaternary structure of MLCP, either by dissociation of the regulatory subunit MYPT1 to PP1c or by disruption of the interactions between the three subunits, decreases MLCP activity toward MLC_{20} dephosphorylation.

PP1c alone is capable of dephosphorylating the MLC_{20}, but its activity is potentiated when complexed to MYPT1. Indeed, MYPT1s targets PP1c to its substrate, MLC_{20}, and confers substrate specificity to the phosphatase. A recent study by crystallography has revealed a basic structure which potentiates the catalytic activity of the holoenzyme (Terrak *et al.*, 2004). Structurally, MYPT1 subunit possesses a C-terminal leucine-zipper domain and a binding site of protein kinase G. Four N-terminal amino acids, KVKF, give a strong affinity for PP1c. MYPT1 also has six repeated domains of ankyrin, which serve as anchoring sites for protein-protein interactions. Several kinases are able to phosphorylate MYPT1 at various sites: Rho kinase, MYPT kinase (also known as Zip-like kinase), integrin-like kinase, myotonic dystrophy kinase (DMPK). PKC also has been shown to phosphorylate ankyrin repeated domains and inhibit the interaction between MYPT1 and PP1c (Toph *et al.*, 2000). In addition to this negative regulation of MLCP, MYPT1 phosphorylation is also associated with an upregulation. Recent studies have reported that protein kinases dependent on cyclic nucleotides such as PKA and PKG are able to phosphorylate MYPT1 on serine 692, 695 and 852. The S695 and S852 residues are adjacent to two residues, T696 and T853, that are phosphorylation sites of MYPT1 by RhoK. Phosphorylation at S695 by PKG or PKA prevents RhoK phosphorylation at S696 and vice versa (Wooldridge *et al.*, 2004), and it is supposed that phosphorylation at S852 has a similar effect on T853 phosphorylation. Thus the phosphorylation of MYPT1 by PKA and PKG protein antagonizes the inhibition of MLCP induced by RhoK (Grassie *et al.*, 2011).

CPI-17 is another regulation pathway of MLCP activity (Ito *et al.*, 2004). CPI-17 is a 17 kDa protein composed of 147 aminoacids that can inhibit the activity of the MLCP holoenzyme as well as that of isolated PP1c. Phosphorylation of CPI-17 at threonine 38 enhances its inhibitory potency about 1000 fold. The proposed mechanism is that CPI-17 binds at the PP1c active site, resulting in formation of a complex of MYPT1/PP1C/CPI-17 (Eto *et al.*, 2004). The major kinase responsible for in *vivo* CPI-17 phosphorylation is PKC, especially PKCα and PKCδ. *In vitro* studies have shown that other kinases can phosphorylate CPI-17 at T38, including Rho kinase, protein kinase N (PKN), MYPT-kinase, integrin-like kinase and PKA (Ito *et al.*, 2004). The scheme of the regulation of MLCP by kinase activity is given in Figure 3.

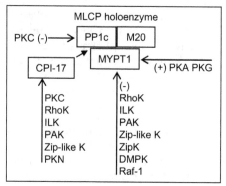

Fig. 3. **Regulation of MLCP activity by protein kinases.** DAP kinase, death-associated protein kinase;; DMPK, myotonic dystrophy protein kinase; ILK, integrin-linked kinase; MLCP, myosin light chain phosphatase; MYPT1, myosin phosphatase target subunit 1 of MLCP; M20, 20 kDa non-catalytic subunit of MLCP; PAK, p21-activated protein kinase; PKA, protein kianse A; PKC, protein kinase C; PKG, protein kinase G; PKN, protein kinase N; PLA2, phospholipase A2; PP1c, catalytic subunit of type1 protein phosphatase; RhoK, Rhokinase; ZipK; zipper-interacting protein kinase (adapted from (Hirano *et al.*, 2003)).

2.4 Thin filament-associated regulation of AM bridge cycling

2.4.1 Caldesmon

Caldesmon is a small protein of 87 kDa associated with the thin filaments of actin and involved in the regulation of smooth muscle contraction. Structurally, caldesmon possesses 4 distinct domains and can bind to actin, tropomyosin, myosin and calmodulin. The fourth domain located at the C-terminus is the most important since it is the binding domain to actin, to myosin and to Ca^{2+}-binding proteins. Caldesmon interacts stoichiometrically with tropomyosin to consolidate the actin filaments (Gusev, 2001), and contributes to the sensitivity of the myofilament to Ca^{2+}. In the absence of Ca^{2+}, the fourth domain interacts closely with actin and inhibits the ATPase activity. $[Ca^{2+}]_i$ increase induces a conformational change of the actin-tropomyosin-caldesmon complex, allowing the ATPase activity of myosin. In addition, it allows an interaction between the thin and thick filaments as it is able to bind to both actin and myosin and this property seems to be involved in the sustained phase of the contractile response (Marston et al., 1991). In vitro studies have shown that caldesmon can be a substrate for several kinases (Kordowska et al., 2006).

2.4.2 Calponin

Calponin is a molecule of approximately 34 kDa related to the skeletal muscle troponin T. Calponin is present in smooth muscle at a concentration almost equivalent to that of tropomyosin. Calponin binding to F-actin disrupts actin-myosin interaction and hence negatively regulates the contraction of smooth muscle. Calponin inhibits the sliding of actin filament on myosin and the ATPase activity of myosin. The inhibitory effect of calponin is suppressed by high $[Ca^{2+}]_i$ values, but only in the presence of calmodulin. Phosphorylation and dephosphorylation of calponin plays a key role in the regulation of smooth muscle contraction. Indeed, the inhibitory effect of calponin is abolished when calponin is phosphorylated by CaM kinase II or PKC, and is restored when calponin is dephosphorylated by phosphatase 2A (Winder et al., 1998; Winder et al., 1990). Based on these properties, 3 physiological roles have been identified for calponin: (1) maintenance of relaxation at rest when $[Ca^{2+}]_i$ is low; (2) Ca^{2+}-dependent contribution, involving calmodulin, to contraction when $[Ca^{2+}]_i$ is increased and, (3) Ca^{2+}-independent contribution to contraction via PKC-dependent phosphorylation.

3. Role of protein kinases in the modulation of excitation-contraction coupling

3.1 General overview

3.1.1 Modulation of Ca^{2+} signalling

Though protein kinases modulate smooth muscle contraction by acting mainly downstream of the Ca^{2+} signal, some of them have been shown to act upstream of the Ca^{2+} signal and regulate the activity of structures involved in Ca^{2+} signal encoding. It has been shown by in vitro studies that PKA can phosphorylate $InsP_3$ receptors on serine 1755 and serine 1589 (Ferris et al., 1991a). Studies using purified receptor protein reconstituted in liposomes have also evidenced that PKC and CaM kinase II can phosphorylate $InsP_3$ receptors (Ferris et al., 1991b). Ryanodine receptors also can be phosphorylated by PKA, PKC and CaM kinase II (Coronado et al., 1994). These phosphorylations may play an important role in smooth muscle physiology since Ca^{2+} release from SR plays a critical role in Ca^{2+} homeodynamics in airway myocyte (Ay et al., 2004; Roux et al., 2004). Indeed, contractile agonists such as acetylcholine act via $InsP_3$ production but also activate PKC (Mbikou et al., 2006). On the other hand, PKA stimulation is a important pathway involved in bronchorelaxation since β_2-adrenergic stimulation, a major pharmacological treatment of asthma, acts mainly via Gs protein stimulation, adenylate cyclase activation, cAMP production and PKA activation (Anderson, 2006; Johnson, 2006). Extracellular ATP has also been shown to induce airway relaxation via PKA activation (Mounkaila *et al.*, 2005). Protein kinase may also modulate Ca^{2+} homeostasis by acting on voltage-dependent Ca^{2+} entry. As explained above, K^+ channels critically contribute to the maintenance of myocyte membrane voltage, and it has been shown in vascular smooth muscle that PKA and PKC can phosphorylate several voltage-dependent K^+ channels, Ca^{2+}-activated K^+ channels and ATP-dependent K^+ channels, thus regulating the electromechanical coupling (Ko *et al.*, 2008). Phosphorylation by PKA activates these K^+ channels and hence tends to hyperpolarize the plasma membrane and limit extracellular Ca^{2+} entry, whereas phosphorylation by PKC has an opposite effect. The effective contribution of these mechanisms in airway in unclear, since it has been shown that in airways activation of Ca^{2+}-activated K^+ channels does not seem to have a significant effect on β-adrenergic relaxation (Corompt *et al.*, 1998).

3.1.2 Modulation of the Ca2+ sensitivity of the contractile apparatus

The signalling pathways capable of modulating the contraction of a given Ca^{2+} signal constitute the modulation of the sensitivity of the contractile apparatus to Ca^{2+}. As detailed previously, the importance of contraction depends on the balance between MLCK and MLCP activity, which is under control of several signalling pathways. In parallel with the canonical MLCK-dependent MLC_{20} phosphorylation, in several smooth muscles, MLC can be directly phosphorylated by Ca^{2+}-independent kinases like ZIP kinases and Integrin-linked kinase (ILK), which can also inhibit MLCP (Deng *et al.*, 2002; Huang *et al.*, 2006; Murthy, 2006; Niiro *et al.*, 2001). Rho kinase, activated by small GTPase (Rho), which seems to be activated both by Ca^{2+}-dependent and Ca^{2+}-independent pathways, can phosphorylate directly MLC and, by inhibition of MLCP, indirectly increase MLC phosphorylation and hence regulate smooth muscle contraction (Bai *et al.*, 2006; Murthy, 2006; Schaafsma *et al.*, 2006). Also, contractile agonists may, in parallel to MLCK activation, inhibit MLCP activity. PKC can inhibit MLCP, and hence modulate contraction (Bai *et al.*, 2006). By contrast, relaxant agonists, such as $β_2$-agonists in airways, may act via MLCP activation and/or MLCK inhibition (Janssen *et al.*, 2004; Johnson, 1998). So, in addition with the Ca^{2+}-calmodulin-MLCK pathway, other Ca^{2+}-dependent and independent enzymatic pathways regulate the contractile apparatus in smooth muscles, which status depends on the balance between MLCP/MLCK phosphorylation and dephosphorylation. Additionally, smooth muscle contraction can be modulated independently from MLC_{20} phosphorylation via caldesmon and calponin. The major kinases involved in smooth muscle contraction are listed below.

3.2 RhoK

Rho-associated kinase (Rho-kinase), originally identified as an effector of the small GTPase Rho, has been shown to play a major role in many processes including cell migration, neuronal polarisation, cytokinesis and cell contraction. It is a serine/threonine kinase structurally related to myotonic dystrophy kinase as there is 72% homology in the kinase domains. Different studies support the idea that Rho-kinase exists as a dimer resulting from parallel association through the coil-coil domain (Shimizu *et al.*, 2003). There are two Rho-kinase members, Rho-kinase I/ROKβ/p160ROCK and Rho-kinase II /ROKα/Rho kinase, which share 65% sequence identity and 95% sequence similarity at the amino acid level (Riento *et al.*, 2003b). The kinase domain is highly conserved between these two proteins (83% identical), suggesting that they may have similar substrate specificity. The consensus phosphorylation sequence for Rho-kinase is R/KXS/T or R/KXXS/T (X is any amino acid). The Rho-kinase protein is composed of three domains: a N-terminal kinase domain, a central coiled-coil domain containing a Rho binding site (RhoBD), and a C-terminal pleckstrin homology-like domain (PH-like domain) containing a Cys-rich region similar to the C1 domain of protein kinase C. The C-terminal region including the RhoBD plus the PH domain has been shown to directly interact with the kinase domain to inhibit its activity (Amano *et al.*, 1999). The interaction between Rho•GTP and the RhoBD releases this autoinhibition and thus activates the kinase (Ishizaki *et al.*, 1996). In addition to Rho other small as GTPases such Rnd3/RhoE, Gem and Rad can bind Rho-kinase outside of the Rho-binding region and inhibit its function (Komander *et al.*, 2008; Riento *et al.*, 2003a). Arachidonic acid has been also shown to activate Rho-kinase via its PH domain (Araki *et al.*, 2001). Although both Rho-kinase I and Rho-kinase II proteins are ubiquitously expressed in most tissues, higher levels of Rho-kinase II are found in brain and muscles whereas higher levels of Rho-kinase I are found in non-neuronal tissues such as liver, lung and testis (Leung *et al.*, 1996; Nakagawa *et al.*, 1996). Functional differences have been reported between Rho-kinase I and Rho-kinase II. For instance Rho-kinase I seems to be essential for the formation of stress fibres, whereas Rho-kinase II seems important for phagocytosis and cell contraction, both being dependent on MLC phosphorylation (Wang *et al.*, 2009; Yoneda *et al.*, 2005). Binding tests revealed that RhoE preferentially binds Rho-kinase I, but not Rho-kinase II, whereas MYPT1 binds only Rho-kinase II (Komander *et al.*, 2008; Wang *et al.*, 2009).

There has been a great deal of interest in the involvement of the Rho/ROCK signalling pathway in excitation–contraction coupling (Iizuka *et al.*, 1999; Mbikou *et al.*, 2011; Yoshii *et al.*, 1999). Activation of the Rho/Rho-kinase pathways is likely due to the stimulation of the G-coupled protein receptor. Stimulation of M_3 muscarinic receptors agonist for instance activates the Gq, G_{12} and G_{13} α-subunits which cause the cascade activation of GEFs, RhoA and Rho-kinase leading to Ca^{2+} sensitization (Hirano *et al.*, 2004; Somlyo *et al.*, 2003). It is well-known that Rho/Rho-Kinase signalling modulates Ca^{2+} sensitivity of the smooth muscle likely either by inhibition of the MLCP activity or by direct phosphorylation of the MLC_{20}. Precisely, Rho-kinase has been shown to phosphorylate the MYPT1 subunit of the MLCP at two inhibitory sites, Thr696 and Thr853, by that causing the dissociation between the MLCP and MLC_{20} (Feng *et al.*, 1999; Kawano *et al.*, 1999; Kimura *et al.*, 1996; Velasco *et al.*, 2002). Phosphorylation of MYPT1 by Rho-kinase alters the MYPT1-PP1c interaction, which decreases MLCP activity toward MLC_{20}. Rho-kinase is also able to directly phosphorylate Ser19 and Thr18 in non-muscle cells (Ueda *et al.*, 2002), but the direct contribution of Rho-kinase to phospho-MLC_{20} levels *in vivo* is not yet proven. Moreover, Rho kinase exerts an inhibitory activity toward PP1c subunit of MLCP upon the phosphorylation of the latter (Eto *et al.*, 1995; Eto *et al.*, 1997). Taken together, the mechanisms triggered by Rho-kinase activity enhance the level of phosphorylated MLC_{20} and, consequently, myosin ATPase activity and therefore contraction. *In vitro* tests have shown that Rho-kinase is also able to phosphorylate CPI17 (MLCP inhibitor protein), thus inducing the inhibition of MLCP and thereby raising the level of phosphor MLC20 (Amano *et al.*, 2010).

3.3 PKC

The protein kinase C (PKC) family is the largest serine/threonine-specific kinase family known. It embraces a large family of enzymes that differ in structure, cofactor requirements and function (Nishizuka, 1995). Depending on their cofactor requirements, the homologous group of PKC can be divided into three groups as follows: the group of conventional (c)PKC isoforms (α, β_I, β_{II} and γ), that require Ca^{2+} and diacylglycerol (DAG) to become activated; the group of novel (n)PKC isoforms (δ, ϵ, ζ, θ and μ) that require only DAG; and group of the atypical (a)PKC isoforms, namely ζ, ι and λ (the mouse homologue of human PKCι), that require neither Ca^{2+} nor DAG. The general structure of a PKC molecule consists of a catalytic domain in N-terminal and a regulatory pseudosubstrate in C-terminal, both framing 3 distinct sites able to bind specifically ATP, Ca^{2+} or phosphatidyl serine. The pseudosubstrate region is a small sequence of amino acids that binds the substrate-binding cavity in the catalytic domain, thus keeping the enzyme inactive. The activity of PKC is controlled by its compartmentalization within the cell. All PKC family members possess a phosphatidylserine binding domain for membrane interaction. The expression and distribution of PKC isoforms is tissue- and species–specific. Some isoforms (e.g. PKCα) are ubiquitously expressed in tissues whereas others seem to be restricted to certain tissues (Webb et al., 2000). In ASM, the protein and mRNA expression of PKC isoforms differs depending on the specie. In human trachealis for instance, there is expression of the conventional α, β_I, and β_{II} PKC isoforms as well as novel (δ, ϵ, ζ, θ) and atypical (ζ) (Webb et al., 1997) while canine trachealis does not express the α isoform and bovine ASM expresses the θ-PKC variants (Webb et al., 2000).

It is well known that the stimulation of a Gq protein-coupled receptor, for instance by cholinergic agonist, induces the elevation of the cytosolic Ca^{2+} concentration and the production of DAG by phospholipase C. Both Ca^{2+} and DAG bind to the C2 and C1 domain, respectively, and recruit PKC to the membrane (Bell et al., 1991; Huang, 1989). This interaction with the membrane results in release of the pseudosubstrate from the catalytic site and activation of the enzyme(Lester et al., 1990).

PKC may modulate the sensitivity of the contractile apparatus to Ca^{2+}, since it can inhibit MLCP activity. However, the effectiveness of PKC contribution to airway contraction remains controversial. PKC has been shown to be involved in force maintenance in human airways (Rossetti et al., 1995). However, other studies do not implicate PKC in agonist-induced Ca^{2+} sensitization and point to other effectors such as the small GTP-binding protein p21rho (Akopov et al., 1998; Itoh et al., 1994; Otto et al., 1996; Yoshii et al., 1999). Our recent studies have shown that PKC contributes to the sustained phase, but not to the initial phase, of cholinergic-induced contraction in rat airways (Mbikou et al., 2006).

3.4 PKA

Cyclic-AMP-dependent protein kinase (PKA) is an ubiquitous mammalian enzyme which catalyzes Ser/Thr phosphorylation in protein substrates that in turn control a wide range of cellular functions including gene regulation, cell cycle, metabolism and cell death (Shabb, 2001). This tetrameric holoenzyme comprises two catalytic (C) subunits that possess kinase activity and two inhibitory regulatory (R) subunits, each including two tandem cAMP binding domains, i.e. CBD-A and CBD-B (Johnson et al., 2001). cAMP is the essential second messenger that activates PKA (Berman et al., 2005; McNicholl et al.). In the absence of cAMP, the R-subunit and the C-subunit create a complex that blocks substrate access and thus prevents the kinase activity. cAMP binding to the R-subunits releases these inhibitory interactions and unleashes the C-subunit, allowing substrate phosphorylation. There are three isoforms of C (Cα, Cβ, and Cγ) and two major isoforms of R (R$_I$ and R$_{II}$) that are further distinguished into subforms (α and β) (Zhao et al., 1998). The physiological importance of these isozyme variations is not fully understood, but anchoring proteins (AKAPs) for R$_{II}$ give it a unique cellular distribution (Scott et al., 1994) R$_I$ and R$_{II}$ show sequence homology in their cAMP-binding and pseudosubstrate domains but differ extensively in their dimerization domains as well as in the sequence connecting the dimerization and pseudosubstrate domains. All known R-subunit isoforms share a common organization that consists of a dimerization domain at the NH$_2$ terminus followed by an autoinhibitor site and two-tandem cAMP-binding domains noted CBD-A and B (Taylor et al., 2005). While the portion of the R subunit COOH-terminal to the inhibition site is responsible for high affinity binding of the C-subunit and cAMP, the remaining NH$_2$ terminus serves as an adaptor for binding to kinase anchoring proteins (Scott et al., 1994) and is responsible for in vivo subcellular localization and targeting of PKA.

The ordered sequential mechanism of PKA activation is described as follows: cAMP binds first to CBD-B, making site CBD-A accessible to a second molecule of cAMP, which in turn causes the release of the active C-subunit (Kim et al., 2007; Su et al., 1995). In other words CBD-B functions as a gatekeeper for CBD-A, whereas the latter acts as the central controlling unit of the PKA system and provides the primary interfaces with the C-subunit (McNicholl et al.). The structures of the R-subunit in its active and inhibited states have also demonstrated that although CBD-A and CBD-B play clearly distinct roles in the activation of PKA, they both share similar allosteric features.

PKA activation is closely dependent on the cytosolic cAMP level which is itself regulated by G proteins via an enzyme named adenylate cyclase. Stimulation of GPCRs, by muscarinic receptor agonist for instance, can rise up or drop off AMPC production depending on the type of GPCRs. Indeed, GPCRs that activate the Gαi subunits inhibit cAMP production whereas GPCRs that activate the Gαs subunits activate cAMP production through this specific sequence: 1) activated Gαs subunit interacts with the adenylate cyclase 2) adenylate cyclase quickly converts ATP into cAMP, 3) AMPc molecule activate the PKA. Activation of PKA is involved in airway smooth muscle relaxation (Zhou et al., 1992), which may involve phosphorylation of a number of effector proteins that cause either reduction of [Ca^{2+}]$_i$ and/or reduction of MLCK sensitivity to Ca^{2+}-calmodulin (de Lanerolle et al., 1991).

Studies have demonstrated that cAMP-dependent signaling pathway activation prevents or reverses the ASM contraction indirectly, via the inhibition of InsP$_3$ receptor (InsP3R) of the sarcoplasmic reticulum, reducing the [Ca^{2+}]$_i$. The InsP3R are responsible for mobilizing Ca^{2+} from sarcoplasmic reticulum in response to agonist binding. PKA has been shown to mediate the phosphorylation of the InsP3R which consequently reduces the ability of Inositol(1,4,5)triphosphate to release Ca^{2+} from membrane vesicles (Schramm et al., 1995). The mechanism by which PKA-induced phosphorylation decreases insP$_3$-induced Ca^{2+} release has not been determined, but known consequences of receptor/channel phosphorylation include altered agonist affinity as well as altered function.

3.5 CaMkinase II

Calcium-calmodulin-dependent protein kinase II (CaMKII) is an oligomeric serine/threonine-specific protein kinase which belongs to a family of enzymes regulated by the calcium-calmodulin complex similarly to the MLCK. Increases in the cytosolic Ca^{2+} concentration following the stimulation modulate the function of many intracellular proteins (Zhou *et al.*, 1994). One of the most important intracellular acceptors of the Ca^{2+} signal is calmodulin (CaM), which exerts a modulating influence on the function of Ca^{2+}/CaM-dependent protein kinases. Among them, the CaMKII shows a broad substrate specificity and has been thought to be a multifunctional protein kinase (Cohen *et al.*, 1992; Colbran *et al.*, 1989a; Colbran *et al.*, 1989b). Four genes encode related but distinct isoforms of CaM kinase II (α, β, γ, and δ). It was originally isolated from brain tissues (Fukunaga *et al.*, 1982) (Goldenring *et al.*, 1983; Kuret *et al.*, 1984) and preparations of CaM kinase II purified from rat forebrain consist of the α (50 kDa) and β (60 kDa) subunits whose cDNAs have been cloned and sequenced (Kolb *et al.*, 1998). Non neuronal tissues express mostly the isoform γ, and δ but in such a low level that it makes difficult the purification of the enzyme; so most biochemical and physical data on the enzyme have been established with CaMkinase II-α and -β/β' from mammalian brain. In smooth muscle, CaM kinase II was first isolated as caldesmon kinase (Ikebe *et al.*, 1990b) with a molecular mass of the major subunit of 56 kDa. Isolated smooth-muscle CaM kinase II has enzymological properties similar to that of brain; however, smooth-muscle CaMKII is a tetramer according to its native molecular mass rather than a decamer or octamer as are the brain enzymes (Zhou *et al.*, 1994).

The CaMKII subunits are thought to assemble to holoenzyme through their C-terminal association domains (Fahrmann *et al.*, 1998). The linear structure of the CaMKII core consists of a catalytic/autoregulatory domain (A) containing a variable region V1, a conserved linker (B), and an association domain that contains two highly conserved sequences (C and D) as well as multiple variable regions (V2-V4). Function of each variable region has been identified as follows: V1 contains insert implicated in SR-membrane targeting; V2 possesses a functional nuclear localization signal, as well as a site of autophosphorylation; Insert X within V3 is rich in proline residues and conforms to a SH_3-binding sequence. The variable regions are diversely expressed in the different subunits. For instance the α and δ_G isoforms are the smallest catalytically competent CaMKII products because they contain the A–D core sequences, but no inserts (Hudmon *et al.*, 2002). The most prominent proteins phosphorylated by the Type II CaM kinase are its own subunits (Bennett *et al.*, 1983; Kennedy *et al.*, 1983; Miller *et al.*, 1985; Miller *et al.*, 1986). Some studies suggest that kinase activity decrease after autophosphorylation (Kuret *et al.*, 1984; LeVine *et al.*, 1985; Yamauchi *et al.*, 1985) while others suggest that it increases (Shields *et al.*, 1984) or becomes autonomous (Saitoh *et al.*, 1985). Despite these discrepancies, the most accepted theory is that the autophosphorylation allows for the activation of the catalytic domain (Hanley *et al.*, 1987). The Ca2+-CaM complex interacts with and promotes autophosphorylation of each subunit of the CaMKII.

The implication of the CAMKII in the artery smooth muscle reactivity has been extensively investigated whereas little is known regarding its role in the airways smooth muscle. In contracted cultured ASM cells from bovine, studies demonstrated that CaMKII is responsible for the phosphorylation of the MLCK (Stull *et al.*, 1993). Biochemistry data showed that MLCK is phosphorylated by CaMKII (Hashimoto *et al.*, 1990; Ikebe *et al.*, 1990a) at a specific serine near the calmodulin-binding domain; and this phosphorylation brings about the reduction of the affinity of MLCK for the $Ca^{2+} \bullet CaM$ complex (Stull *et al.*, 1990). In another hand the phosphorylation of MLCK by CaMKII have been shown to decrease the

Ca^{2+} sensitivity of MLC_{20} phosphorylation (Tansey *et al.*, 1994; Tansey *et al.*, 1992). Taken together, these findings would suggest that the net effect of the CaMKII is the relaxation of the ASM. However, enzymatic tests revealed two key elements which exclude this hypothesis: 1) the rate of phosphorylation of MLCK likely by the CaMKII is first of all slower than the rates of increase in cytosolic Ca^{2+} concentrations, 2) and also slower than the rate of phosphorylation of the MLC in intact tracheal smooth muscle cells in culture (Tansey *et al.*, 1994). Therefore, the CaMKII activity is thought not to affect significantly the reactivity of the bovine ASM. Moreover, other studies showed that the CAMKII do not play a role in the profile of the contractile response upon stimulation in intact ASM from rat or cow (Liu *et al.*, 2005; Mbikou *et al.*, 2011; Sakurada *et al.*, 2003).

3.6 PI3K

The phosphatidy inositol 3-kinases (PI3K) superfamily draws together all the enzymes capable of phosphorylating specifically the hydroxyl group of a membrane phospholipid called phosphatidyl inositol (PtdIns). Cloning approaches revealed the existence of eight distinct PI3K genes expressing eight isoforms in human and mouse genomes. Based on their domain structure, lipid substrate specificity and associated regulatory subunits, these isoforms have been divided into three main classes as follows: class I including p110α, p110β, p110δ and p110γ; class II including PI3K-C2α, PI3K-C2β and PI3K-C2γ, and the class III consisting of the sole enzyme Vps34. The PI3K phosphorylates the PtdIns into four possible products, PtdIns(3)phosphate, PtdIns(3,4)biphosphate, PtdIns(3,5) biphosphate and PtdIns(3,4,5)triphosphate, which are involved in a wide range of cellular functions, including cell growth, proliferation, motility, differentiation, survival and intracellular trafficking (Fry, 2001; Fry, 1994; Katso *et al.*, 2001; Rameh *et al.*, 1999).

Purified PI3K is a heterodimer of 85 and 110 kDa subunits (Carpenter *et al.*, 1990; Fry *et al.*, 1992; Morgan *et al.*, 1990; Shibasaki *et al.*, 1991). Analysis of the primary sequence of p85 reveals a multidomain protein which contains a number of non-catalytic domains, a Src homology region 3(SH3), and a region with significant sequence similarity to the product of the breakpoint cluster region gene BCR (Otsu *et al.*, 1991). In vascular smooth muscle, PI3K appears to play a role in the regulation of contraction as experiments showed that the specific isoform PI3K-C2α is necessary for Rho/RhoKinase-dependent MLCP inhibition and consequently for the MLC20 phosphorylation and the contraction (Yoshioka *et al.*, 2007). However, in the ASM, the PI3K does not regulate the agonist-induced contractile response (Mbikou *et al.*, 2006).

4. Theoretical modelling of PK and ASMC contraction

4.1 Interest and general priniples

Since the contractile pattern of SMC in response to contractile agonists not only depends on the structure of the regulatory network that modulates the contractile apparatus but also on the dynamics of the reactions, understanding of the mechanisms underlying this contractile profile is quite impossible by a non-mathematical intuitive approach. In the following sections, we will present the concepts of our recent mathematical modelling of isometric contraction and force development in airway smooth muscle cells based on the 4-state latch bridge model (Hai *et al.*, 1988a) and upgraded by the unique description of MLCK and

MLCP regulatory pathways (Fajmut *et al.*, 2008; Fajmut *et al.*, 2005a; Fajmut *et al.*, 2005b; Fajmut *et al.*, 2005c; Mbikou *et al.*, 2011; Mbikou *et al.*, 2006).

The latch state was first described by the mathematical model by Hai and Murphy in 1988 (Hai *et al.*, 1988a). They introduced the model of isometric contraction based on the 4-state kinetic scheme of actomyosin crossbridges, in which to actin bound and unbound to myosin (phosphorylated and dephosphorylated) represent four different states. Even nowadays, that model represents the reference in modelling of smooth muscle contraction.

Huxley (Huxley, 1957) pioneered the modelling of smooth muscle contraction. His model from 1957 was based on the sliding filament theory. Until the occurrence of Hai and Murphy's model in 1988 (Hai *et al.*, 1988b), relatively small number of models of smooth muscle contraction was developed compared to striated muscles. In 1986 Gestrelius and Börgström (Gestrelius *et al.*, 1986) introduced new concepts in modelling. They first considered viscoelastic properties of the filaments and the cytoskeleton. The later model and Huxley's model enabled the study of nonisometric contraction and took into account mostly the dynamics and mechanics of the filaments and myosin crossbridges. Although the original Hai and Murphy's model (Hai *et al.*, 1988a) enabled only the prediction of isometric force its advantage was in considering the regulatory mechanisms that drive smooth muscle contraction. Several authors have later upgraded it to enable the prediction of nonisometric contraction. In 1997 Yu et al. (Yu *et al.*, 1997) expanded it to simulate the nonisometric contractions of smooth muscles, added length dynamics and assumed length-dependent bonding and unbonding rates to be distributed according to the Gaussian distribution. In 1999 Fredberg et al. (Fredberg *et al.*, 1999) and Mijailovich et al. (Mijailovich *et al.*, 2000) integrated the latch regulation scheme of Hai and Murphy with Huxley's sliding filament model of muscle contraction for the studies of the effects of length fluctuations on the dynamically evolving cross-bridge distributions, simulating those that occur in airway smooth muscle during breathing. The later model has been recently upgraded by simple description of Ca^{2+}-dependent regulation of MLCK activity (Bates *et al.*, 2009; Wang *et al.*, 2008). Our approach in the modelling of smooth muscle contraction is to upgrade the 4-state kinetic description of Hai and Murphy (Hai *et al.*, 1988a) with the signalling pathways that regulate the MLC phosphorylation and dephosphorylation. This was for a long time the missing part in the modelling of smooth muscle contraction.

Development of force in smooth muscles is achieved by interactions between myosin cross-bridges and actin filaments. To describe theses interactions Hai and Murphy proposed a kinetic scheme shown in Figure 1 (Hai *et al.*, 1988a). The scheme was later upgraded by Rembold and Murphy (Rembold *et al.*, 1990) with the consideration of the attachment of dephosphorylated myosin to actin with a very slow rate. Many authors ignore this interaction, since the rate of attachment of dephosphorylated myosin is 150-fold lower than that of phosphorylated myosin (0.002 s^{-1} and 0.3 s^{-1}, respectively) (Trybus, 1996). In our studies we consider it in the modelling (Fajmut *et al.*, 2008; Fajmut *et al.*, 2005a; Mbikou *et al.*, 2011; Mbikou *et al.*, 2006).

The four different states of the myosin cross-bridges are presented in Figure 1. : A+M - detached, dephosphorylated, A+MP - detached, phosphorylated, AMP - attached, phosphorylated, and AM - attached, dephosphorylated, the last one termed also a latch bridge. The corresponding reaction velocities of phosphorylation/dephosphorylation and attachment/detachment of myosin cross bridges, the relevant variables of mathematical modelling, are indicated together with sites of action of enzymes MLCK and MLCP responsible for phosphorylation and dephosphorylation of MLCs, respectively. In the model, the magnitude of stress in smooth muscles is proportional to the concentration of myosin cross-bridges associated with actin filaments (AMP and AM), whereby myosin cross-bridges in the state AMP generate stress and cross-bridges in the state AM maintain stress in smooth muscles. v are either the velocities of crossbridge attachment/detachment or the velocities of phosphorylation/ dephosphorylation, which are regulated by MLCK and MLCP.

Hai and Murphy's model considered a very simple semi-theoretical description of the Ca^{2+}-dependent MLCK activation, and thus first coupled Ca^{2+} signalling pathway with the contraction. In this chapter we will present an upgrade from that model in the sense of purely theoretical and more detailed description of Ca^{2+}-dependent MLCK activation and functioning as well as MLCP regulation of force development. Our new concepts will be compared with the old ones. As a basic model scheme, we take the four-state model, in which we incorporate explicitly the description of both enzymatic reactions. The activity of MLCK is under the influence of transmitting Ca^{2+} signal and, hence, Ca^{2+} directly affects the force development. On the other hand, the action of MLCP is considered either independent of other signalling pathways or being under influence of Rho-kinase (RhoK), which phosphorylates MLCP and regulates its activity and catalytic properties.

4.2 Modelling of MLCK activity

According to the generally accepted view, MLCK is activated by the Ca^{2+}-CaM complex (Kamm et al., 2001). In general, an increase in $[Ca^{2+}]_i$ is considered to initiate the binding of four Ca^{2+} ions to CaM, which is finally followed by association of complex Ca_4CaM with MLCK (Dabrowska et al., 1982; Smith et al., 2000). In this chapter we will present three different approaches to deal with interactions between Ca^{2+}, CaM and MLCK: a three-state model proposed by Kato et al. in 1984 (Kato et al., 1984), our eight-state model proposed by Fajmut et al. in 2005 (Fajmut et al., 2005b) and a semi-theoretical approach proposed by Rembold and Murphy in 1990 (Rembold et al., 1990). The kinetic scheme of a three-state model is presented in Figure 4.

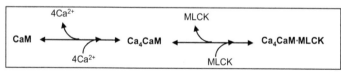

Fig. 4. **The three-state kinetic scheme of MLCK activation proposed by Kato et al.** (Kato et al., 1984).

It assumes four independent and equivalent Ca^{2+} binding sites on CaM and considers a minimum number of possible states for CaM in the activation of MLCK. According to the kinetic scheme, MLCK is activated in two steps. The first step assumes simultaneous binding of four Ca^{2+} ions to CaM. In the second step, the intermediate complex of CaM with four Ca^{2+} ions bound (Ca_4CaM) is associated with MLCK. The final product Ca_4CaM-MLCK represents the active form of MLCK in the sense that it is able to phosphorylate MLCs. It should be noted, that Kato's kinetic scheme (Kato et al., 1984) (Figure 4) reflects a somewhat older and simplified view on the interactions between Ca^{2+}, CaM and MLCK. Newer experimental studies show that not only Ca_4CaM complex but also Ca_2CaM complexes and Ca^{2+}-free CaM interact with MLCK (Bayley *et al.*, 1996; Johnson *et al.*, 1996). It has been also shown that CaM and MLCK can interact without the presence of Ca^{2+} (Wilson *et al.*, 2002) and that the velocity of Ca^{2+} binding to CaM is different with respect to N and C terminals on CaM (Brown *et al.*, 1997; Johnson *et al.*, 1996; Minowa *et al.*, 1984; Persechini *et al.*, 2000).

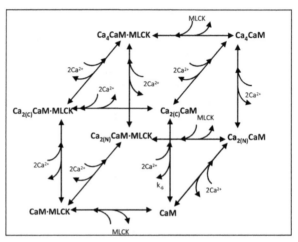

Fig. 5. **The eight-state kinetic scheme of MLCK activation proposed by Fajmut et al.** (Fajmut *et al.*, 2005a).

On the basis of these experimental findings we proposed more detailed and profound kinetic model of MLCK activation that considers either six (Fajmut *et al.*, 2005c) or eight (Fajmut *et al.*, 2005a) different states of CaM with respect to binding sites for Ca^{2+} on C and N terminals and a binding site for MLCK. The corresponding eight-state kinetic scheme is presented in Figure 5.

Binding reactions and reaction components involved can easily be recognised from the scheme. The complex $Ca_4CaM \cdot MLCK$ represents the active form of MLCK and can be applied in the 4-state kinetic scheme of force development as indicated in Figure 1.

In contrast to theoretical models of MLCK activation described here, a semi-theoretical approach was proposed by Hai and Murphy (Hai *et al.*, 1988a). This approach was based on the expression fitted to the experimentally determined dependence of MLC phosphorylation on $[Ca^{2+}]_i$. The corresponding expression was built into the mathematical model of force development proposed by them (Hai *et al.*, 1988a). The predictions of all abovementioned models for the steady state relative amount of the active MLCK (A), i.e., the ratio between $Ca_4CaM \cdot MLCK$ and total MLCK versus $[Ca^{2+}]_i$, are presented in Figure 6.

From the comparison of the model results (lines on the left panel) with the measurements (open circles) (Geguchadze *et al.*, 2004) in Figure 6 one can conclude that predictions of the eight-state model (Fajmut *et al.*, 2005a) (full line), most properly describes the process of MLCK activation. The other two models give either too sensitive (dotted line) (Rembold *et al.*, 1990) or too insensitive (dashed line) (Kato *et al.*, 1984) responses to $[Ca^{2+}]$ and they also

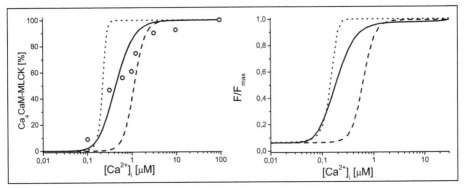

Fig. 6. **Model prediction of MLCK activation.** *Left panel:* Relative amount of the active MLCK (*A*) in dependence on [Ca^{2+}] according to: semi-theoretical approach by Rembold and Murphy (Rembold *et al.*, 1990) (dotted line); eight-state model by Fajmut et al. (Fajmut *et al.*, 2005a) (full line); three-state model by Kato et al. (Kato *et al.*, 1984) (dashed line). Experimental results of Geguchadze et al. (Geguchadze *et al.*, 2004) (open circles). *Right panel:* The corresponding steady state force dependencies on [Ca^{2+}] according to: Rembold and Murphy (Rembold *et al.*, 1990) (curve 1); Fajmut et al. (Fajmut *et al.*, 2005a) (curve 2); Kato et al. (Kato *et al.*, 1984) (curve 3).

do not have strong support by other experimental evidence (Gallagher *et al.*, 1993; Geguchadze *et al.*, 2004). The corresponding steady state force dependency on [Ca^{2+}] (right panel) shows similar behaviour. Rembold and Murphy's model gives sensitive and Kato's model gives insensitive response to [Ca^{2+}] and again don't have a strong support by experiments (Sieck *et al.*, 1998). These findings speak in favour of the necessity for more complex description of interactions between Ca^{2+}, CaM and MLCK in Ca^{2+}-signal transduction pathway. Another important property, which has not been implicated in this steady state analysis, is time dependency. In the past the interactions between Ca^{2+}, CaM and MLCK were considered as very fast (Kasturi *et al.*, 1993; Torok *et al.*, 1994) and thus modelled as being in the steady state. The reason for that were the results obtained in *in vitro* experiments, which exhibited very fast kinetics between Ca^{2+}, CaM and MLCK. However, experiments from 2001 performed by Wilson et al. (Wilson *et al.*, 2002) revealed that this might not be the case *in vivo*. Their experiments showed that at low [Ca^{2+}]$_i$ MLCK is present in complexes with Ca^{2+}-free CaM as well as with Ca$_2$CaM, in which Ca^{2+} is bound to the C-terminal of CaM (Wilson *et al.*, 2002). It has been suggested that this is a consequence of the increased affinity of CaM for Ca^{2+} by the presence of MLCK in the complex (Wilson *et al.*, 2002). In accordance with these findings the Ca^{2+}-CaM activation of MLCK appears at sufficiently high Ca^{2+} levels, when in addition to the C-terminal binding sites, the N-terminal binding sites for Ca^{2+} on CaM are saturated. However, the transition from CaM-MLCK and Ca$_2$CaM ·MLCK complexes is much slower compared to the transition from Ca^{2+}- and MLCK- free CaM state. Our model simulations show for typical physiological total amounts of CaM and MLCK (10 μM and 2 μM, respectively) in smooth muscle cells that for [Ca^{2+}]$_i$ = 0.3 μM approximately one third of MLCK is in complexes with Ca$_2$CaM, one third is Ca^{2+}-CaM-free and one third is in active form. Moreover, the half saturation time for achieving the final active MLCK form for [Ca^{2+}]$_i$ = 0.1 μM is 1.5 s and for [Ca^{2+}]$_i$ = 0.5 μM it is 0.5 s. These results of the eight-state model show that the processes of activation/deactivation of MLCK are not as fast as proposed by earlier models (Kato *et al.*, 1984) and some *in vitro* experiments on isolated CaM and MLCK (Kasturi *et al.*, 1993; Torok *et al.*, 1994). Moreover, the half-saturation time of MLCK activation/deactivation is of the same order of magnitude as the typical periods of oscillatory Ca^{2+} signals in smooth muscle cells (Mbikou *et al.*, 2006; Perez *et al.*, 2005), thus the processes of MLCK activation significantly contribute to decoding of oscillatory Ca^{2+} signal into a rather steady developed force already at the cellular level (Fajmut *et al.*, 2008; Fajmut *et al.*, 2005b; Mbikou *et al.*, 2011; Mbikou *et al.*, 2006) and add a small delay in force development after $[Ca^{2+}]_i$ increase.

4.3 Modelling of the MLCK/MLP balance and Ca^{2+}-contraction coupling

In our models (Fajmut *et al.*, 2008; Fajmut *et al.*, 2005b; Mbikou *et al.*, 2011; Mbikou *et al.*, 2006) we showed that the transduction of the Ca^{2+} signal from its appearance in the cytosol as a time-dependent variation of concentration to the development of force in smooth muscle cells is decoded mainly by the interactions between Ca^{2+}, CaM and MLCK, and is further translated to force by the balance between the phosphorylation and dephosphorylation of MLC, whereby both processes are regulated by MLCK and MLCP, respectively. The abundance of actomyosin crossbridges either phosphorylated or in the latch state is reflected in the magnitude of developed force. Our model results point out that a complete description of MLCK activation by Ca^{2+}-CaM is necessary for the relevant prediction of Ca^{2+}-contraction coupling and that 4-state latch bridge model upgraded with Ca^{2+}-CaM-dependent MLCK activation well describes the fast phase (first few minutes) of force development in the isometric contraction of rat tracheal rings (Fajmut *et al.*, 2008; Mbikou *et al.*, 2011; Mbikou *et al.*, 2006).

Essentially, the last and the most elaborated version of the model describing Ca^{2+}-contraction coupling consists of three parts (Mbikou *et al.*, 2011). The first one describes the activation of MLCK by Ca^{2+}-CaM complexes, considers binding of MLCK to Ca^{2+}-free CaM as well as to various Ca^{2+}-CaM complexes, predicts the concentration of the active form of MLCK, i.e. the complex $Ca_4CaM \cdot MLCK$, in dependence of $[Ca^{2+}]_i$. The kinetic scheme of these interactions is presented in Figure 5.

The second part describes the regulation of MLCP activity and represents an essential upgrade from the original description of Hai and Murphy (Hai *et al.*, 1988a), in which dephosphorylation velocity was taken as a linear function of AMP and MP. In our first models (Fajmut *et al.*, 2008; Fajmut *et al.*, 2005b; Mbikou *et al.*, 2006) we treated dephosphorylation process with two parallel enzymatic reactions of Michaelis-Menten type with AMP and MP as the substrates for MLCP and with MLCP·MP and MLCP·AMP as the intermediate complexes. All these models considered constant total amount of the enzyme and one Ca^{2+} independent catalytic activity of MLCP. In accordance with recent experimental results indicating that the activity of MLCP is under the influence of Rho-Kinase (RhoK) (Somlyo *et al.*, 2000), we modelled RhoK-dependent MLCP regulation (Mbikou *et al.*, 2011). RhoK phosphorylates MLCP and thus modifies its enzymatic properties. It decreases the rate constant of enzyme-substrate breakdown (k_{cat}) and increases the Michaelis constant (K_M) (Feng *et al.*, 1999; Ichikawa *et al.*, 1996; Lukas, 2004). The mechanisms by which RhoK is itself activated have not yet been fully determined. However, according to our present experimental results, RhoK is likely to be activated by the Ca^{2+} signal (Mbikou *et al.*, 2011). Thus, we consider the transition of RhoK from inactive state into active state to be Ca^{2+}-dependent. On these statements, RhoK activity has been modelled according to the kinetic scheme, represented in Figure 7 (Mbikou *et al.*, 2011).

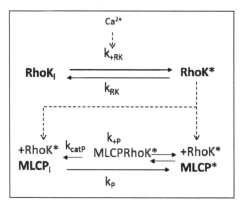

Fig. 7. **Kinetic scheme of RhoK dependent MLCP phosphorylation and partial inhibition proposed by Mbikou et al. (Mbikou *et al.*, 2011).** RhoK$_i$, inactive RhoK; RhoK*, active Rhok. MLCP$_i$, inactive MLCP; MLCP*, active MLCP.

RhoK can be either in an inactive (RhoK$_i$) or active (RhoK*) state. k$_{+RK}$ and k$_{-RK}$ are the corresponding on- and off- rate constants. k$_{+RK}$ depends on the Ca^{2+} response, that is the [Ca^{2+}]$_i$ above baseline. Modelling of MLCP phosphorylation by RhoK* is based on the steady-state Michaelis-Menten enzyme kinetics, whereby MLCP*RhoK* is the intermediate complex and k$_{catP}$ is the rate constant for the breakdown of this intermediate complex into product, i.e. MLCP$_i$. k$_{+P}$ and k$_{-P}$ are the corresponding overall rate constants for MLCP phosphorylation and dephosphorylation, respectively. The values of the rate constant of MLCP-substrate breakdown (k$_{cat}$) and the Michaelis constant of MLCP (K$_M$) depend on the net effective form of MLCP, which is dependent on the fraction of unphosphorylated and inhibited MLCP ([MLCP]*/[MLCP]$_i$).

The third part of the modelling represents the well-known 4-state actomyosin latch bridge model (Hai *et al.*, 1988a). Links between all three parts of the model are the active form of MLCK and the net effective form of MLCP, which both modulate the rate of MLC phosphorylation and dephosphorylation.

In our experimental and theoretical study (Mbikou *et al.*, 2006), the version of the model without RhoK-dependent regulation of MLCP was first developed and analysed. The model was applied to the studies of the effect of different calcium signals to the amplitude and the velocity of force developed in airway smooth muscles. It was shown that the velocity and magnitude of the force that develops in several seconds after cholinergic stimulation are determined by the following signal parameters: the amplitude and the frequency of the oscillating Ca^{2+} signal as well as the plateau - but not the peak - in the biphasic Ca^{2+} signal, which comprises a peak followed by a decline to a plateau phase (Mbikou *et al.*, 2006). On the other hand, the increased frequency of oscillating Ca^{2+} signal is translated into the increase of force magnitude (Fajmut *et al.*, 2008; Mbikou *et al.*, 2006). One main physiological implication of that model (Mbikou *et al.*, 2006) was the prediction of the temporal delay of force generation with respect to Ca^{2+} transient. Figure 8 (left panel) presents the time dependent relative MLC phosphorylation (p) (full line) and relative MLCK activity (A) (dotted line) after biphasic $[Ca^{2+}]_i$ with a peak and a plateau with the characteristic values (peak: 0.6 µM, plateau: 0.2 µM, baseline: 0.15 µM) as well as the time dependent relative force development (F) as predicted by the model (Mbikou et al., 2006) (right panel).

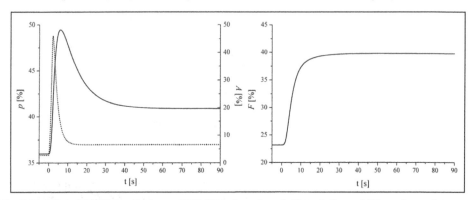

Fig. 8. **Predicted MLCK activity and MLC20 phosphorylation.** *Left panel:* Time dependent relative MLC phosphorylation (p) (full line) and relative MLCK activity (A) (dotted line) after biphasic $[Ca^{2+}]_i$ with a peak and plateau as predicted by the model (Mbikou et al., 2006). *Right panel:* Time dependent relative force development (F) as predicted by the model (Mbikou et al., 2006). F is defined relatively to the force obtained at supramaximal steady $[Ca^{2+}]_i$.

This time-delay originates also from the process of MLCK activation/inactivation. Namely, the process of Ca^{2+}-CaM-dependent MLCK activation contributes significantly to the time delay, in contrast with other studies (Kasturi et al., 1993; Torok et al., 1994), hypothesizing that the process of MLCK activation is extremely fast and is not likely to contribute more than a few milliseconds to the overall delay in force development (Sieck et al., 1998). Slow force generation in our model is also a consequence of the slow-rate of MLCK activation/inactivation kinetics. Additionally, the kinetics of formation of actomyosin cross bridges explains the delay between MLC phosphorylation and force development, and the Hill-shaped time course of isometric contraction.

In another purely theoretical study (Fajmut et al., 2008) we confirmed that upon biphasic Ca^{2+}-signal transduction through the system, MLCK controls amplitude more than duration, whereas MLCP tends to control both. These general characteristic regulatory properties of kinases and phosphatases were previously described by Heinrich et al. and Hornberg et al. (Heinrich et al., 2002; Hornberg et al., 2005) in other signalling processes.

In our most recent work (Mbikou et al., 2011), the model was applied to the studies of RhoK contribution to the early phase of the Ca^{2+}-contraction coupling in airway smooth muscle. For this purpose, the simulation of RhoK inhibitor Y27632 was simulated by the model. Theoretical results of early stress development agreed with experimental results, which showed an evident drop in the stress development after RhoK inhibition in the early phase of contraction, whereby the shape and the characteristic time of stress development did not change significantly. The model further predicted that maximal RhoK activation and subsequent MLCP inactivation occur in less than 10 s, i.e., before the short time maximal contraction is achieved (Mbikou et al., 2011).

An essential property of our models coupling Ca^{2+} and contraction is the explicit consideration of MLC phosphorylation and dephosphorylation steps described by Michaelis-Menten kinetics. This permits simulations of the effects that variations in enzyme contents and their catalytic properties exert on their velocities of phosphorylation/dephosphorylation, signal transduction and development of force. In this sense our models provide an upgrade with respect to the original Hai and Murphy's model as well as other models describing smooth muscle contraction.

4.4 Modelling of thin filament-associated regulation of contraction

An additional upgrade to the original Hai and Murphy's model has been proposed by Hai and Kim in 2005 (Fajmut *et al.*, 2005c) to address some experimental data that could not be explained by the four-state model. Because in phorbol ester-induced force development constant myosin phosphorylation could not be explained by the original 4-state model, the authors proposed and postulated a thin-filament-regulated latch-bridge model that includes two latch-bridge 4-state cycles, one of which is identical to the original Hai and Murphy's model, and the other one is the ultraslow 4-state cycle with lower cross-bridge cycling rates (Hai *et al.*, 2005). The model is able to fit phorbol ester-induced contractions at constant myosin phosphorylation. This was achieved by shifting cross bridges from the regular to the ultraslow cross-bridge cycle. It was also proposed that PKC activation leads to the thin-filament-based inhibition of actomyosin ATPase activity in ultraslow cycle, however, authors did not specify the target, the signalling pathway and the mechanism of this regulation. They hypothesized about calponin and caldesmon – the thin-filament-based regulatory proteins – being the candidates for the inhibition of actomyosin ATPase activity caused by PKC, because both proteins can exist in the unphosphorylated and phosphorylated form (Gerthoffer *et al.*, 1994).

In the early 1990s it was believed that the actin bound proteins, calponin and caldesmon, have large modulatory role in the latch state, however, in the late 1990s, the discovery of the significant regulatory role of myosin light chain phosphatase in smooth muscle contraction draw attention away from the thin-filament-based regulatory proteins (Paul, 2009). There is evidence that regulation of the response to a given $[Ca^{2+}]_i$, that is, regulation of the 'Ca^{2+} sensitivity' via modulating phosphatase activity, is as important as regulation of $[Ca^{2+}]_i$ in the control of contractility (Paul, 2009). PKC-mediated phosphorylation of CPI-17 has been postulated as a mechanism of PKC-mediated inhibition of MLCP (Hirano *et al.*, 2003). However, according to Hai and Kim (Hai *et al.*, 2005) this mechanism can potentially explain only the initial increase in the MLC phosphorylation but not the continued force development after myosin phosphorylation has already reached steady state.

In our experimental studies (Mbikou *et al.*, 2011; Mbikou *et al.*, 2006) we also observed the biphasic force development similar to that observed by phorbol ester-induced contraction (Hai *et al.*, 2005). The analysis of the time course of isometric contraction and MLC phosphorylation showed that the contractile response of rat tracheal rings to cholinergic stimulation developed in two distinct phases (Mbikou *et al.*, 2006). The first, short-time contractile response, which represents 70 % of the total contraction, was associated with a fast $[Ca^{2+}]_i$ peak followed by a plateau with, in some cases, with superimposed $[Ca^{2+}]_i$ oscillations, in correlation with fast and transient MLC phosphorylation, and Hill-shaped force development (Mbikou *et al.*, 2006) as shown in our model simulations presented in

Figure 8. The first fast phase was then followed by the second slow phase in which progressive increase of force reached plateau after 30 minutes (Mbikou et al., 2006). Our model (Mbikou et al., 2006) of force development generated by Ca^{2+}-dependent MLCK activation properly predicted the short-time contractile response. Simulations with the existing model showed that the long-time contractile response might be explained either by a long-term increase in oscillation frequency or $[Ca^{2+}]_i$ plateau. However, recordings of $[Ca^{2+}]_i$ responses to cholinergic stimulation for several minutes did not support such a hypothesis (Perez et al., 2005). A possible explanation was, in parallel with the activation of MLCK, the inactivation of MLCP, which may be due to the action of PKC or RhoK (Mizuno et al., 2008). For RhoK we have recently shown that it is implicated only in the fast phase of force development (Mbikou et al., 2011). However, incorporation of progressive slow inactivation of MLCP in our modelling predicts a time course of isometric contraction similar to the experimental one, and explains the second increase in MLC phosphorylation and a slow phase in force development (Mbikou et al., 2006). But, it does not explain a decrease in MLC20 phosphorylation associated with maximal force observed after 30 minutes (Mbikou et al., 2006).

5. Conclusion

In airway smooth muscle as in other smooth muscles, actin-myosin cross bridge cycling critically depends on the phosphorylation of MLC_{20} and hence on MLCK/MLCP balance. MLCK, which activity is modulated by the Ca^{2+} signal through the formation of the Ca^{2+}-calmodulin-MLCK complex, is the most important kinase in airway myocyte contraction and the contractile properties of airway smooth muscle cells are lost when MLCK is inactivated or deleted. However, though activation of MLCK is indispensable, contraction of airway smooth muscle, both in its amplitude and time-course, is modulated by a network of kinases that can act upstream the Ca^{2+} signal, modulating the Ca^{2+} signal itself, or downstream, modulating the sensitivity of the contractile apparatus to Ca^{2+}. The main targets of the protein kinases acting on the decoding of the Ca^{2+} signal are MLCP and MLCK, though direct MLC_{20} phosphorylation, in parallel to MLCK, may be possible. Indeed, MLCP and MLCK have several sites of phosphorylation and their enzymatic activity depends on whether these sites are phosphorylated or not. These phosphorylations may up- or downregulate MLCP and MLCK activity. Stimulation of airways by contractile agonists such as acetylcholine activates kinases such as Rho kinase and PKC that inhibit MLCP activity and hence increase the sensitivity of the contractile apparatus to Ca^{2+}. By contrast, β_2-adrenergic stimulation, a major relaxant pathway, activates PKA which inhibits MLCK and favours MLCP activity. Additionally, contraction may be modulated by phosphorylation of caldesmon and calponin, proteins associated with the thin filament of actin. The MLCK/MLCP balance is hence embedded in a network of protein kinases. The resultant contractile behaviour of the SMC depends on the dynamics of the reaction of this regulatory network, and mathematical modelling is essential to decipher how the different protein kinases determine the time-dependent variation of the contractile status of the airway smooth muscle cell.

6. Acknowledgements

This work was supported by a Proteus Hubert-Curien partnership. The authors gratefully acknowledge a grant from the Cultural Service of the French Embassy in Slovenia and the French Institute Charles Nodier for the stay of A. Fajmut in Bordeaux.

7. References

Akopov SE, Zhang L, Pearce WJ (1998). Regulation of Ca2+ sensitization by PKC and rho proteins in ovine cerebral arteries: effects of artery size and age. *Am J Physiol* 275(3 Pt 2): H930-939.

Amano M, Chihara K, Nakamura N, Kaneko T, Matsuura Y, Kaibuchi K (1999). The COOH terminus of Rho-kinase negatively regulates rho-kinase activity. *J Biol Chem* 274(45): 32418-32424.

Amano M, Nakayama M, Kaibuchi K (2010). Rho-kinase/ROCK: A key regulator of the cytoskeleton and cell polarity. *Cytoskeleton (Hoboken)* 67(9): 545-554.

Anderson GP (2006). Current issues with beta2-adrenoceptor agonists: pharmacology and molecular and cellular mechanisms. *Clin Rev Allergy Immunol* 31(2-3): 119-130.

Araki S, Ito M, Kureishi Y, Feng J, Machida H, Isaka N, *et al.* (2001). Arachidonic acid-induced Ca2+ sensitization of smooth muscle contraction through activation of Rho-kinase. *Pflugers Arch* 441(5): 596-603.

Ay B, Prakash YS, Pabelick CM, Sieck GC (2004). Store-operated Ca2+ entry in porcine airway smooth muscle. *Am J Physiol Lung Cell Mol Physiol* 286(5): L909-917.

Bai Y, Sanderson MJ (2006). Modulation of the Ca2+ sensitivity of airway smooth muscle cells in murine lung slices. *Am J Physiol Lung Cell Mol Physiol.*

Barany K, Barany M (1996). Myosin Light Chains. In: Barany M (ed)^(eds). *Biochemistry of Smooth Muscle Contraction*, edn. San Diego: Academic Press. p^pp 21-35.

Bates JHT, Bullimore SR, Politi AZ, Sneyd J, Anafi RC, Lauzon A-M (2009). Transient oscillatory force-length behavior of activated airway smooth muscle. *American Journal of Physiology - Lung Cellular and Molecular Physiology* 297(2): L362-L372.

Bayley PM, Findlay WA, Martin SR (1996). Target recognition by calmodulin: Dissecting the kinetics and affinity of interaction using short peptide sequences. *Protein Sci.* 5: 1215-1228.

Bell RM, Burns DJ (1991). Lipid activation of protein kinase C. *J Biol Chem* 266(8): 4661-4664.

Bennett MK, Erondu NE, Kennedy MB (1983). Purification and characterization of a calmodulin-dependent protein kinase that is highly concentrated in brain. *J Biol Chem* 258(20): 12735-12744.

Bergner A, Sanderson MJ (2002). Acetylcholine-induced calcium signaling and contraction of airway smooth muscle cells in lung slices. *J Gen Physiol* 119(2): 187-198.

Berman HM, Ten Eyck LF, Goodsell DS, Haste NM, Kornev A, Taylor SS (2005). The cAMP binding domain: an ancient signaling module. *Proc Natl Acad Sci U S A* 102(1): 45-50.

Brown SE, Martin SR, Bayley PM (1997). Kinetic control of the dissociation pathway of calmodulin-peptide complexes. *J. Biol. Chem.* 272(6): 3389-3397.

Carpenter CL, Duckworth BC, Auger KR, Cohen B, Schaffhausen BS, Cantley LC (1990). Purification and characterization of phosphoinositide 3-kinase from rat liver. *J Biol Chem* 265(32): 19704-19711.

Cohen A, Boeijinga JK, van Haard PM, Schoemaker RC, van Vliet-Verbeek A (1992). Gastrointestinal blood loss after non-steroidal anti-inflammatory drugs. Measurement by selective determination of faecal porphyrins. *Br J Clin Pharmacol* 33(1): 33-38.

Colbran RJ, Schworer CM, Hashimoto Y, Fong YL, Rich DP, Smith MK, *et al.* (1989a). Calcium/calmodulin-dependent protein kinase II. *Biochem J* 258(2): 313-325.

Colbran RJ, Smith MK, Schworer CM, Fong YL, Soderling TR (1989b). Regulatory domain of calcium/calmodulin-dependent protein kinase II. Mechanism of inhibition and regulation by phosphorylation. *J Biol Chem* 264(9): 4800-4804.

Corompt E, Bessard G, Lantuejoul S, Naline E, Advenier C, Devillier P (1998). Inhibitory effects of large Ca2+-activated K+ channel blockers on beta-adrenergic- and NO-donor-mediated relaxations of human and guinea-pig airway smooth muscles. *Naunyn Schmiedebergs Arch Pharmacol* 357(1): 77-86.

Coronado R, Morrissette J, Sukhareva M, Vaughan DM (1994). Structure and function of ryanodine receptors. *Am J Physiol* 266(6 Pt 1): C1485-1504.

Dabrowska R, Hinkins S, Walsh MP, Hartshorne DJ (1982). The binding of smooth-muscle myosin light chain kinase to actin. *Biochem. Biophys. Res. Commun.* 107: 1524-1531.

de Lanerolle P, Paul RJ (1991). Myosin phosphorylation/dephosphorylation and regulation of airway smooth muscle contractility. *Am J Physiol* 261(2 Pt 1): L1-14.

Deng JT, Sutherland C, Brautigan DL, Eto M, Walsh MP (2002). Phosphorylation of the myosin phosphatase inhibitors, CPI-17 and PHI-1, by integrin-linked kinase. *Biochem J* 367(Pt 2): 517-524.

Deng JT, Van Lierop JE, Sutherland C, Walsh MP (2001). Ca2+-independent smooth muscle contraction. a novel function for integrin-linked kinase. *J Biol Chem* 276(19): 16365-16373.

Eto M, Kitazawa T, Brautigan DL (2004). Phosphoprotein inhibitor CPI-17 specificity depends on allosteric regulation of protein phosphatase-1 by regulatory subunits. *Proc Natl Acad Sci U S A* 101(24): 8888-8893.

Eto M, Ohmori T, Suzuki M, Furuya K, Morita F (1995). A novel protein phosphatase-1 inhibitory protein potentiated by protein kinase C. Isolation from porcine aorta media and characterization. *J Biochem* 118(6): 1104-1107.

Eto M, Senba S, Morita F, Yazawa M (1997). Molecular cloning of a novel phosphorylation-dependent inhibitory protein of protein phosphatase-1 (CPI17) in smooth muscle: its specific localization in smooth muscle. *FEBS Lett* 410(2-3): 356-360.

Fahrmann M, Mohlig M, Schatz H, Pfeiffer A (1998). Purification and characterization of a Ca2+/calmodulin-dependent protein kinase II from hog gastric mucosa using a protein-protein affinity chromatographic technique. *Eur J Biochem* 255(2): 516-525.

Fajmut A, Brumen M (2008). MLC-kinase/phosphatase control of Ca2+ signal transduction in airway smooth muscles. *J Theor Biol* 252(3): 474-481.

Fajmut A, Brumen M, Schuster S (2005a). Theoretical model of the interactions between Ca2+, calmodulin and myosin light chain kinase. *FEBS Lett.* 579: 4361-4366.

Fajmut A, Dobovisek A, Brumen M (2005b). Mathematical modeling of the relation between myosin phosphorylation and stress development in smooth muscles. *J. Chem. Inf. Model.* 45(6): 1610-1615.

Fajmut A, Jagodic M, Brumen M (2005c). Mathematical modeling of the myosin light chain kinase activation. *J. Chem. Inf. Model.* 45(6): 1605-1609.

Feng J, Ito M, Ichikawa K, Isaka N, Nishikawa M, Hartshorne DJ, *et al.* (1999). Inhibitory phosphorylation site for Rho-associated kinase on smooth muscle myosin phosphatase. *J Biol Chem* 274(52): 37385-37390.

Ferris CD, Cameron AM, Bredt DS, Huganir RL, Snyder SH (1991a). Inositol 1,4,5-trisphosphate receptor is phosphorylated by cyclic AMP-dependent protein kinase at serines 1755 and 1589. *Biochem Biophys Res Commun* 175(1): 192-198.

Ferris CD, Huganir RL, Bredt DS, Cameron AM, Snyder SH (1991b). Inositol trisphosphate receptor: phosphorylation by protein kinase C and calcium calmodulin-dependent protein kinases in reconstituted lipid vesicles. *Proc Natl Acad Sci U S A* 88(6): 2232-2235.

Fredberg JJ, Inoyue DS, Mijalovich SM, Butler JP (1999). Perturbed equilibrium of myosin binding in airway smooth muscle and its implications in bronchospasm. *Am. J. Respir. Crit. Care Med.* 159(3): 959-967.

Fry MJ (2001). Phosphoinositide 3-kinase signalling in breast cancer: how big a role might it play? *Breast Cancer Res* 3(5): 304-312.

Fry MJ (1994). Structure, regulation and function of phosphoinositide 3-kinases. *Biochim Biophys Acta* 1226(3): 237-268.

Fry MJ, Panayotou G, Dhand R, Ruiz-Larrea F, Gout I, Nguyen O, *et al.* (1992). Purification and characterization of a phosphatidylinositol 3-kinase complex from bovine brain by using phosphopeptide affinity columns. *Biochem J* 288 (Pt 2): 383-393.

Fukunaga K, Yamamoto H, Matsui K, Higashi K, Miyamoto E (1982). Purification and characterization of a Ca2+- and calmodulin-dependent protein kinase from rat brain. *J Neurochem* 39(6): 1607-1617.

Gallagher P, Herring B, Trafny A, Sowadski J, Stull J (1993). A molecular mechanism for autoinhibition of myosin light chain kinases. *J. Biol. Chem.* 268(35): 26578-26582.

Geguchadze R, Zhi G, Lau KS, Isotani E, Persechini A, Kamm KE, *et al.* (2004). Quantitative measurements of Ca2+/calmodulin binding and activation of myosin light chain kinase in cells. *FEBS Lett.* 557(1-3): 121-124.

Gerthoffer W, Pohl J (1994). Caldesmon and calponin phosphorylation in regulation of smooth muscle contraction. *Can J Physiol Pharmacol* 72(11): 1410-1414.

Gestrelius S, Borgstrom P (1986). A dynamic model of smooth muscle contraction. *Biophys J* 50(1): 157-169.

Goldenring JR, Gonzalez B, McGuire JS, Jr., DeLorenzo RJ (1983). Purification and characterization of a calmodulin-dependent kinase from rat brain cytosol able to phosphorylate tubulin and microtubule-associated proteins. *J Biol Chem* 258(20): 12632-12640.

Grassie ME, Moffat LD, Walsh MP, MacDonald JA (2011). The myosin phosphatase targeting protein (MYPT) family: a regulated mechanism for achieving substrate specificity of the catalytic subunit of protein phosphatase type 1delta. *Arch Biochem Biophys* 510(2): 147-159.

Gunter TE, Buntinas L, Sparagna G, Eliseev R, Gunter K (2000). Mitochondrial calcium transport: mechanisms and functions. *Cell Calcium* 28(5-6): 285-296.

Gusev NB (2001). Some properties of caldesmon and calponin and the participation of these proteins in regulation of smooth muscle contraction and cytoskeleton formation. *Biochemistry (Mosc)* 66(10): 1112-1121.

Hai C-M, Kim HR (2005). An expanded latch-bridge model of protein kinase C-mediated smooth muscle contraction. *J. Appl. Physiol.* 98(4): 1356-1365.

Hai CM, Murphy RA (1988a). Cross-bridge phosphorylation and regulation of latch state in smooth muscle. *Am J Physiol Cell Physiol* 254(1): C99-106.

Hai CM, Murphy RA (1988b). Regulation of shortening velocity by cross-bridge phosphorylation in smooth muscle. *Am J Physiol* 255(1 Pt 1): C86-94.

Hanley RM, Means AR, Ono T, Kemp BE, Burgin KE, Waxham N, *et al.* (1987). Functional analysis of a complementary DNA for the 50-kilodalton subunit of calmodulin kinase II. *Science* 237(4812): 293-297.

Harnett KM, Biancani P (2003). Calcium-dependent and calcium-independent contractions in smooth muscles. *Am J Med* 115 Suppl 3A: 24S-30S.

Hartshorne DJ, Hirano K (1999). Interactions of protein phosphatase type 1, with a focus on myosin phosphatase. *Mol Cell Biochem* 190(1-2): 79-84.

Hartshorne DJ, Ito M, Erdodi F (1998). Myosin light chain phosphatase: subunit composition, interactions and regulation. *J Muscle Res Cell Motil* 19(4): 325-341.

Hashimoto Y, Soderling TR (1990). Phosphorylation of smooth muscle myosin light chain kinase by Ca2+/calmodulin-dependent protein kinase II: comparative study of the phosphorylation sites. *Arch Biochem Biophys* 278(1): 41-45.

Heinrich R, Neel B, Rapoport T (2002). Mathematical models of protein kinase signal transduction. *Mol Cell* 9(5): 957-970.

Hirano K, Derkach DN, Hirano M, Nishimura J, Kanaide H (2003). Protein kinase network in the regulation of phosphorylation and dephosphorylation of smooth muscle myosin light chain. *Mol Cell Biochem* 248(1-2): 105-114.

Hirano K, Hirano M, Kanaide H (2004). Regulation of myosin phosphorylation and myofilament Ca2+ sensitivity in vascular smooth muscle. *J Smooth Muscle Res* 40(6): 219-236.

Hong F, Haldeman BD, Jackson D, Carter M, Baker JE, Cremo CR (2011). Biochemistry of smooth muscle myosin light chain kinase. *Arch Biochem Biophys* 510(2): 135-146.

Hornberg JJ, Bruggeman FJ, Binder B, Geest CR, de Vaate AJMB, Lankelma J, et al. (2005). Principles behind the multifarious control of signal transduction. ERK phosphorylation and kinase/phosphatase control. *FEBS Journal* 272(1): 244-258.

Huang J, Mahavadi S, Sriwai W, Hu W, Murthy KS (2006). Gi-coupled receptors mediate phosphorylation of CPI-17 and MLC20 via preferential activation of the PI3K/ILK pathway. *Biochem J* 396(1): 193-200.

Huang KP (1989). The mechanism of protein kinase C activation. *Trends Neurosci* 12(11): 425-432.

Hudmon A, Schulman H (2002). Neuronal CA2+/calmodulin-dependent protein kinase II: the role of structure and autoregulation in cellular function. *Annu Rev Biochem* 71: 473-510.

Huxley AF (1957). Muscle structure and theories of contraction. *Prog Biophys Biophys Chem* 7: 255-318.

Hyvelin JM, Martin C, Roux E, Marthan R, Savineau JP (2000a). Human isolated bronchial smooth muscle contains functional ryanodine/caffeine-sensitive Ca-release channels. *Am J Respir Crit Care Med* 162(2 Pt 1): 687-694.

Hyvelin JM, Roux E, Prevost MC, Savineau JP, Marthan R (2000b). Cellular mechanisms of acrolein-induced alteration in calcium signaling in airway smooth muscle. *Toxicol Appl Pharmacol* 164(2): 176-183.

Ichikawa K, Ito M, Hartshorne DJ (1996). Phosphorylation of the large subunit of myosin phosphatase and inhibition of phosphatase activity. *J Biol Chem* 271(9): 4733-4740.

Iizuka K, Yoshii A, Samizo K, Tsukagoshi H, Ishizuka T, Dobashi K, et al. (1999). A major role for the rho-associated coiled coil forming protein kinase in G-protein-mediated Ca2+ sensitization through inhibition of myosin phosphatase in rabbit trachea. *Br J Pharmacol* 128(4): 925-933.

Ikebe M, Reardon S (1990a). Phosphorylation of smooth myosin light chain kinase by smooth muscle Ca2+/calmodulin-dependent multifunctional protein kinase. *J Biol Chem* 265(16): 8975-8978.

Ikebe M, Reardon S, Scott-Woo GC, Zhou Z, Koda Y (1990b). Purification and characterization of calmodulin-dependent multifunctional protein kinase from smooth muscle: isolation of caldesmon kinase. *Biochemistry* 29(51): 11242-11248.

Ishizaki T, Maekawa M, Fujisawa K, Okawa K, Iwamatsu A, Fujita A, et al. (1996). The small GTP-binding protein Rho binds to and activates a 160 kDa Ser/Thr protein kinase homologous to myotonic dystrophy kinase. *EMBO J* 15(8): 1885-1893.

Ito M, Nakano T, Erdodi F, Hartshorne DJ (2004). Myosin phosphatase: structure, regulation and function. *Mol Cell Biochem* 259(1-2): 197-209.

Itoh T, Suzuki A, Watanabe Y (1994). Effect of a peptide inhibitor of protein kinase C on G-protein-mediated increase in myofilament Ca(2+)-sensitivity in rabbit arterial skinned muscle. *Br J Pharmacol* 111(1): 311-317.

Janssen LJ, Tazzeo T, Zuo J (2004). Enhanced myosin phosphatase and Ca(2+)-uptake mediate adrenergic relaxation of airway smooth muscle. *Am J Respir Cell Mol Biol* 30(4): 548-554.

Johnson DA, Akamine P, Radzio-Andzelm E, Madhusudan M, Taylor SS (2001). Dynamics of cAMP-dependent protein kinase. *Chem Rev* 101(8): 2243-2270.

Johnson JD, Snyder C, Walsh M, Flynn M (1996). Effects of myosin light chain kinase and peptides on Ca2+ exchange with the N- and C-terminal Ca2+ binding sites of calmodulin. *J. Biol. Chem.* 271(2): 761-767.

Johnson M (1998). The beta-adrenoceptor. *Am J Respir Crit Care Med* 158(5 Pt 3): S146-153.

Johnson M (2006). Molecular mechanisms of beta(2)-adrenergic receptor function, response, and regulation. *J Allergy Clin Immunol* 117(1): 18-24; quiz 25.

Kajita J, Yamaguchi H (1993). Calcium mobilization by muscarinic cholinergic stimulation in bovine single airway smooth muscle. *Am J Physiol* 264(5 Pt 1): L496-503.

Kamm KE, Stull JT (2001). Dedicated myosin light chain kinases with diverse cellular functions. *J. Biol. Chem.* 276(7): 4527-4530.

Kasturi R, Vasulka C, Johnson J (1993). Ca2+, caldesmon, and myosin light chain kinase exchange with calmodulin. *J. Biol. Chem.* 268(11): 7958-7964.

Kato S, Osa T, Ogasawara T (1984). Kinetic model for isometric contraction in smooth muscle on the basis of myosin phosphorylation hypothesis. *Biophys. J.* 46(1): 35-44.

Katso R, Okkenhaug K, Ahmadi K, White S, Timms J, Waterfield MD (2001). Cellular function of phosphoinositide 3-kinases: implications for development, homeostasis, and cancer. *Annu Rev Cell Dev Biol* 17: 615-675.

Kawano Y, Fukata Y, Oshiro N, Amano M, Nakamura T, Ito M, et al. (1999). Phosphorylation of myosin-binding subunit (MBS) of myosin phosphatase by Rho-kinase in vivo. *J Cell Biol* 147(5): 1023-1038.

Kennedy MB, McGuinness T, Greengard P (1983). A calcium/calmodulin-dependent protein kinase from mammalian brain that phosphorylates Synapsin I: partial purification and characterization. *J Neurosci* 3(4): 818-831.

Kim C, Cheng CY, Saldanha SA, Taylor SS (2007). PKA-I holoenzyme structure reveals a mechanism for cAMP-dependent activation. *Cell* 130(6): 1032-1043.

Kimura K, Ito M, Amano M, Chihara K, Fukata Y, Nakafuku M, et al. (1996). Regulation of myosin phosphatase by Rho and Rho-associated kinase (Rho-kinase). *Science* 273(5272): 245-248.

Ko EA, Han J, Jung ID, Park WS (2008). Physiological roles of K+ channels in vascular smooth muscle cells. *J Smooth Muscle Res* 44(2): 65-81.

Kolb SJ, Hudmon A, Ginsberg TR, Waxham MN (1998). Identification of domains essential for the assembly of calcium/calmodulin-dependent protein kinase II holoenzymes. *J Biol Chem* 273(47): 31555-31564.

Komander D, Garg R, Wan PT, Ridley AJ, Barford D (2008). Mechanism of multi-site phosphorylation from a ROCK-I:RhoE complex structure. *EMBO J* 27(23): 3175-3185.

Kordowska J, Huang R, Wang CL (2006). Phosphorylation of caldesmon during smooth muscle contraction and cell migration or proliferation. *J Biomed Sci* 13(2): 159-172.

Kuo KH, Herrera AM, Seow CY (2003). Ultrastructure of airway smooth muscle. *Respir Physiol Neurobiol* 137(2-3): 197-208.

Kuret J, Schulman H (1984). Purification and characterization of a Ca2+/calmodulin-dependent protein kinase from rat brain. *Biochemistry* 23(23): 5495-5504.

Lester DS, Doll L, Brumfeld V, Miller IR (1990). Lipid dependence of surface conformations of protein kinase C. *Biochim Biophys Acta* 1039(1): 33-41.

Leung T, Chen XQ, Manser E, Lim L (1996). The p160 RhoA-binding kinase ROK alpha is a member of a kinase family and is involved in the reorganization of the cytoskeleton. *Mol Cell Biol* 16(10): 5313-5327.

LeVine H, 3rd, Sahyoun NE, Cuatrecasas P (1985). Calmodulin binding to the cytoskeletal neuronal calmodulin-dependent protein kinase is regulated by autophosphorylation. *Proc Natl Acad Sci U S A* 82(2): 287-291.

Liu C, Zuo J, Pertens E, Helli PB, Janssen LJ (2005). Regulation of Rho/ROCK signaling in airway smooth muscle by membrane potential and [Ca2+]i. *Am J Physiol Lung Cell Mol Physiol* 289(4): L574-582.

Liu X, Farley JM (1996). Frequency modulation of acetylcholine-induced Ca(++)-dependent Cl- current oscillations are mediated by 1, 4, 5-trisphosphate in tracheal myocytes. *J Pharmacol Exp Ther* 277: 796-804.

Lukas TJ (2004). A signal transduction pathway model prototype I: from agonist to cellular endpoint. *Biophys. J.* 87(3): 1406-1416.

Marston SB, Redwood CS (1991). The molecular anatomy of caldesmon. *Biochem J* 279 (Pt 1): 1-16.

Marthan R (2004). Store-operated calcium entry and intracellular calcium release channels in airway smooth muscle. *Am J Physiol Lung Cell Mol Physiol* 286(5): L907-908.

Mbikou P, Fajmut A, Brumen M, Roux E (2011). Contribution of Rho kinase to the early phase of the calcium-contraction coupling in airway smooth muscle. *Exp Physiol* 96(2): 240-258.

Mbikou P, Fajmut A, Brumen M, Roux E (2006). Theoretical and experimental investigation of calcium-contraction coupling in airway smooth muscle. *Cell Biochem Biophys* 46(3): 233-252.

McNicholl ET, Das R, SilDas S, Taylor SS, Melacini G Communication between tandem cAMP binding domains in the regulatory subunit of protein kinase A-Ialpha as revealed by domain-silencing mutations. *J Biol Chem* 285(20): 15523-15537.

Mijailovich SM, Butler JP, Fredberg JJ (2000). Perturbed equilibria of myosin binding in airway smooth muscle: bond-length distributions, mechanics, and ATP metabolism. *Biophys J* 79(5): 2667-2681.

Miller SG, Kennedy MB (1985). Distinct forebrain and cerebellar isozymes of type II Ca2+/calmodulin-dependent protein kinase associate differently with the postsynaptic density fraction. *J Biol Chem* 260(15): 9039-9046.

Miller SG, Kennedy MB (1986). Regulation of brain type II Ca2+/calmodulin-dependent protein kinase by autophosphorylation: a Ca2+-triggered molecular switch. *Cell* 44(6): 861-870.

Minowa O, Yagi K (1984). Calcium binding to tryptic fragments of calmodulin. *J. Biochem. (Tokyo)* 56: 1175-1182.

Mizuno Y, Isotani E, Huang J, Ding H, Stull JT, Kamm KE (2008). Myosin light chain kinase activation and calcium sensitization in smooth muscle in vivo. *Am J Physiol Cell Physiol* 295(2): C358-364.

Morgan SJ, Smith AD, Parker PJ (1990). Purification and characterization of bovine brain type I phosphatidylinositol kinase. *Eur J Biochem* 191(3): 761-767.

Mounkaila B, Marthan R, Roux E (2005). Biphasic effect of extracellular ATP on human and rat airways is due to multiple P2 purinoceptor activation. *Respir Res* 6(1): 143.

Murthy KS (2006). Signaling for contraction and relaxation in smooth muscle of the gut. *Annu Rev Physiol* 68: 345-374.

Nakagawa O, Fujisawa K, Ishizaki T, Saito Y, Nakao K, Narumiya S (1996). ROCK-I and ROCK-II, two isoforms of Rho-associated coiled-coil forming protein serine/threonine kinase in mice. *FEBS Lett* 392(2): 189-193.

Niiro N, Ikebe M (2001). Zipper-interacting protein kinase induces Ca(2+)-free smooth muscle contraction via myosin light chain phosphorylation. *J Biol Chem* 276(31): 29567-29574.

Nishizuka Y (1995). Protein kinase C and lipid signaling for sustained cellular responses. *FASEB J* 9(7): 484-496.

Otsu M, Hiles I, Gout I, Fry MJ, Ruiz-Larrea F, Panayotou G, *et al.* (1991). Characterization of two 85 kd proteins that associate with receptor tyrosine kinases, middle-T/pp60c-src complexes, and PI3-kinase. *Cell* 65(1): 91-104.

Otto B, Steusloff A, Just I, Aktories K, Pfitzer G (1996). Role of Rho proteins in carbachol-induced contractions in intact and permeabilized guinea-pig intestinal smooth muscle. *J Physiol* 496 (Pt 2): 317-329.

Paul RJ (2009). Regulation of smooth muscle contraction/relaxation: paradigm shifts and quantifying arrows. *The Journal of Physiology* 587(14): 3413-3414.

Perez JF, Sanderson MJ (2005). The frequency of calcium oscillations induced by 5-HT, ACH, and KCl determine the contraction of smooth muscle cells of intrapulmonary bronchioles. *J Gen Physiol* 125(6): 535-553.

Persechini A, Yano K, Stemmer PM (2000). Ca2+ binding and energy coupling in the calmodulin-myosin light chain kinase complex. *J. Biol. Chem.* 275(6): 4199-4204.

Prakash YS, Kannan MS, Walseth TF, Sieck GC (1998). Role of cyclic ADP-ribose in the regulation of [Ca2+]i in porcine tracheal smooth muscle. *Am J Physiol* 274(6 Pt 1): C1653-1660.

Prakash YS, Pabelick CM, Kannan MS, Sieck GC (2000). Spatial and temporal aspects of ACh-induced [Ca2+]i oscillations in porcine tracheal smooth muscle. *Cell Calcium* 27(3): 153-162.

Rameh LE, Cantley LC (1999). The role of phosphoinositide 3-kinase lipid products in cell function. *J Biol Chem* 274(13): 8347-8350.

Rembold CM, Murphy RA (1990). Latch-bridge model in smooth-muscle - [Ca-2+]i can quantitatively predict stress. *Am. J. Physiol.* 259(2): C251-C257.

Riento K, Guasch RM, Garg R, Jin B, Ridley AJ (2003a). RhoE binds to ROCK I and inhibits downstream signaling. *Mol Cell Biol* 23(12): 4219-4229.

Riento K, Ridley AJ (2003b). Rocks: multifunctional kinases in cell behaviour. *Nat Rev Mol Cell Biol* 4(6): 446-456.

Rossetti M, Savineau JP, Crevel H, Marthan R (1995). Role of protein kinase C in nonsensitized and passively sensitized human isolated bronchial smooth muscle. *Am J Physiol* 268(6 Pt 1): L966-971.

Roux E, Hyvelin JM, Savineau JP, Marthan R (1998). Calcium signaling in airway smooth muscle cells is altered by in vitro exposure to the aldehyde acrolein. *Am J Respir Cell Mol Biol* 19(3): 437-444.

Roux E, Marhl M (2004). Role of sarcoplasmic reticulum and mitochondria in ca(2+) removal in airway myocytes. *Biophys J* 86(4): 2583-2595.

Roux E, Noble PJ, Noble D, Marhl M (2006). Modelling of calcium handling in airway myocytes. *Prog Biophys Mol Biol* 90(1-3): 64-87.

Saitoh T, Schwartz JH (1985). Phosphorylation-dependent subcellular translocation of a Ca2+/calmodulin-dependent protein kinase produces an autonomous enzyme in Aplysia neurons. *J Cell Biol* 100(3): 835-842.

Sakurada S, Takuwa N, Sugimoto N, Wang Y, Seto M, Sasaki Y, *et al.* (2003). Ca2+-dependent activation of Rho and Rho kinase in membrane depolarization-induced and receptor stimulation-induced vascular smooth muscle contraction. *Circ Res* 93(6): 548-556.

Sanders KM (2001). Invited review: mechanisms of calcium handling in smooth muscles. *J. Appl. Physiol.* 91(3): 1438-1449.

Schaafsma D, Boterman M, de Jong AM, Hovens I, Penninks JM, Nelemans SA, *et al.* (2006). Differential Rho-kinase dependency of full and partial muscarinic receptor agonists in airway smooth muscle contraction. *Br J Pharmacol* 147(7): 737-743.

Schramm CM, Chuang ST, Grunstein MM (1995). cAMP generation inhibits inositol 1,4,5-trisphosphate binding in rabbit tracheal smooth muscle. *Am J Physiol* 269(5 Pt 1): L715-719.

Scott JD, McCartney S (1994). Localization of A-kinase through anchoring proteins. *Mol Endocrinol* 8(1): 5-11.

Shabb JB (2001). Physiological substrates of cAMP-dependent protein kinase. *Chem Rev* 101(8): 2381-2411.

Shah V, Bharadwaj S, Kaibuchi K, Prasad GL (2001). Cytoskeletal organization in tropomyosin-mediated reversion of ras-transformation: Evidence for Rho kinase pathway. *Oncogene* 20(17): 2112-2121.

Shibasaki F, Homma Y, Takenawa T (1991). Two types of phosphatidylinositol 3-kinase from bovine thymus. Monomer and heterodimer form. *J Biol Chem* 266(13): 8108-8114.

Shields SM, Vernon PJ, Kelly PT (1984). Autophosphorylation of calmodulin-kinase II in synaptic junctions modulates endogenous kinase activity. *J Neurochem* 43(6): 1599-1609.

Shimizu T, Ihara K, Maesaki R, Amano M, Kaibuchi K, Hakoshima T (2003). Parallel coiled-coil association of the RhoA-binding domain in Rho-kinase. *J Biol Chem* 278(46): 46046-46051.

Sieck GC, Han Y-S, Prakash YS, Jones KA (1998). Cross-bridge cycling kinetics, actomyosin ATPase activity and myosin heavy chain isoforms in skeletal and smooth respiratory muscles. *Comp. Biochem. Physiol. Part B: Biochem. Mol. Biol.* 119(3): 435-450.

Smith L, Stull JT (2000). Myosin light chain kinase binding to actin filaments. *FEBS Lett.* 480: 298-300.

Somlyo AP, Somlyo AV (2003). Ca2+ sensitivity of smooth muscle and nonmuscle myosin II: modulated by G proteins, kinases, and myosin phosphatase. *Physiol Rev* 83(4): 1325-1358.

Somlyo AP, Somlyo AV (1994). Signal transduction and regulation in smooth muscle. *Nature* 372(6503): 231-236.

Somlyo AP, Somlyo AV (2000). Signal transduction by G-proteins, rho-kinase and protein phosphatase to smooth muscle and non-muscle myosin II. *J Physiol* 522 Pt 2: 177-185.

Somlyo AP, Somlyo AV, Kitazawa T, Bond M, Shuman H, Kowarski D (1983). Ultrastructure, function and composition of smooth muscle. *Ann Biomed Eng* 11(6): 579-588.

Stull JT, Hsu LC, Tansey MG, Kamm KE (1990). Myosin light chain kinase phosphorylation in tracheal smooth muscle. *J Biol Chem* 265(27): 16683-16690.

Stull JT, Tansey MG, Tang DC, Word RA, Kamm KE (1993). Phosphorylation of myosin light chain kinase: a cellular mechanism for Ca2+ desensitization. *Mol Cell Biochem* 127-128: 229-237.

Su Y, Dostmann WR, Herberg FW, Durick K, Xuong NH, Ten Eyck L, *et al.* (1995). Regulatory subunit of protein kinase A: structure of deletion mutant with cAMP binding domains. *Science* 269(5225): 807-813.

Tansey MG, Luby-Phelps K, Kamm KE, Stull JT (1994). Ca(2+)-dependent phosphorylation of myosin light chain kinase decreases the Ca2+ sensitivity of light chain phosphorylation within smooth muscle cells. *J Biol Chem* 269(13): 9912-9920.

Tansey MG, Word RA, Hidaka H, Singer HA, Schworer CM, Kamm KE, *et al.* (1992). Phosphorylation of myosin light chain kinase by the multifunctional calmodulin-dependent protein kinase II in smooth muscle cells. *J Biol Chem* 267(18): 12511-12516.

Taylor SS, Kim C, Vigil D, Haste NM, Yang J, Wu J, *et al.* (2005). Dynamics of signaling by PKA. *Biochim Biophys Acta* 1754(1-2): 25-37.

Terrak M, Kerff F, Langsetmo K, Tao T, Dominguez R (2004). Structural basis of protein phosphatase 1 regulation. *Nature* 429(6993): 780-784.

Toph A, Kiss E, Gergely P, Walsh MP, Hartshorne DJ, Erdodi F (2000). phosphorylation of MYPT1 by protein kinase C attenuates interaction with PP1 catalytic subunit and 20 kDa light chain of myosin. *FEBS Lett* 484: 113-117.

Torok K, Trentham DR (1994). Mechanism of 2-chloro-(epsilon-amino-Lys75)-[6-[4-(N,N-diethylamino)phenyl]-1,3,5-triazin-4-yl]calmodulin interactions with smooth muscle myosin light chain kinase and derived peptides. *Biochem.* 33: 12807-12820.

Trybus KM (1996). In: Barany M (ed)^(eds). *Biochemistry of smooth muscle contraction*, edn. San Diego, CA: Academic Press. p^pp 37-46.

Ueda K, Murata-Hori M, Tatsuka M, Hosoya H (2002). Rho-kinase contributes to diphosphorylation of myosin II regulatory light chain in nonmuscle cells. *Oncogene* 21(38): 5852-5860.

Velasco G, Armstrong C, Morrice N, Frame S, Cohen P (2002). Phosphorylation of the regulatory subunit of smooth muscle protein phosphatase 1M at Thr850 induces its dissociation from myosin. *FEBS Lett* 527(1-3): 101-104.

Wang I, Politi A, Tania N, Bai Y, Sanderson M, Sneyd J (2008). A mathematical model of airway and pulmonary arteriole smooth muscle. *Biophys J* 94(6): 2053-2064.

Wang Y, Zheng XR, Riddick N, Bryden M, Baur W, Zhang X, *et al.* (2009). ROCK isoform regulation of myosin phosphatase and contractility in vascular smooth muscle cells. *Circ Res* 104(4): 531-540.

Webb BL, Hirst SJ, Giembycz MA (2000). Protein kinase C isoenzymes: a review of their structure, regulation and role in regulating airways smooth muscle tone and mitogenesis. *Br J Pharmacol* 130(7): 1433-1452.

Webb BL, Lindsay MA, Seybold J, Brand NJ, Yacoub MH, Haddad EB, *et al.* (1997). Identification of the protein kinase C isoenzymes in human lung and airways smooth muscle at the protein and mRNA level. *Biochem Pharmacol* 54(1): 199-205.

Wilson DP, Sutherland C, Walsh MP (2002). Ca2+ activation of smooth muscle contraction. Evidence for the involvement of calmodulin that is bound to the triton-insoluble fraction even in the absence of Ca2+. *J. Biol. Chem.* 277(3): 2186-2192.

Winder SJ, Allen BG, Clement-Chomienne O, Walsh MP (1998). Regulation of smooth muscle actin-myosin interaction and force by calponin. *Acta Physiol Scand* 164(4): 415-426.

Winder SJ, Walsh MP (1990). Smooth muscle calponin. Inhibition of actomyosin MgATPase and regulation by phosphorylation. *J Biol Chem* 265(17): 10148-10155.

Wingard CJ, Nowocin JM, Murphy RA (2001). Cross-bridge regulation by Ca(2+)-dependent phosphorylation in amphibian smooth muscle. *Am J Physiol Regul Integr Comp Physiol* 281(6): R1769-1777.

Wooldridge AA, MacDonald JA, Erdodi F, Ma C, Borman MA, Hartshorne DJ, *et al.* (2004). Smooth muscle phosphatase is regulated in vivo by exclusion of phosphorylation of threonine 696 of MYPT1 by phosphorylation of Serine 695 in response to cyclic nucleotides. *J Biol Chem* 279(33): 34496-34504.

Yamauchi T, Fujisawa H (1985). Self-regulation of calmodulin-dependent protein kinase II and glycogen synthase kinase by autophosphorylation. *Biochem Biophys Res Commun* 129(1): 213-219.

Yoneda A, Multhaupt HA, Couchman JR (2005). The Rho kinases I and II regulate different aspects of myosin II activity. *J Cell Biol* 170(3): 443-453.

Yoshii A, Iizuka K, Dobashi K, Horie T, Harada T, Nakazawa T, *et al.* (1999). Relaxation of contracted rabbit tracheal and human bronchial smooth muscle by Y-27632 through inhibition of Ca2+ sensitization. *Am J Respir Cell Mol Biol* 20(6): 1190-1200.

Yoshioka K, Sugimoto N, Takuwa N, Takuwa Y (2007). Essential role for class II phosphoinositide 3-kinase alpha-isoform in Ca2+-induced, Rho- and Rho kinase-dependent regulation of myosin phosphatase and contraction in isolated vascular smooth muscle cells. *Mol Pharmacol* 71(3): 912-920.

Yu SN, Crago PE, Chiel HJ (1997). A nonisometric kinetic model for smooth muscle. *Am J Physiol Cell Physiol* 272(3): C1025-1039.

Yuen SL, Ogut O, Brozovich FV (2009). Nonmuscle myosin is regulated during smooth muscle contraction. *Am J Physiol Heart Circ Physiol* 297(1): H191-199.

Zhang WC, Peng YJ, Zhang GS, He WQ, Qiao YN, Dong YY, *et al.* (2010). Myosin light chain kinase is necessary for tonic airway smooth muscle contraction. *J Biol Chem* 285(8): 5522-5531.

Zhao J, Hoye E, Boylan S, Walsh DA, Trewhella J (1998). Quaternary structures of a catalytic subunit-regulatory subunit dimeric complex and the holoenzyme of the cAMP-dependent protein kinase by neutron contrast variation. *J Biol Chem* 273(46): 30448-30459.

Zhou HL, Newsholme SJ, Torphy TJ (1992). Agonist-related differences in the relationship between cAMP content and protein kinase activity in canine trachealis. *J Pharmacol Exp Ther* 261(3): 1260-1267.

Zhou ZL, Ikebe M (1994). New isoforms of Ca2+/calmodulin-dependent protein kinase II in smooth muscle. *Biochem J* 299 (Pt 2): 489-495.

cGMP-Dependent Protein Kinase in the Regulation of Cardiovascular Functions

Yuansheng Gao, Dou Dou, Xue Qin, Hui Qi and Lei Ying
[1]Department of Physiology and Pathophysiology,
Peking University Health Science Center,
[2]Key Laboratory of Molecular Cardiovascular Science (Peking University),
Ministry of Education,
China

1. Introduction

Cyclic GMP-dependent protein kinase (PKG) was discovered in 1970 in lobster muscle (Kuo & Greengard, 1970). It is a serine/threonine protein kinase specifically activated by cyclic guanosine monophosphate (cGMP). PKG is a ubiquitous intracellular second messenger mediating the biological effects of cGMP elevating agents including nitric oxide (NO), natriuretic peptides, and guanylin (an intestinal peptide involved in intestinal fluid regulation). It is now well recognized that PKG plays a central role in a broad range of physiological processes, such as contractility and proliferation of smooth muscle and cardiac myocytes, platelet aggregation, synaptic plasticity and learning, behavior, intestinal chloride reabsorption, renin secretion, and endochondral ossification (Francis et al., 2010; Hofmann et al., 2009; Lincoln et al., 2001; Lohmann & Walter, 2005). This chapter will focus on the role of PKG in the regulation of cardiovascular functions under physiological and pathophysiological conditions.

2. PKG structure and tissue distribution

In mammalian cells PKG exists as two types, PKG-I and PKG-II, respectively. They are encoded by two separated genes *prkg1* and *prkg2*. The human *prkg1* gene is located on chromosome 10 at p11.2 - q11.2 and has 15 exons. The NH$_2$ terminus (the first 100 residues) of PKG I is encoded by two alternative exons that produce the isoforms PKG Iα and PKG Iβ. The human *prkg2* gene is located on chromosome 4 at q13.1 - q21.1 and has 19 exons. PKG-I and PKG-II is composed of two identical subunits of the homodimer about 75-80 kDa and 84-86 kDa, respectively and shares common structural features. Each subunit of the enzyme consists of a regulatory domain and a catalytic domain. The regulatory domain is composed of an N-terminal domain and a cGMP binding domain. The N-terminal domain mediates homodimerization, suppression of the kinase activity in the absence of cGMP, and interactions with other proteins including protein substrates. The cGMP binding domain contains a high and a low cGMP affinity binding sites. The two cGMP-binding sites interact allosterically. Binding of cGMP releases the inhibition of the catalytic center by the N-terminal autoinhibitory/pseudosubstrate domain and allows the phosphorylation of target proteins. The catalytic domain contains a MgATP and a target protein-binding site, which catalyze the phosphate transfer from ATP to the hydroxyl group of a serine/threonine side chain of the target protein. When stimulated with cGMP, the phosphotransferase activity increases by 3- to 10-fold (Francis et al., 2010; Hofmann et al., 2009).

PKG-I is predominantly localized in the cytoplasm (except in the platelets where it is with the membrane). PKG-II is anchored to the plasma membrane by N-terminal myristoylation. In general, PKG-I and PKG-II are expressed in different cell types. PKG-I exists at high concentrations in all types of smooth muscle cells (~0.1μM) including vascular smooth muscle cells and at lower levels in vascular endothelium and cardiomyocytes. The enzyme has also been detected in other cell types such as fibroblasts, certain types of renal cells and leukocytes, and in specific regions of the nervous system. Platelets express predominantly PKG Iβ while both PKG Iα and PKG Iβ isoforms are present in smooth muscle, including uterus, blood vessels, intestines, and trachea. PKG-II is expressed in several brain nuclei, intestinal mucosa, kidney, chondrocytes and the lung but not in cardiac and vascular myocytes (Francis et al., 2010; Hofmann et al., 2009).

Existing research results show that PKG-I is the major type of the enzymes in the cardiovascular system involved in the regulation of vascular tone, regulation of vascular smooth muscle cells and myocardial cells proliferation and phenotypic modulation, and inhibiting platelet aggregation. Both PKG-Iα and PKG-Iβ can be specially activated by cGMP, with the former is about 10 times more sensitive to cGMP than the latter. PKG can also be activated by cAMP, although more than 100 times less potent than cGMP. The main role of PKG-II is phosphorylation in the intestinal mucosa of cystic fibrosis transmembrane conductance regulator, regulation of intestinal chloride ion/fluid secretion, inhibition of renin secretion in the kidney, and the regulation of bone tissue and bone endochondral bone growth (Francis et al., 2010; Hofmann et al., 2009; Lincoln et al., 2001; Lohmann & Walter, 2005).

3. PKG function in the cardiovascular system

3.1 Blood vessels

3.1.1 Vasodilatation

PKG is involved in vasodilatation caused by cGMP elevating agents including endothelium-derived NO, ANP, CNP, and exogenous nitrovasodilators (Gao, 2010; Hofmann et al., 2009). In certain vessel types such as ovine perinatal pulmonary artery and vein (Dhanakoti et al., 2000; Gao et al., 1999) as well as porcine coronary artery and vein (Qin et al., 2007; Qi et al., 2007) relaxation caused by nitrovasodilators is primarily mediated by PKG. Studies show that the expression and activity of PKG can be modulated by physiological variables such as oxygenation (Gao et al., 2003).

Activation of calcium activated potassium (BK) channels has been implicated as a mechanism for PKG-mediated relaxation of vascular smooth muscle in a number of vessel types including cerebral artery (Robertson et al., 1993), coronary artery (White et al., 2000), and pulmonary artery (Barman et al., 2003), which leads to increased membrane polarization and thus decreased Ca^{2+} influx and vasodilatation. PKG may stimulate BK channels by direct phosphorylation of the α-subunit at serine 1072 (Fukao et al., 1999) or through activation of protein phosphatase 2A (Zhou et al., 1996). In ovine basilar arterial smooth muscle cells PKG has been shown to play a larger role in the regulation of BK activity in fetal than in adult myocytes, indicating a developmental changes in the role of PKG (Lin et al., 2005).

PKG may also modulate Ca^{2+} release from the inositol-trisphosphate receptor (IP_3R) of the sarcoplasmic/endoplasmic reticulum (SERCA) through phosphorylation of IP_3R-associated cGMP kinase substrate (IRAG), a 125-kDa protein that resides in the SERCA membrane in a trimeric complex with PKG Iβ and IP_3R. In aortic smooth muscle cells of mice expressing a mutated IRAG protein that does not interact with the IP_3R the inhibition of cGMP on hormone-induced increases in $[Ca^{2+}]_i$ and contractility are blunted (Geiselhöringer et al., 2004). NO-, ANP-, and cGMP-dependent relaxation of aortic vessels is also attenuated in IRAG-knockout mice (Desch et al., 2010).

Increasing evidence has pointed to Ca^{2+} desensitization through interference with RhoA and Rho kinase (ROK) signaling as a key mechanism for PKG-mediated vasodilatation (A.P. Somlyo & A.V. Somlyo, 2003). PKG may phosphorylate RhoA at Ser188, resulting in increased extraction of Rho A from cell membranes and thus reduced activation of this small GTPase protein and attenuated vasocontractility (Loirand et al., 2006). PKG may suppress the inhibitory effect of ROK on myosin light chain phosphatase (MLCP) by phosphorylation of the regulatory subunit of MLCP, myosin phosphatase targeting subunit (MYPT1), at Ser695 and Ser852, which leads to decreased phosphorylation of MYPT1 at Thr696 and Thr853 by ROK, increased activity of MLCP, decreased phosphorylation of myosin light chain (MLC), and diminished vasoreactivity (Wooldridge et al., 2004; Gao et al., 2007 &2008). The effect of PKG on MLCP requires its binding to the leucine zipper domain in the C-terminal of MYPT1. The expression of the leucine zipper domain in MYPT1 is modulated by various physiological and pathophyiological conditions (Chen et al., 2006; Dou et al., 2010; Payne et al., 2006), which may alter the action of PKG on MLCP. Studies also show that PKG/MYPT1 signaling plays a greater role in mediating relaxation of proximal arteries induced by NO than that of distal arteries in coronary vasculature (Ying et al., 2011).

A number of PKG substrates not mentioned above may also be targeted by PKG and involved in PKG-mediated vasodilatation, such as phosphodiesterase 5 (PDE5), phospholamban, and RGS (regulator of G-protein signaling) proteins (Schlossmann & Desch, 2009). It is worth noting that cGMP may affect vasodilatation by PKG-dependent and independent mechanism. Global PKG-knockout causes only a slight hypertension in young mice whereas in the adult the basal blood pressure of the PKG-knockout mice is not different from the control (Pfeifer et al. 1998), indicating other mechanisms may take place to compensate the lose of PKG in maintaining a normal blood pressure.

3.1.2 Phenotype modulation and antiproliferation action

Vascular smooth muscle cells (VSMCs) exist in either a differentiated, contractile or a dedifferentiated, synthetic phenotype. A normal PKG activity appears critical to maintain vascular smooth muscle cells in a contractile and differentiated state. Repetitively passaged VSMCs of the rat aorta do not express PKG and exist in the synthetic phenotype. Transfection of PKG Iα cDNA induces a morphologic change of VSMCs consistent with the contractile phenotype, which is prevented by the inhibition of PKG (Dey et al., 2005).

Myocardin is a smooth muscle and cardiac muscle-specific transcriptional coactivator of serum response factor (SRF) while E26-like protein-1 (Elk-1) is a SRF/myocardin transcription antagonist. PKG-I has been shown to decrease Elk-1 activity by sumo modification of Elk-1, thereby increasing myocardin-SRF activity on SMC-specific gene expression and keeping the cells in a contractile phenotype (Choi et al., 2010). In VSMCs of ovine fetal pulmonary veins hypoxia-induced reduction in PKG protein expression is closely correlated with the repressed expression of VSMC phenotype markers, along with a reduced expression of myocardin and increased expression of Elk-1. It is postulated that the increased expression of Elk-1 resulting from the downregulation of PKG under hypoxia displaces myocardin from SRF and thereby leads to suppression of SMC marker genes and activation of expressions of genes related with the synthetic phenotype (Zhou et al., 2007 & 2009). The PKG-dependent modulation of phenotypes of VSMCs appears to need the cysteine-rich LIM-only protein CRP4 to act as a scaffolding protein that promotes cooperation between SRF and other transcription factors and cofactors since PKG stimulation of the SM-α-actin promoter is suppressed when CRP4 is deficient in PKG binding (Zhang et al., 2007).

PKG has been reported to exert anti- and pro-atherogenic effects in vascular smooth muscle. In coronary and cerebral arterial smooth muscle cells (El-Mowafy et al., 2008; Luo et al., 2011) the proliferation induced by vascular mitogens was inhibited by the cGMP elevating agent or PKG I transfection. However, 8-Br-cGMP stimulated proliferation of aortic SMCs from the wide-type mice but not from PKG I-deficient mice (Wolfsgruber et al., 2003). The contradictory effects may have in part resulted from differences in PKG activation levels (i.e., basal activation vs. hyperactivation). For instance, PKG at low activation levels prevents apoptosis whereas high-level activation causes apoptosis of aortic SMCs of the mice (Wong & Fiscus, 2010).

3.2 Heart

3.2.1 Cardiac contractility

A critical role for PKG in the negative inotropic effect caused by NO and cGMP has been demonstrated in myocardial preparations from PKG I-deficient juvenile mice, from the cardiomyocyte-specific knockout adult mice (Wegener et al., 2002), and in rat ventricular myocytes with the PKG inhibitor (Layland et al., 2002). The Ca^{2+} current of rat ventricular cells is inhibited by cGMP and a catalytically active fragment of PKG (Mery et al., 1991). In murine cardiac myocytes overexpressing PKG I the basal and stimulated activities of L-type Ca^{2+} channels are inhibited by NO and the cGMP analog (Schroder et al. 2003). Hence, PKG may exert its negative inotropic effect action by reducing $[Ca^{2+}]_i$ through inhibiting the activity of Ca^{2+} channels. Recent studies suggest that PKG I-mediated inhibition of L-type Ca^{2+} channels of cardiac myocytes may result from the phosphorylation of Ca_v 1.2 α_{1c} and β_2 subunits (Yang et al., 2007).

PKG may also reduce $[Ca^{2+}]_i$ of cardiomyocytes through phosphorylation of phospholamban, which leads to an increased activity of the sarcoplasmic reticulum Ca^{2+}-ATPase (SERCA) and thereby an increased Ca^{2+} uptake from the cytosol. Indeed, phospholamban of the rabbit cardiac myocytes is phosphorylated by cGMP in a manner sensitive to the inhibition of PKG. Moreover, the inhibitory effect of contractility of the cardiac myocytes caused by cGMP is prevented by the inhibition of PKG or SERCA (Zhang et al., 2005). C-type natriuretic peptide and the cGMP analog have been found to cause positive inotropic and lusitropic responses of murine hearts, which are associated with an increased phosphorylation of phospholamban (Wollert et al., 2003). These observations are in vary with those obtained in studies by Zhang et al. discussed above. The underlying reasons for the different inotropic effects remain to be determined (Wollert et al., 2003).

In intact cardiomyocytes of the rat, the negative inotropic and relaxant effects of DEA/NO, an NO donor, occur without significant changes in the amplitude or kinetics of the intracellular Ca^{2+} transient. The effect is diminished in the presence of the inhibitor of soluble guanylyl cyclase (sGC) or PKG, indicating a PKG-dependent Ca^{2+} desensitization of the myofilaments. Meanwhile, hearts treated with DEA/NO showed a significant increase in troponin I phosphorylation (Layland et al., 2002). The PKG may reduce the Ca^{2+} sensitivity of cardiac myofilaments through phosphorylation of cardiac troponin I (cTnI) at the same sites (Ser23/24) as those phosphorylated by PKA (Layland et al., 2005). Studies suggest that cardiac Troponin T may serve as an anchoring protein for PKG to facilitate preferential and rapid cTnI phosphorylation (Yuasa et al., 1999).

3.2.2 Antihypertrophy

An increased left ventricular mass has been recognized as an independent risk factor that correlates closely with cardiovascular risk and has strong prognostic implications. In the mice administration of sildenafil, which elevates cGMP level by inhibiting PDE5, suppresses the development of cardiac hypertrophy caused by chronic pressure overload and can even reverse pre-established cardiac enlargement. These effects are associated with an increased activity of PKG I (Takimoto et al. 2005). Mice with myocyte-specific PDE5 gene overexpression develop more severe cardiac hypertrophy and PKG activation is inhibited as compared to controls in response to pressure overload. Under such a cardiomyopathic state, the suppression of PDE5 expression/activity in myocytes enhanced PKG activity and reversed all previously amplified maladaptive responses (Zhang et al., 2010). In contrast to many studies which indicate an antihypertrophic role for PKG, Lukowski et al. have found that total PKG I-knockout and myocyte-specific rescue of PKG expression (in the context of global gene silencing) did not affect isoproterenol and stress-induced development of cardiac hypertrophy in mice (Lukowski et al., 2010). It is suspected that the lack of differences between controls and PKG I-deficient mice may be in part due to that PKG I-targeted cascades have not been activated under the experimental conditions (Kass & Takimoto, 2010).

RGSs are GTPase-accelerating proteins that promote GTP hydrolysis by the alpha subunit of heterotrimeric G proteins, thereby accelerating signal termination in response to GPCR stimulation (Schlossmann & Desch, 2009). Among more than 30 RGS proteins, RGS4 is richly expressed in murine coronary myocytes. In cultured cardiac myocytes, atrial natriuretic peptide stimulated PKG-dependent phosphorylation of RGS4 and association of RGS4 with the alpha subunit of Gq protein. Mice lacking guanylyl cyclase-A (GC-A), a natriuretic peptide receptor, have pressure-independent cardiac hypertrophy, reduced expression and phosphorylation of RGS4 in the hearts compared with wild-type mice. The RGS4 overexpression in GC-A-KO mice reduced cardiac hypertrophy and suppressed the augmented cardiac expressions of hypertrophy-related genes. These results suggest that GC-A activation may counteract cardiac hypertrophy via RGS4 in a PKG-dependent mechanism (Tokudome et al., 2008). It appears that ANP-cGMP-PKG-RGS signaling is involved in β-adrenergic but not angiotensin II (Ang II)–induced (Gs vs. Gaq mediated) cardiomyocyte hypertrophy. ANP attenuated Ang II-stimulated Ca^{2+} currents of cardiomycytes but had no effect on isoproterenol stimulation. The effect of ANP on Ang II stimulation was eliminated in cardiomyocytes of mice deficient in GC-A, in PKG I, or in RGS2. Furthermore, cardiac hypertrophy induced by Ang II but not by β- adrenoreceptor was exacerbated in mice with cardiomyocyte-restricted GC-A deletion (Klaiber et al., 2010).

Multiple subclasses of transient receptor potential (TRP) channels are expressed in the heart. These channels, especially the TRPC subclass, have been implicated being involved in the regulation of the cardiac hypertrophic response, most likely coordinating signaling within local domains or through direct interaction with Ca^{2+}-dependent regulatory proteins. Overexpression of TRPC6 in mice lacking GC-A exacerbated cardiac hypertrophy while the blockade of TRPC channels attenuated the cardiac hypertrophy. ANP inhibited agonist-evoked Ca^{2+} influx of murine cardiomyocytes. The inhibitory effects of ANP were abolished by PKG inhibitors or by substituting an alanine for threonine 69 in TRPC6, suggesting that PKG-dependent phosphorylation of TRPC6 at threonine 69 is a critical target of antihypertrophic effects elicited by ANP (Kinoshita et al., 2010)

3.2.3 Cardioprotective action against ischemia-reperfusion injury

In isolated murine heart and cardiomyocytes elevation of cGMP by the activators of soluble or particulate guanylyl cyclase, by the inhibitor of PDE5, or by the cGMP analog elicits potent protection against myocardial ischemia-reperfusion injury and reduces cardiomyocyte necrosis and apoptosis. These effects are accompanied by an increased PKG activity and attenuated by PKG inhibitors or by selective knockdown of PKG in cardiomyocytes (Das, 2008 et al.; Gorbe et al., 2010). It is generally recognized that ischemia/reperfusion injury arises primarily from the opening of the mitochondrial permeability transition pore (mPTP) in the first minutes of reperfusion. cGMP-PKG signaling may prevent opening of mPTP via activation of the mitochondrial K_{ATP} channels, direct phosphorylation of an unknown protein on the mitochondrial outer membrane, and upregulation of the antiapoptotic protein Bcl-2 (Costa et al.,2008; Deschepper, 2010). Glycogen synthase kinase 3β (GSK-3β) plays a central role in transferring cardio protective signals downstream to target(s) that act at or in proximity to the mPTP. Phosphorylation and inhibition of GSK-3β has also been demonstrated being involved in PKG-mediated cardioprotective action (Das et al., 2008; Juhaszova et al., 2009; Xi et al., 2010).

3.3 eNOS activity and endothelial permeability

PKG I has been detected within a range of 0.15 to 0.5 µg/mg cellular protein in adult artery and vein endothelial cells (ECs) and in microvascular ECs (Diwan et al., 1994; Draijer et al., 1995; MacMillan-Crow et al., 1994). There are only limited studies on the role of PKG in the regulation of endothelial function, which is related to eNOS activity and endothelial permeability (Butt et al., 2000; Draijer et al., 1995; Moldobaeva et al., 2006; Rentsendorj et al., 2008). Studies using recombinant human eNOS suggest that the enzyme can be phosphorylated at Ser^{1177}, Ser^{633}, Thr^{495} and activated by PKG II in a manner independent of Ca^{2+} and calmodulin (Butt et al., 2000). Cyclic GMP analog inhibits an increase in $[Ca^{2+}]_i$ and endothelial permeability caused by thrombin in cultured ECs expressing PKG I but not those lacking PKG expressing (Draijer et al., 1995). In human pulmonary artery endothelial cells infected with adenovirus encoding PKG Iβ the cGMP analog prevents the increase in endothelial permeability caused by H_2O_2. The barrier protection effect was not affected by inhibition of the expression of VASP, a PKG substrate (Moldobaeva et al., 2006; Rentsendorj et al., 2008).

3.4 Anti-platelet aggregation action

Substantial evidence supports a critical role for PKG in mediating the anti-platelet aggregation action caused by cGMP elevating agents such as endothelium-derived NO (EDNO) and exogenous nitrovasodilators (Walter & Gambaryan, 2009; Dangel et al., 2010). PKG Iβ is the predominate isoform of the enzyme in platelets. The concentration of PKG Iβ in human platelets is 3.65μM, which is higher than that of any other cell type examined (Antl et al., 2007; Eigenthaler et al., 1992). In PKG-deficient murine platelets the inhibition of the cGMP analog on granule secretion, aggregation and adhesion is severely affected (Massberg et al. 1999; Schinner et al., 2011). The effect of PKG may be in part mediated by IRAG. IRAG is abundantly expressed in platelets and constitutively formed in a macrocomplex with PKGIβ and the $InsP_3R$. PKGIβ phosphorylates IRAG at Ser664 and Ser677 in intact platelets, resulting in attenuated release of Ca^{2+} from the sarcoplasmic reticulum evoked by IP_3. Targeted deletion of the IRAG-InsP3RI interaction in $IRAG^{\Delta 12/\Delta 12}$ mutant mice causes a loss of NO/cGMP-dependent inhibition of $[Ca^{2+}]_i$ increase and platelet aggregation. The preventive effect of NO on arterial thrombosis in the injured carotid artery was observed in wide-type platelets but not in $IRAG^{\Delta 12/\Delta 12}$ mutants (Antl et al. 2007).

Vasodilator-stimulated phosphoprotein (VASP) belongs to the Ena-VASP protein family. It is associated with filamentous actin formation and may play a widespread role in cell adhesion and motility. In VASP-deficient mice, the inhibitory effect of NO on platelet adhesion is impaired. Under physiologic conditions, platelet adhesion to endothelial cells was enhanced in VASP null mutants. Under pathophyiological conditions, the loss of VASP augments platelet adhesion to the postischemic intestinal microvasculature, to the atherosclerotic endothelium of ApoE-deficient mice, and to the subendothelial matrix of blood vessels (Massberg et al. 2004). In VASP-deficient mice, although cGMP-mediated inhibition of platelet aggregation is impaired, cGMP-dependent inhibition of agonist-induced increases in cytosolic calcium concentrations and granule secretion is preserved (Aszódi et al., 1999).

Although it is a currently prevailing concept that PKG signaling inhibits platelet function, some studies show that activation of NO-cGMP-PKG pathway promotes platelet aggregation (Blackmore, 2011; Li et al., 2003; Zhang et al., 2011). In PKG knockout mice platelet responses to von Willebrand factor (vWF) or low doses of thrombin are impaired and bleeding time is prolonged. Human platelet aggregation induced by these agents is also diminished by PKG inhibitors but enhanced by cGMP (Li et al., 2003). A defect in platelet aggregation in response to low doses of collagen or thrombin also occurs in platelet-specific sGC-deficient mice (Zhang et al., 2011). It appears that cGMP at low concentrations promotes while at higher concentrations inhibits platelet aggregation (Blackmore et al., 2011; Li et al., 2003).

4. PKG and cardiovascular diseases

4.1 Hypertension

Global deletion of eNOS (Huang et al., 1995), sGC (Friebe et al., 2007), or PKG I (Pfeifer et al. 1998) results in hypertension in mice. About 80% of the mice that are deficient in PKG I died at age of 8-week. Those lived to adulthood showed no significant difference in blood pressure from the wild type animals, indicating compensatory mechanisms are functioning (Pfeifer et al. 1998). Loss of PKG I abolishes NO- and cGMP-dependent relaxations of smooth muscle (Pfeifer et al. 1998). In mice with a selective mutation in the N-terminal protein interaction domain of PKG Iα also results in reduced vasodilator response to EDNO and cGMP and increased systemic blood pressure, suggesting that the hypertension results from a diminished response of blood vessels to cGMP (Michael et al., 2008). Vascular reconstitution of PKG Iα or PKG Iβ in PKG I-deficient mice restores the diminished vasodilatation to NO and cGMP and normalizes the elevated blood pressure, (Weber et al., 2007). In spontaneously hypertensive rats (SHR) cardiomyocytes PKG-I expression is decreased, making the NO/cGMP-dependent regulation on calcium transient in cardiomyocytes weakened, and promoting cardiac hypertrophy (Mazzetti et al., 2001).

Abnormality in the renin-angiotensin-aldosterone system is an important etiologic event in the development of hypertension. Renal renin mRNA levels under stimulatory (low-salt diet plus ramipril) and inhibitory (high-salt diet) conditions were elevated in PKG II deficient mice. The deletion of PKG II abolishes the attenuation of forskolin-stimulated renin secretion caused by 8-Br-cGMP in cultured renal juxtaglomerular cells. Activation of PKG by 8-Br-cGMP decreased renin secretion from the isolated perfused rat kidney of the wild-type mice but not that of PKG II-/- mice. These findings suggest that PKG II exerts an inhibitory effect on renin secretion (Wagner et al., 1998). Mice deficient in PKG II display no elevated blood pressure, suggesting that PKG II is not critically involved in the regulation of overall systemic blood pressure (Hofmann et al., 2009).

4.2 Atherosclerosis

In an animal model of late-stage atherosclerosis obtained by feeding 8-week-old rabbits with hypercholesterol diet for 50 weeks the protein levels of sGC and PKG I of the aorta were reduced. These changes were most prominent in the neointimal layer. Phosphorylation of VASP at Ser239, a specific indicator of PKG activity, was also reduced. The preferential down-regulation of cGMP/PKG signaling in neointima suggests a direct connection of these changes to neointimal proliferation and vascular dysfunction occurred in atherosclerosis (Melichar et al., 2004). It seems that the decreased PKG expression occurred only at late-stage atherosclerosis, as the protein level of PKG was unaltered in Watanabe heritable hyperlipidemic rabbits of three month old (Warnholtz et al., 2002). Thrombospondin-1 and osteopontin are extracellular matrix (ECM) proteins involved in the development of atherosclerosis. PKG may exert its anti- atherosclerotic effect in part through these two ECM proteins, since their expression could be marked reduced by PKG I (Dey et al., 1998). Interestingly, postnatal ablation of PKG I selectively in the VSMCs of mice reduced atherosclerotic lesion area, which would suggest that smooth muscle PKG I promotes atherogenesis (Wolfsgruber et al., 2003).

4.3 Diabetic vascular disease

High glucose exposure has been found to reduce the protein and mRNA levels of PKG I as well as PKG I activity in cultured rat VSMCs. PKG I protein levels were decreased in femoral arteries from diabetic mice. Glucose-mediated decrease in PKG I levels was inhibited by the superoxide scavenger or NAD(P)H oxidase inhibitors. High glucose exposure increased the protein levels and phosphorylated levels of p47phox (an NADPH oxidase subunit) in VSMCs, associated with increased superoxide production. The suppressed PKG expression and increased superoxide production were prevented by transfection of cells with siRNA-p47phox, suggesting that NADPH oxidase-derived superoxide may mediate the high glucose-induced downregulation of PKG occurred in diabetic blood vessels (Liu et al., 2007). Studies also show that activation of PKG by expression of constitutively active PKG suppressed high glucose-induced VSMC proliferation and inhibited first gap phase (G1) to synthesis phase (S) phase progression of the cell cycle. These changes were accompanied with reduced glucose-induced cyclin E expression and cyclin E-cyclin-dependent kinase 2 activity as well as inhibition of glucose-induced phosphorylation of retinoblastoma protein (Rb) and p27 degradation. It suggests that PKG may inhibit VSMC proliferation through attenuation of cyclin E expression and increase in p27 protein stability, which leads to decreased CDK 2 activity and reduced Rb phosphorylation, thereby resulting in cell cycle arrest and cell growth inhibition (Wang & Li, 2009). Increased activity of transforming growth factor-β (TGF-β) is implicated in the development of diabetic macrovascular fibroproliferative remodeling. High glucose was found to stimulate the expression of thrombospondin1 (TSP1), a major activator of transforming growth factor-β (TGF-β), and to stimulate TGF-β activation in primary murine aortic SMCs. These effects were inhibited by overexpression of constitutively active PKG. Since PKG is downregulated in diabetic vasculature, it is likely that the downregulation of PKG action may relieve its suppression on TSP1 expression and TGF-β activity, thereby leading to augmented vascular remodeling in diabetes (Wang et al., 2010).

4.4 Pulmonary arterial hypertension

PKG expression and/or activity are/is reduced in animal models of pulmonary arterial hypertension (PAH) induced by ligation of the ductus arteriosus of fetal lambs (Resnik et al., 2006) and in caveolin-1 (Cav-1) knockout mice (Zhao et al., 2009). Cav-1, a 21-kDa integral membrane protein, is an intracellular physiological inhibitor of eNOS activity. Mice deficient in Cav-1 led to chronic eNOS activation and PAH. Activation of eNOS in Cav-1-/- lungs resulted in an impaired PKG activity through tyrosine nitration, probably at Tyr345 or Tyr549 of the catalytic domain of human PKG Iα. The PAH phenotype in Cav-1-/- lungs could be rescued by overexpression of PKG 1α. The treatment of these mice with either a superoxide scavenger or an eNOS inhibitor reverses their pulmonary vascular pathology and PAH phenotype, suggesting that an increased peroxynitrite formed from chronic overproduction of NO and superoxide may result in tyrosine nitration and loss of activity of PKG. Clinically, lung tissues from patients with idiopathic PAH have been found to display reduced Cav-1 expression, increased eNOS activation, and PKG nitration (Zhao et al., 2009). In ovine fetal pulmonary veins hypoxic exposure also causes peroxynitrite-mediated PKG nitration, reduced PKG activity, and suppressed dilator response to 8-Br-cGMP (Negash et al., 2007).

An upregulated ROK activity is implicated in a number of cardiovascular diseases including PAH (Satoh et al., 2011). ROK augments vasoconstriction primarily by inhibiting MLCP activity through phosphorylation of the regulatory subunit MYPT1 at Thr696 and Thr853, which leads to increased Ca^{2+}-sensitization of smooth muscle. The effect of ROK can be counteracted by the stimulatory action of PKG through phosphorylation of MYPT1 at Ser695 and Ser852. Pulmonary arteries from fetuses exposed to chronic intrauterine hypoxia (CH) displayed thickening vessel walls and diminished relaxant response to 8-Br-cGMP, two important characteristics of newborn PAH (Bixby et al., 2007; Gao et al., 2007). Rp-8-Br-PET-cGMPS, a specific PKG inhibitor, attenuated relaxation to 8-Br-cGMP in control vessels to a greater extent than in CH vessels while Y-27632, a ROCK inhibitor, potentiated 8-Br-cGMP-induced relaxation of CH vessels and had only a minor effect in control vessels. The specific activity of PKG was decreased while ROK activity was increased in CH vessels as compared with the controls. The phosphorylation of MYPT1 at Thr696 and Thr853 was inhibited by 8-Br-cGMP to a lesser extent in CH vessels than in controls. The difference was eliminated by Y-27632. These data indicate that the attenuated PKG-mediated relaxation in pulmonary arteries exposed to chronic hypoxia in utero is due to inhibition of PKG activity and due to enhanced ROCK activity. Increased ROCK activity may inhibit PKG action through increased phosphorylation of MYPT1 at Thr696 and Thr853 (Gao et al., 2007). In contrast to pulmonary arteries, relaxation of pulmonary veins of fetuses exposed to (CH) displayed no changes in the thickness of vessel walls and relaxant response to 8-Br-cGMP. In these veins phosphorylation of MYPT1 at Thr696 by ROK and at Ser695 by PKG was diminished as compared with control veins, suggesting that CH attenuates both PKG action and ROK action on MYPT1, resulting in an unaltered response to cGMP (Gao et al., 2008).

Bone morphogenetic proteins (BMPs) are members of the TGF-β superfamily. Mutations in the BMP type II receptor (BMPR-II) are responsible for the majority of cases of heritable PAH. Dysfunction in BMP signaling is implicated in idiopathic PAH and in a number of experimental models of PAH (Toshner et al., 2010). Studies found that PKG I may regulate the activation of BMP receptor and receptor-regulated Smad, a key mediator for BMP signaling, at the plasma membrane and regulate the expression of BMP target genes in the nucleus. These mechanisms may enable PKG I to compensate for the aberrant cellular responses to BMP caused by mutations in BMPRII found in PAH patients. Indeed, the overexpression of PKG I restores normal BMP responsiveness in cells expressing signaling deficient PAH mutant receptors such as the mutant BMPRII-Q657ins16 (Schwappacher et al., 2009; Thomson et al., 2000).

4.5 Nitrate tolerance

Nitroglycerine (NTG) is a widely used vasodilator in the treatment of angina pectoris and acute heart failure. It is converted inside the cell to NO or an NO-related intermediate and causes vasodilatation in a cGMP-dependent fashion. The effectiveness of NTG is often diminished when it is continuously used for a period of time, termed nitrate tolerance. The underlying mechanisms include an increased production of reactive oxygen species (ROS), impairment of biotransformation of NTG by aldehyde dehydrogenase, desensitization of sGC, upregulation of phosphodiesterases, and downregulation PKG activity (Münzel et al., 2005). In human arteries and veins, nitrate tolerance is associated with decreased PKG activity (Schulz et al., 2002). In the arteries of rats and rabbits, nitrate tolerance induced by low-dose NTG is associated with decreased PKG activity, while the tolerance induced by high-dose NTG is associated with decreased PKG protein level and activity (Mülsch *et al.*, 2001). In porcine coronary arteries nitrate tolerance induced by NTG at low concentrations is prevented by the scavenger of ROS. However the tolerance induced by NTG at higher concentrations is not affected by the scavenger of ROS and shows cross-tolerance to the NO donor and 8-Br-cGMP. Meanwhile, the protein and mRNA levels of PKG are reduced. It seems that the tolerance induced by NTG at higher concentrations may be due to suppression of PKG expression resulting from sustained activation of the enzyme (Dou et al., 2008). A diminished expression and activity of PKG was also observed in pulmonary veins of newborn lambs after prolonged exposure to the NO donor (Gao et al., 2004).

Activation of MLCP is a key mechanism for vasodilatation induced by nitrovasodilators such as NTG and NO. MLCP is a heterotrimer, composed of a catalytic subunit PP1cδ, a regulatory subunit MYPT1, and a subunit with unknown function. The regulatory subunit MYPT1 exists as isoform either with or without leucine zipper domain in its C-terminal [MYPT1 (LZ+) and MYPT1 (LZ−), respectively]. The presence of leucine zipper is necessary for PKG binding to MYPT1 and for PKG-mediated stimulatory effect on MLCP. Studies consistently demonstrate that the expression of MYPT1 (LZ+) determines the sensitivity to cGMP-mediated vasodilatation (Lee et al., 2007; A.P. Somlyo & A.V. Somlyo, 2003). Nitrate tolerance induced under *in vitro* conditions in porcine coronary arteries and induced under *in vivo* preparations in murine aorta show a decreased protein levels of MYPT1 (LZ+) but not of PP1cδ. The decrease in the MYPT1 (LZ+) protein level of coronary artery can also be induced by the NO donor and 8-Br-cGMP in a manner sensitive to the inhibitors of sGC and PKG, respectively. The tolerance to NTG in porcine coronary artery and mouse aorta is ameliorated by proteasome inhibitors. Therefore a downregulation of MYPT1 (LZ+) caused by increased proteasome-dependent degradation may contribute to development of nitrate tolerance (Dou et al., 2010).

5. Conclusion

Overwhelming evidence, obtained by genetic manipulation and pharmacological tools, under both *in vivo* and *in vitro* conditions, suggests that PKG is the primary enzyme in mediating vasodilatation, antiproliferation of vascular smooth muscle, and anti-platelet aggregation action induced by endogenous and exogenous nitrovasodilators via cGMP elevation (Francis et al., 2010; Gao, 2010; Hofmann et al., 2009; Walter & Gambaryan, 2009). Studies also support a barrier protection effect in the vascular endothelium (Moldobaeva et al., 2006; Rentsendorj et al., 2008). Increasing evidence also suggests that PKG exerts negative inotropic and antihypertrophic actions in the heart (Takimoto et al. 2005; Yang et al., 2007; Zhang et al., 2010) as well as a cardioprotective action against ischemia-reperfusion injury (Das et al., 2008; Juhaszova et al., 2009; Xi et al., 2010). Despite substantial progress has been made in elucidating the role of PKG in the regulation of cardiovascular functions there are many aspects remain to be explored. For instance, the developing and ageing aspects for the role of PKG, the gender difference, and the heterogeneity in the role of PKG in different vasculatures. Also, the roles of many PKG substrates in the regulation of cardiovascular activities remain to be defined (Schlossmann & Desch, 2009).

Dysfunction in NO-cGMP signaling is a common initiator and independent predictor of cardiovascular events (Vanhoutte et al., 2009). An impaired PKG action has been implicated in various cardiovascular disorders such as hypertension, atherosclerosis, diabetic vascular disease, pulmonary arterial hypertension, and nitrate tolerance (Francis et al., 2010; Gao, 2010; Hofmann et al., 2009). Cardiovascular alterations are a long-term process comprising functional and structural changes with remarkable complexities, which undoubtedly make the dissection of the role of PKG rather challenging. However, a better understanding of its role and the underlying mechanism will be of great therapeutic significance.

6. References

Antl, M.;, von Bruhl, M.L.; Eiglsperger, C.; Werner, M.; Konrad, I.; Kocher, T.; Wilm, M.; Hofmann, F.; Massberg, S.; Schlossmann, J. (2007). IRAG mediates NO/cGMP-dependent inhibition of platelet aggregation and thrombus formation. *Blood* Vol.109, No.2, (January 2007), pp. 552-559.

Aszódi, A.; Pfeifer, A.; Ahmad, M.; Glauner, M.; Zhou, X.H.; Ny, L.; Andersson, K.E.; Kehrel, B.; Offermanns, S.; Fässler, R. (1999). The vasodilator- stimulated phosphoprotein (VASP) is involved in cGMP- and cAMP-mediated inhibition of agonist- induced platelet aggregation, but is dispensable for smooth muscle function. *EMBO J.* (January 1999), Vol.18, No.1, pp. 37-48.

Barman, S.A.; Zhu, S.; Han, G.; White, R.E. (2003). cAMP activates BK_{Ca} channels in pulmonary arterial smooth muscle via cGMP-dependent protein kinase. *Am. J. Physiol. Lung Cell Mol. Physiol.,* (June 2003), Vol.284, No.6, pp. L1004-L1011.

Bixby, C.E.; Ibe, B.O.; Abdallah, M.F.; Zhou, W.; Hislop, A.A.; Longo, L.D., Raj, J.U. (2007). Role of platelet-activating factor in pulmonary vascular remodeling associated with chronic high altitude hypoxia in ovine fetal lambs. *Am. J. Physiol. Lung Cell Mol. Physiol.,* (December 2007), Vol.293, No. 6, pp.L1475-82.

Blackmore, P.F. (2011). Biphasic effects of nitric oxide on calcium influx in human platelets. *Thromb. Res.,* (January 2011), Vol.127, No.1 pp.e8-e14.

Butt, E.; Bernhardt, M.; Smolenski, A.; Kotsonis, P; Frohlich, L.G.; Sickmann, A.; Meyer, H.E.; Lohmann, S.M.; Schmidt, H.H. (2000). Endothelial nitric-oxide synthase (type III) is activated and becomes calcium independent upon phosphorylation by cyclic nucleotide-dependent protein kinases. *J. Biol. Chem.,* (February 2000), Vol.275, No.7, pp.5179-5187.

Chen, F.C.; Ogut, O.; Rhee, A.Y.; Hoit, B.D.; Brozovich, F.V. (2006). Captopril prevents myosin light chain phosphatase isoform switching to preserve normal cGMP-mediated vasodilatation. *J. Mol. Cell Cardiol.,* (September 2006), Vol.41, No.3, pp.488-495.

Choi, C.; Sellak, H.; Brown, F.M.; Lincoln, T.M. (2010). cGMP-dependent protein kinase and the regulation of vascular smooth muscle cell gene expression: possible involvement of Elk-1 sumoylation. *Am. J. Physiol. Heart Circ. Physiol.,* (November 2010), Vol.299, No. 5, pp.H1660-H1670.

Costa, A.D.; Pierre, S.V.; Cohen, M.V.; Downey, J.M.; Garlid, K.D. (2008). cGMP signalling in pre- and post-conditioning: the role of mitochondria. *Cardiovasc. Res.,* (January 2008), Vol.77, No.2, pp.344-352.

Dangel, O.; Mergia, E.; Karlisch, K.; Groneberg, D.; Koesling, D.; Friebe, A. (2010). Nitric oxide-sensitive guanylyl cyclase is the only nitric oxide receptor mediating platelet inhibition. *J. Thromb. Haemost.,* (June 2010), Vol.8, No. 6, pp.1343-1352.

Das, A.; Xi, L.; Kukreja, R.C. (2008). Protein Kinase G-dependent Cardioprotective Mechanism of Phosphodiesterase-5 Inhibition Involves Phosphorylation of ERK and GSK3β. *J. Biol. Chem.,* (October 2008), Vol.283, No. 43, pp.29572-29585.

Desch, M.; Sigl, K.; Hieke, B.; Salb, K.; Kees, F.; Bernhard, D.; Jochim, A.; Spiessberger, B.; Höcherl, K.; Feil, R.; Feil, S.; Lukowski, R.; Wegener, J.W.; Hofmann, F.;

Schlossmann, J. (2010). IRAG determines nitric oxide- and atrial natriuretic peptide-mediated smooth muscle relaxation. *Cardiovasc. Res.*, (June 2010), Vol.86, No.3, pp.496-505.

Deschepper, C.F. (2010). Cardioprotective actions of cyclic GMP: lessons from genetic animal models. *Hypertension*, (February 2010), Vol.55, No. 2, pp.453-458.

Dey, N.B.; Boerth, N.J.; Murphy-Ullrich, J.E.; Chang, P.L.; Prince, C.W.; Lincoln, T.M. (1998). Cyclic GMP-dependent protein kinase inhibits osteopontin and thrombospondin production in rat aortic smooth muscle cells. *Circ. Res.*, (Feb 1998), Vol.82, No.2, pp.139-146.

Dey, N.B.; Foley, K.F.; Lincoln, T.M.; Dostmann, W.R. (2005). Inhibition of cGMP-dependent protein kinase reverses phenotypic modulation of vascular smooth muscle cells. *J. Cardiovasc. Pharmacol.*, (May 2005), Vol.45, No.5, pp404-413.

Dhanakoti, S.; Gao, Y.; Nguyen, M.Q.; Raj, J.U. (2000). Involvement of cGMP-dependent protein kinase in the relaxation of ovine pulmonary arteries to cGMP and cAMP. *J. Appl. Physiol.*, (May 2000), Vol.88, No.5, pp.1637-1642.

Diwan, A.H.; Thompson, W.J.; Lee, A.K.; Strada, S.J. (1994). Cyclic GMP-dependent protein kinase activity in rat pulmonary microvascular endothelial cells. *Biochem. Biophys. Res. Commun.*, (July 1994), Vol.202, No.2, pp.728-735.

Dou, D.; Ma, H.; Zheng, X.; Ying, L.; Guo, Y.; Yu, X.; Gao, Y. (2010). Degradation of leucine zipper-positive isoform of MYPT1 may contribute to development of nitrate tolerance. *Cardiovasc. Res.*, (April 2010), Vol.86, No.1, pp.151-159.

Dou, D.; Zheng, X.; Qin, X.; Qi, H.; Liu, L.; Raj, J.U.; Gao, Y. (2008). Role of cGMP-dependent protein kinase in development of tolerance to nitroglycerine in porcine coronary arteries. *Br. J. Pharmacol.*, (February 2008), Vol.153, No.3, pp.497-507.

Draijer, R.; Vaandrager, A.B.; Nolte, C.; de Jonge, H.R.; Walter, U.; van Hinsbergh, V.W. (1995). Expression of cGMP-dependent protein kinase I and phosphorylation of its substrate, vasodilator-stimulated phosphoprotein, in human endothelial cells of different origin. *Circ. Res.*, (November 1995), Vol.77, No.5, pp.897-905.

Eigenthaler, M.; Nolte, C.; Halbrugge, M.; Walter, U. (1992). Concentration and regulation of cyclic nucleotides, cyclic-nucleotide-dependent protein kinases and one of their major substrates in human platelets. Estimating the rate of cAMP-regulated and cGMP-regulated protein phosphorylation in intact cells. *Eur. J. Biochem.*, (April 1992), Vol.205, No. 2, pp.:471-481.

El-Mowafy, A.M.; Alkhalaf, M.; El-Kashef, H.A. (2008). Resveratrol reverses hydrogen peroxide-induced proliferative effects in human coronary smooth muscle cells: a novel signaling mechanism.*Arch. Med. Res.*, (February 2008), Vol.39, No.2, pp.155-161.

Francis, S.H.; Busch, J.L.; Corbin, J.D.; Sibley, D. (2010). cGMP-dependent protein kinases and cGMP phosphodiesterases in nitric oxide and cGMP action. *Pharmacol. Rev.*, (September 2010), Vol.62, No.3, pp.525-563.

Friebe, A.; Mergia, E.; Dangel, O.; Lange, A.; Koesling, D. (2007). Fatal gastrointestinal obstruction and hypertension in mice lacking nitric oxide-sensitive guanylyl cyclase. *Proc. Natl. Acad. Sci. USA*, (May 2007), Vol.104, No.18, pp.7699-7704.

Fukao, M.; Mason, H.S.; Britton, F.C.; Kenyon, J.L.; Horowitz, B.; Keef, K.D. (1999). Cyclic GMP-dependent protein kinase activates cloned BKCa channels expressed in mammalian cells by direct phosphorylation at serine 1072. *J. Biol. Chem.*, (April 1999), Vol.274, No.16, pp.10927-10935.

Gao, Y. (2010). The multiple actions of NO. Pflugers Arch.- Eur. J. Physiol., (May 2010), Vol.459, No.6, pp.829-839.

Gao, Y.; Dhanakoti, S.; Tolsa, J.-F.; Raj, J.U. (1999). Role of protein kinase G in nitric oxide and cGMP-induced relaxation of newborn ovine pulmonary veins. *J. Appl. Physiol.* (September 1999), Vol.87, No.3, pp.993-998.

Gao, Y.; Dhanakoti, S.; Trevino, E.M.; Sander, F.C.; Portuga, A.M.; Raj, J.U. (2003). Effect of oxygen on cyclic GMP-dependent protein kinase-mediated relaxation in ovine fetal pulmonary arteries and veins. *Am. J. Physiol. Lung Cell Mol. Physiol.*, (September 2003), Vol.285, No.3, pp.L611-L618.

Gao, Y.; Dhanakoti, S; Trevino, E.M.; Wang, X.; Sander, F.C.; Portuga, A.D.; Raj, J.U. (2004). Role of cGMP-dependent protein kinase in development of tolerance to nitric oxide in pulmonary veins of newborn lambs. *Am J Physiol Lung Cell Mol Physiol.*, (April 2004), Vol.286, No.4, ppL786–L792.

Gao, Y.; Portugal, A.D.; Negash, S.; Zhou, W.; Longo, L.D.; Usha Raj, J. (2007). Role of Rho kinases in PKG-mediated relaxation of pulmonary arteries of fetal lambs exposed to chronic high altitude hypoxia. *Am J Physiol Lung Cell Mol Physiol.*, (March 2007), Vol.292, No.3, pp.L678-L684.

Gao, Y.; Portugal, A.D.; Liu, J.; Negash, S.; Zhou, W.; Tian, J.; Xiang, R.; Longo, L.D.; Raj, J.U. (2008). Preservation of cGMP-induced relaxation of pulmonary veins of fetal lambs exposed to chronic high altitude hypoxia: role of PKG and Rho kinase. *Am. J. Physiol. Lung Cell Mol. Physiol.*, (November 2008), Vol.295, No.5, pp.L889-L896.

Geiselhöringer, A.; Werner, M.; Sigl, K.; Smital, P.; Wörner, R.; Acheo, L.; Stieber, J.; Weinmeister, P.; Feil, R.; Feil, S.; Wegener, J.; Hofmann, F.; Schlossmann, J. (2004). IRAG is essential for relaxation of receptor-triggered smooth muscle contraction by cGMP kinase. *EMBO J.*, (October 2004), Vol.23, No.21, pp.4222-4231.

Gorbe, A.; Giricz, Z.; Szunyog, A.; Csont, T.; Burley, D.S.; Baxter, G.F.; Ferdinandy, P. (2010). Role of cGMP-PKG signaling in the protection of neonatal rat cardiac myocytes subjected to simulated ischemia/reoxygenation. *Basic. Res. Cardiol.*, (September 2010), Vol.105, No.5, pp.643-650.

Hofmann, F.; Bernhard, D.; Lukowski, R.; Weinmeister, P. (2009). cGMP regulated protein kinases (cGK). *Handb. Exp. Pharmacol.*, Vol.191, 137-162.

Huang, P.L.; Huang, Z.; Mashimo, H.; Bloch, K.D.; Moskowitz, M.A.; Bevan, J.A.; Fishman, M.C. (1995). Hypertension in mice lacking the gene for endothelial nitric oxide synthase. *Nature*, (September 1995), Vol.377, No.6546, pp.239-242.

Juhaszova, M.; Zorov, D.B.; Yaniv, Y.; Nuss, H.B.; Wang ,S.; Sollott, S.J. (2009). Role of Glycogen Synthase Kinase-3β in Cardioprotection. *Circ. Res.*, (June 2009), Vol.104, No.11, pp.1240-1252

Kass, D.A.; Takimoto, E. (2010). Regulation and role of myocyte cyclic GMP-dependent protein kinase-I. *Proc. Natl. Acad. Sci. USA.* (June 2010), Vol.107, No.24, pp.E98.

Kinoshita, H.; Kuwahara, K.; Nishida, M.; Jian, Z.; Rong, X.; Kiyonaka, S.; Kuwabara, Y.; Kurose, H.; Inoue, R.; Mori, Y.; Li, Y.; Nakagawa, Y.; Usami, S.; Fujiwara, M.; Yamada, Y.; Minami, T.; Ueshima, K.; Nakao, K. (2010). Inhibition of TRPC6 channel activity contributes to the antihypertrophic effects of natriuretic peptides-guanylyl cyclase-A signaling in the heart. *Circ. Res.*, (June 2010), Vol.106, No.12, pp.1849-1860.

Klaiber, M.; Kruse, M.; Völker, K.; Schröter, J.; Feil, R.; Freichel, M.; Gerling, A.; Feil, S.; Dietrich, A.; Londoño, J.E.; Baba, H.A.; Abramowitz, J.; Birnbaumer, L.; Penninger, J.M.; Pongs, O.; Kuhn, M. (2010). Novel insights into the mechanisms mediating the local antihypertrophic effects of cardiac atrial natriuretic peptide: role of cGMP-dependent protein kinase and RGS2. *Basic Res. Cardiol.*, (September 2010), Vol.105, No.5, pp.583–595.

Kuo, J.F.; Greengard, P. (1970). Cyclic nucleotide-dependent protein kinases. VI. Isolation and partial purification of a protein kinase activated by guanosine 3',5'-monophosphate. *J. Biol. Chem.*, (May 1970), Vol.245, No.10, pp.:2493-248.

Layland, J.; Li, J.M.; Shah, A.M. (2002). Role of cyclic GMP-dependent protein kinase in the contractile respose to exogenous nitric oxide in rat cardiac myocytes. *J. Physiol.*, (April 2002), Vol.540(Pt. 2), pp.457-467.

Layland, J.; Solaro, R.J.; Shah, A.M. (2005). Regulation of cardiac contractile function by troponin I phosphorylation. *Cardiovasc. Res.*, (April 2005), Vol.66, No.1, pp.12–21.

Lee, E.; Hayes, D.B.; Langsetmo, K.; Sundberg, E.J; Tao, T.C. (2007). Interactions between the leucine-zipper motif of cGMP-dependent protein kinase and the C-terminal region of the targeting subunit of myosin light chain phosphatase. *J. Mol. Biol.*, (November 2007), Vol.373, No.5, pp.1198-1212.

Li, Z.; Xi, X.; Gu, M.; Feil, R.; Ye, R.D.; Eigenthaler, M.; Hofmann, F.; Du, X. (2003). A stimulatory role for cGMP-dependent protein kinase in platelet activation. *Cell*, (January 2003), Vol.112, No.1, pp.77-86.

Lin, M.T.; Longo, L.D.; Pearce, W.J.; Hessinger, D.A. (2005). Ca^{2+}-activated K^+ channel-associated phosphatase and kinase activities during development *Am. J. Physiol. Heart Circ. Physiol.*, (July 2005), Vol.289, No.1, H414-H425.

Lincoln ,T.M.; Dey, N.; Sellak, H. (2001). cGMP-dependent protein kinase signaling mechanisms in smooth muscle: from the regulation of tone to gene expression. *J. Appl. Physiol.*, (September 2001), Vol.91, No.3, pp.1421-1430.

Liu, S.; Ma, X.; Gong, M.; Shi, L.; Lincoln, T.; Wang, S. (2007). Glucose down-regulation of cGMP-dependent protein kinase I expression in vascular smooth muscle cells involves NAD(P)H oxidase-derived reactive oxygen species. *Free Radic. Biol. Med.*, (March 2007), Vol.42, No.6, pp.852-863.

Lohmann, S.M.; Walter, U. (2005).Tracking functions of cGMP-dependent protein kinases (cGK). *Front. Biosci.*, (May 2005), Vol.10, pp.1313-1328.

Loirand, G.; Guilluy, C.; Pacaud, P. (2006). Regulation of Rho proteins by phosphorylation in the cardiovascular system. *Trends. Cardiovasc. Med.*, (August 2006), Vol.16, No.6, pp.199-204.

Lukowski, R.; Rybalkin, S.D.; Loga, F.; Leiss, V.; Beavo, J.A.; Hofmann, F. (2010). Cardiac hypertrophy is not amplified by deletion of cGMP-dependent protein kinase I in cardiomyocytes. *Proc. Natl. Acad. Sci. USA*, (March 2010), Vol.107, No.12, pp.5646-5651.

Luo, C.; Yi, B.; Chen, Z.; Tang, W.; Chen, Y.; Hu, R.; Liu, Z.; Feng, H.; Zhang, J.H. (2011). PKGIα inhibits the proliferation of cerebral arterial smooth muscle cell induced by oxyhemoglobin after subarachnoid hemorrhage. *Acta. Neurochir. Suppl.*, Vol.110(Pt 1), pp.167-171.

MacMillan-Crow, L.A.; Murphy-Ullrich, J.E.; Lincoln, T.M. (1994). Identification and possible localization of cGMP-dependent protein kinase in bovine aortic endothelial cells. *Biochem. Biophys. Res. Commun.*, (June 1994), Vol.201, No.2, pp.531-537.

Massberg, S.; Gruner, S.; Konrad, I.; Garcia Arguinzonis, M.I.; Eigenthaler, M.; Hemler, K.; Kersting, J.; Schulz, C.; Muller, I.; Besta, F.; Nieswandt, B.; Heinzmann, U.; Walter, U.; Gawaz, M. (2004). Enhanced in vivo platelet adhesion in vasodilator-stimulated phosphoprotein (VASP)-deficient mice. *Blood*, (January 2004), Vol.103, No.1, pp.136–142.

Massberg, S.; Sausbier, M.; Klatt, P.; Bauer, M.; Pfeifer, A.; Siess, W.; Fassler, R.; Ruth, P.; Krombach, F.; Hofmann, F. (1999). Increased adhesion and aggregation of platelets lacking cyclic guanosine 3': 5'- monophosphate kinase I. *J. Exp. Med.*, (April 1999), Vol.189, No.8, pp.1255–1264.

Mazzetti, L.; Ruocco, C.; Giovannelli, L.; Ciuffi, M.; Franchi-Micheli, S.; Marra, F.; Zilletti, L.; Failli, P. (2001). Guanosine 3': 5'-cyclic monophosphate-dependent pathway alterations in ventricular cardiomyocytes of spontaneously hypertensive rats. *Br. J. Pharmacol.*, (October 2001), Vol.134, No.3, pp.596-602.

Melichar, V.O.; Behr-Roussel, D.; Zabel, U.; Uttenthal, L.O.; Rodrigo, J.; Rupin, A.; Verbeuren ,T.J.; Kumar, H.S.A.; Schmidt, H.H. (2004). Reduced cGMP signaling associated with neointimal proliferation and vascular dysfunction in late-stage atherosclerosis. *Proc. Natl. Acad. Sci. USA*, (November 2004), Vol.101, No.47, pp.16671-16676.

Mery, P.F.; Lohmann, S.M.; Walter, U.; Fischmeister, R. (1991). Ca^{2+} current is regulated by cyclic GMP-dependent protein kinase in mammalian cardiac myocytes. *Proc. Natl. Acad. Sci. USA*, (February 1991), Vol.88, No.4, pp.1197–1201.

Michael, S.K.; Surks, H.K.; Wang, Y.; Zhu, Y.; Blanton, R.; Jamnongjit, M.; Aronovitz, M.; Baur, W.; Ohtani, K.; Wilkerson, M.K.; Bonev, A.D.; Nelson, M.T.; Karas, R.H.; Mendelsohn, M.E. (2008). High blood pressure arising from a defect in vascular function. *Proc Natl Acad Sci USA*, (May 2008), Vol.105, No.18, 6702-6707.

Moldobaeva, A.; Welsh-Servinsky, L.E.; Shimoda, L.A.; Stephens, R.S.; Verin, A.D.; Tuder, R.M.; Pearse, D.B. (2006). Role of protein kinase G in barrier-protective effects of cGMP in human pulmonary artery endothelial cells. *Am. J. Physiol. Lung Cell Mol. Physiol.*, (May 2006), Vol.290, No.5, pp.L919–L930.

Mülsch, A.; Oelze, M.; Klöss, S.; Mollnau, H.; Töpfer, A.; Smolenski, A.; Walter, U.; Stasch, J.P.; Warnholtz, A.; Hink, U.; Meinertz, T.; Münzel, T. (2001). Effects of in vivo nitroglycerin treatment on activity and expression of the guanylyl cyclase and cGMP-dependent protein kinase and their downstream target vasodilator-stimulated phosphoprotein in aorta. *Circulation*, (May 2001), Vol.103, No.17, pp.2188-2194.

Münzel, T.; Daiber, A.; Mülsch, A. (2005). Explaining the phenomenon of nitrate tolerance. *Circ. Res.*, (September 2005), Vol.97, No.7, pp.618-628.

Negash, S.; Gao, Y.; Zhou, W.; Liu, J.; Chinta, S.; Raj. J.U. (2007). Regulation of cGMP-dependent protein kinase-mediated vasodilation by hypoxia-induced reactive species in ovine fetal pulmonary veins. *Am. J. Physiol. Lung Cell Mol. Physiol.*, (October 2007), Vol.293, No.4, pp.L1012-L1020.

Payne, M.C.; Zhang, H.Y.; Prosdocimo, T.; Joyce, K.M.; Koga, Y.; Ikebe, M.; Fisher, S.A. (2006). Myosin phosphatase isoform switching in vascular smooth muscle development. *J. Mol. Cell Cardiol.*, (February 2006), Vol.40, No.2, pp.274-282.

Pfeifer, A.; Klatt, P.; Massberg, S.; Ny, L.; Sausbier, M.; Hirneiss, C.; Wang, G.X.; Korth, M.; Aszodi, A.; Andersson, K.E.; Krombach, F.; Mayerhofer, A.; Ruth, P.; Fassler, R.; Hofmann, F. (1998). Defective smooth muscle regulation in cGMP kinase I-deficient mice. *Embo. J.*, (June 1998), Vol.17, No.11, pp.3045-P3051.

Qi, H.; Zheng, X.; Qin, X.; Dou, D.; Xu, H.; Raj, J.U.; Gao, Y. (2007). PKG regulates the basal tension and plays a major role in nitrovasodilator-induced relaxation of porcine coronary veins. *Br. J. Pharmacol.*, (December 2007), Vol.152, No.7, pp.1060-1069.

Qin, X.; Zheng, X.; Qi, H.; Dou, D.; Raj, J.U.; Gao, Y. (2007). cGMP-dependent protein kinase in regulation of basal tone and in nitroglycerin and nitric oxide induced relaxation in porcine coronary artery. *Pflügers Archiv.*, (September 2007), Vol.454, No.6, pp.913-923.

Rentsendorj, O.; Mirzapoiazova, T.; Adyshev, D.; Servinsky, L.E.; Renné, T.; Verin, A.D.; Pearse, D.B. (2008). Role of vasodilator-stimulated phosphoprotein in cGMP-mediated protection of human pulmonary artery endothelial barrier function. *Am. J. Physiol. Lung Cell Mol. Physiol.*, (April 2008), Vol.294, No.4, pp.L686-L697.

Resnik, E.; Herron, J.; Keck, M.; Sukovich, D.; Linden, B.; Cornfield. D.N. (2006). Chronic intrauterine pulmonary hypertension selectively modifies pulmonary artery smooth muscle cell gene expression. *Am. J. Physiol. Lung Cell Mol. Physiol.*, (March 2006), Vol.290, No.3, pp.L426-L433.

Robertson, B.E.; Schubert, R.; Hescheler, J.; Nelson, M.T. (1993). cGMP-dependent protein kinase activates Ca-activated K channels in cerebral artery smooth muscle cells. *Am. J. Physiol.*, (July 1993), Vol.265(1 Pt 1), C299- C303.

Satoh, K.; Fukumoto, Y.; Shimokawa, H. (2011). Rho-kinase: important new therapeutic target in cardiovascular diseases. *Am. J. Physiol. Heart Circ. Physiol.*, (August 2011), Vol.301, No.2, pp.H287-H296.

Schinner, E.; Salb, K.; Schlossmann, J. (2011). Signaling via IRAG is essential for NO/cGMP-dependent inhibition of platelet activation. *Platelets*, Vol.22, No.3, pp.217-227.

Schlossmann, J.; Desch, M. (2009). cGK substrates. *Handb. Exp. Pharmacol.*, Vol.191, pp.163-193.

Schroder, F.; Klein, G.; Fiedler, B.; Bastein, M.; Schnasse, N.; Hillmer, A.; Ames, S.; Gambaryan, S.; Drexler, H.; Walter, U.; Lohmann, S.M.; Wollert, K.C. (2003). Single L-type Ca^{2+} channel regulation by cGMP-dependent protein kinase type I in adult cardiomyocytes from PKG I transgenic mice. *Cardiovasc. Res.*, (November 2003), Vol.60, No.2, pp.268-277.

Schulz, E.; Tsilimingas, N.; Rinze, R.; Reiter, B.; Wendt, M.; Oelze, M.; Woelken-Weckmüller. S.; Walter, U.; Reichenspurner, H.; Meinertz, T.; Münzel, T. (2002). Functional and biochemical analysis of endothelial (dys)function and NO/cGMP signaling in human blood vessels with and without nitroglycerin pretreatment. *Circulation*, (March 2002), Vol.105, No.10, pp.1170-1175.

Schwappacher, R.; Weiske, J.; Heining, E.; Ezerski, V.; Marom, B.; Henis, Y.I.; Huber, O.; Knaus, P. (2009). Novel crosstalk to BMP signalling: cGMP-dependent kinase I modulates BMP receptor and Smad activity. *EMBO. J.* (June 2009), Vol.28, No.11, pp.1537-1550

Somlyo, A.P.; Somlyo, A.V. (2003). Ca^{2+} sensitivity of smooth muscle and nonmuscle myosin II: modulated by G proteins, kinases, and myosin phosphatase. *Physiol. Rev.,* (October 2003), Vol.83, No.4, pp.1325-1358.

Takimoto, E.; Champion, H.C.; Li, M.; Belardi, D.; Ren, S.; Rodriguez, E.R.; Bedja, D.; Gabrielson, K.L.; Wang, Y.; Kass, D.A. (2005). Chronic inhibition of cyclic GMP phosphodiesterase 5A prevents and reverses cardiac hypertrophy. *Nat. Med.,* (February 2005), Vol.11, No.2, pp:214–222.

Thomson, J.R.; Machado, R.D.; Pauciulo, M.W.; Morgan, N.V.; Humbert, M.; Elliott, G.C.; Ward, K.; Yacoub, M.; Mikhail, G.; Rogers, P.; Newman, J.; Wheeler, L.; Higenbottam, T.; Gibbs, J.S.; Egan, J.; Crozier, A.; Peacock, A.; Allcock, R.; Corris, P.; Loyd, J.E.; Trembath, R.C.; Nichols, W.C. (2000). Sporadic primary pulmonary hypertension is associated with germline mutations of the gene encoding BMPR-II, a receptor member of the TGF-beta family. *J. Med. Genet.,* (October 2000), Vol.37, No.10, pp.741-745.

Tokudome, T.; Kishimoto, I.; Horio, T.; Arai, Y.; Schwenke, D.O.; Hino, J.; Okano, I.; Kawano, Y.; Kohno, M.; Miyazato, M.; Nakao, K.; Kangawa, K. (2008). Regulator of G-protein signaling subtype 4 mediates antihypertrophic effect of locally secreted natriuretic peptides in the heart. *Circulation,* (May 2008), Vol.117, No.18, pp.2329-2339

Toshner, M.; Tajsic, T.; Morrell, N.W. (2010).Pulmonary hypertension: advances in pathogenesis and treatment. *Br. Med. Bull.,* Vol.94, pp.21–32.

Vanhoutte, P.M.; Shimokawa, H.; Tang, E.H.; Feletou, M. (2009). Endothelial dysfunction and vascular disease. Acta. Physiol. (Oxf.). (Jun 2009), Vol196, No.2, pp.193-222.

Wagner, C.; Pfeifer, A.; Ruth, P.; Hofmann, F.; Kurtz, A. (1998). Role of cGMP-kinase II in the control of renin secretion and renin expression. *J. Clin. Invest.,* (October 1998),Vol.102, No.8, pp.1576–1582.

Walter U, Gambaryan S. (2009). cGMP and cGMP-dependent protein kinase in platelets and blood cells. *Handb. Exp. Pharmacol.,* Vol.191, pp.533-548.

Wang, S.; Li, Y. (2009). Expression of constitutively active cGMP-dependent protein kinase inhibits glucose-induced vascular smooth muscle cell proliferation. *Am. J. Physiol. Heart Circ. Physiol.,* (December 2009), Vol.297, No.6, pp.H2075-H2083.

Wang, S.; Lincoln, T.M.; Murphy-Ullrich, J.E. (2010). Glucose downregulation of PKG-I protein mediates increased thrombospondin1-dependent TGF-β activity in vascular smooth muscle cells. *Am. J. Physiol. Cell Physiol.,* (May 2010), Vol.298, No.5, pp.C1188-C1197.

Warnholtz, A.; Mollnau, H.; Heitzer, T.; Kontush, A.; Möller-Bertram, T.; Lavall, D.; Giaid, A.; Beisiegel, U.; Marklund, S.L.; Walter, U.; Meinertz, T.; Munzel, T. (2002). Adverse effects of nitroglycerin treatment on endothelial function, vascular nitrotyrosine levels and cGMP-dependent protein kinase activity in hyperlipidemic Watanabe rabbits. *J. Am. Coll. Cardiol.,* (October 2002), Vol.40, No.7, pp.1356-1363.

Weber, S.; Bernhard, D.; Lukowski, R.; Weinmeister, P.; Worner, R.; Wegener, J.W.; Valtcheva. N.; Feil, S.; Schlossmann, J.; Hofmann, F.; Feil, R. (2007). Rescue of cGMP kinase I knockout mice by smooth muscle specific expression of either isozyme. *Circ. Res.,* (November 2007), Vol.101, No.11, pp.1096–1103.

Wegener, J.W.; Nawrath, H.; Wolfsgruber, W.; Kuhbandner, S.; Werner, C.; Hofmann, F.; Feil, R. (2002). cGMP-dependent protein kinase I mediates the negative inotropic effect of cGMP in the murine myocardium. *Circ. Res.,* (January 2002), Vol.90, No.1, pp18–20.

White, R.E.; Kryman, J.P.; El-Mowafy, A.M.; Han, G.; Carrier, G.O. (2000). cAMP-Dependent Vasodilators Cross-Activate the cGMP-Dependent Protein Kinase to Stimulate BKCa Channel Activity in Coronary Artery Smooth Muscle Cells *Circ. Res.*, (April 2000), Vol.86, No.8, pp. 897-905.

Wolfsgruber, W.; Feil, S.; Brummer, S.; Kuppinger, O.; Hofmann, F.; Feil, R. (2003). A proatherogenic role for cGMP-dependent protein kinase in vascular smooth muscle cells. *Proc. Natl. Acad. Sci. USA*, (November 2003), Vol.100, No. 23, pp.13519-13524.

Wollert, K.C.; Yurukova, S.; Kilic, A.; Begrow, F.; Fiedler, B.; Gambaryan, S.; Walter, U.; Lohmann, S.M.; Kuhn, M. (2003). Increased effects of C-type natriuretic peptide on contractility and calcium regulation in murine hearts overexpressing cyclic GMP-dependent protein kinase I. *Br. J. Pharmacol.*, (December 2003) Vol.140, No.7, pp.1227–1236.

Wong, J.C.; Fiscus, R.R. (2010). Protein kinase G activity prevents pathological-level nitric oxide-induced apoptosis and promotes DNA synthesis/cell proliferation in vascular smooth muscle cells. *Cardiovasc. Pathol.*, (November-December 2010), Vol.19, No.6, pp.e221-e231.

Wooldridge, A.A.; MacDonald, J.A.; Erdodi, F.; Ma, C.; Borman, M.A.; Hartshorne, D.J.; Haystead, T.A.J. (2004). Smooth muscle phosphatase is regulated in vivo by exclusion of phosphorylation of threonine 696 of MYPT1 by phosphorylation of serine 695 in response to cyclic nucleotides. *J. Biol. Chem.*, (August 2004), Vol.279, No.33, pp.34496–34504.

Xi, J.; Tian, W.; Zhang, L.; Jin, Y.; Xu, Z. (2010). Morphine prevents the mitochondrial permeability transition pore opening through NO/cGMP/PKG/Zn^{2+}/GSK-3beta signal pathway in cardiomyocytes. *Am. J. Physiol. Heart Circ. Physiol.*, (February 2010), Vol.298, No.2, pp.H601-H607.

Yang ,L.; Liu, G.; Zakharov, S.I.; Bellinger, A.M.; Mongillo, M.; Marx, S.O. (2007). Protein kinase G phosphorylates Ca_v 1.2 α_{1c} and β_2 subunits. *Circ. Res.*, (August 2007), Vol.101, No.5, pp.465-474.

Ying, L.; Xu, X.; Liu, J.; Dou, D.; Yu, X.; Ye L.; He, Q.; Gao, Y. (2011). Heterogeneity in relaxation of different sized porcine coronary arteries to nitrovasodilators: Role of PKG and MYPT1. *Pflügers Archiv. – Eur. J. Physiol.*, (February 2012), Vol. 463, No. 2, pp.257-268.

Yuasa, K.; Michibata, H.; Omori, K.; Yanaka, N. (1999).A novel interaction of cGMP-dependent protein kinase I with troponin T. *J. Biol. Chem.*, (December 1999), Vol.274, No.52, pp.37429-37434.

Zhang, G.; Xiang, B.; Dong, A.; Skoda, R.C.; Daugustherty, A.; Smyth, S.S.; Du, X.; Li, Z. (2011). Biphasic roles for soluble guanylyl cyclase in platelet activation. *Blood*, (September 2011), Vol. 118, No.13, pp.3670-3690.

Zhang, M.; Takimoto, E.; Hsu, S.; Lee, D.I.; Nagayama, T.; Danner, T.; Koitabashi, N.; Barth, A.S.; Bedja, D.; Gabrielson, K.L.; Wang, Y.; Kass, D.A. (2010). Myocardial remodeling is controlled by myocyte-targeted gene regulation of phosphodiesterase type 5. *J. Am. Coll. Cardiol.*, (December 2010), Vol.56, No.24, pp.2021-2030.

Zhang, Q.; Scholz, P.M.; He, Y.; Tse, J.; Weiss, H.R. (2005). Cyclic GMP signaling and regulation of SERCA activity during cardiac myocyte contraction. *Cell Calcium*, (March 2005), Vol.37, No.3, pp.259-266.

Zhang, T.; Zhuang, S.; Casteel, D.E.; Looney, D.J.; Boss, G.R.; Pilz, R.B. (2007). A Cysteine-rich LIM-only Protein Mediates Regulation of Smooth Muscle-specific Gene Expression by cGMP-dependent Protein Kinase. *J. Biol. Chem.*, (November 2007), Vol.282, No.46, pp.33367-33380.

Zhao, Y.Y.; Zhao, Y.D.; Mirza, M.K.; Huang, J.H.; Potula, H.H.; Vogel, S.M.; Brovkovych, V.; Yuan, J.X.; Wharton, J.; Malik, A.B. (2009). Persistent eNOS activation secondary to caveolin-1 deficiency induces pulmonary hypertension in mice and humans through PKG nitration. J Clin Invest., (July 2009),Vol.119, No. 7, pp.2009-2018.

Zhou, W.; Dasgupta, C.; Negash, S.; Raj, J.U. (2007). Modulation of pulmonary vascular smooth muscle cell phenotype in hypoxia: role of cGMP-dependent protein kinase. *Am. J. Physiol. Lung Cell Mol. Physiol.*, (June 2007), Vol.292, No.6, pp.L1459-L1466.

Zhou, W.; Negash, S.; Liu, J.; Raj, J.U. (2009). Modulation of pulmonary vascular smooth muscle cell phenotype in hypoxia: role of cGMP-dependent protein kinase and myocardin. *Am. J. Physiol. Lung Cell Mol. Physiol.*, (May 2009), Vol.296, No.5, pp.L780-L789.

Zhou, X.B.; Ruth, P.; Schlossmann, J.; Hofmann, F.; Korth, M. (1996). Protein phosphatase 2A is essential for the activation of Ca2+-activated K+ currents by cGMP-dependent protein kinase in tracheal smooth muscle and Chinese hamster ovary cells. J. Biol. Chem., (August 1996), Vol.271, No.33, pp.19760-19767.

5

Regulation of Na$^+$/H$^+$ Exchanger Isoform 3 by Protein Kinase A in the Renal Proximal Tubule

Adriana Castello Costa Girardi[1] and Luciene Regina Carraro-Lacroix[2]
[1]*University of São Paulo Medical School*
[2]*Hospital for Sick Children*
[1]*Brazil*
[2]*Canada*

1. Introduction

One of the major functions of the kidneys is to maintain the volume and composition of the body fluids constant despite wide variation in the daily intake of water and solutes. To accomplish this task, the activities of a number of transport proteins along the nephron are tightly regulated.

The nephron is the functional unit of the kidneys. Each human kidney contains approximately 1.2 millions of nephrons. At the beginning of each nephron, in the glomerulus, the blood is filtered: cells and most proteins are retained, whereas water and small solutes pass from the glomerular capillaries to the Bowman's capsule. As the glomerular filtrate leaves Bowman's capsule and enters the renal tubule, it flows sequentially through the proximal tubule, the loop of Henle, the distal tubule, and the collecting duct. Along this course, greater part of the glomerular filtrate is transported across and between the tubule cells and reenters the blood (reabsorption), whereas some is secreted from the blood into the luminal fluid (secretion). The formation of urine involves the sum of these three major processes: ultrafiltration of plasma by the glomerulus, reabsorption of water and solutes from the ultrafiltrate, and secretion of solutes into the tubular fluid. Although 180 liters of plasma is filtered by the human glomeruli each day, less than 1% of water, sodium chloride and variable amounts of other solutes are excreted in the urine. By the processes of reabsorption and secretion the renal tubule modulates the volume and composition of the urine. Consequently, the tubules precisely control the volume, composition, and pH of the body fluids.

The renal proximal tubule is responsible for reabsorption of the majority of the filtered sodium, bicarbonate, chloride and water. Na$^+$/H$^+$ exchange is the predominant mechanism for absorption of Na$^+$ and secretion of H$^+$ across the apical membrane of proximal tubule cells (Alpern, 1990). Apical membrane Na$^+$/H$^+$ exchange also has a major role in mediating chloride reabsorption in the proximal tubule through its combined activity with a Cl$^-$/base exchanger and by creating an increase in luminal chloride concentration that favors the diffusion of the anion from the tubular lumen to the blood (Warnock and Yee, 1981; Aronson and Giebisch, 1997). The sodium/proton exchanger isoform 3 (NHE3) represents the major topic of this chapter; therefore the properties of the NHE family will be briefly discussed bellow.

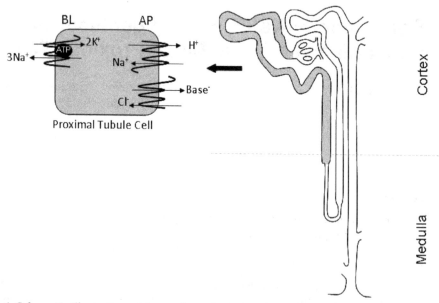

Fig. 1. Schematic illustration of the nephron depicting the most important transport mechanisms involved with NaCl reabsorption in proximal tubule. The inset shows Na^+/K^+-ATPase, Na^+/H^+ exchanger and $Cl^-/Base$ exchanger localization in proximal tubule. BL = basolateral; AP = apical.

The mammalian Na^+/H^+ exchanger (NHE) gene family (SLC9) consists of secondary active transporters that mediate the electroneutral exchange of intracellular protons for extracellular sodium (Aronson, 1985). The transport activity of this protein is crucial to regulation of intracellular pH and cellular volume. In polarized epithelia, Na^+/H^+ exchangers are also involved in transepithelial NaHCO3 and NaCl transport.

All members of the NHE family share a common structural feature. They consist of two major portions, an N-terminal transmembrane domain and a large cytoplasmic C-terminal domain. The N-terminal portion of all known isoforms is predicted to span the plasma membrane twelve times. This domain is responsible for the Na^+/H^+ exchange transport function (Pouyssegur, 1994). The C-terminal portion is mainly hydrophilic and it is the portion through which the activity of the exchanger is regulated.

Thus far, five sodium proton exchangers (NHE1, NHE2, NHE3, NHE4 and NHE8) have been identified in plasma membrane of renal tubular cells (Biemesderfer et al., 1992; Biemesderfer et al., 1993a; Amemiya et al., 1995; Chambrey et al., 1997; Chambrey et al., 1998). Of these, NHE3, the most abundant NHE isoform in renal tissue, is confined to the apical membrane of proximal tubule and thin and thick ascending limb. Several lines of evidence strongly support the conclusion that NHE3 is the principal NHE isoform responsible for apical membrane Na^+/H^+ exchange in the proximal tubule. First, studies

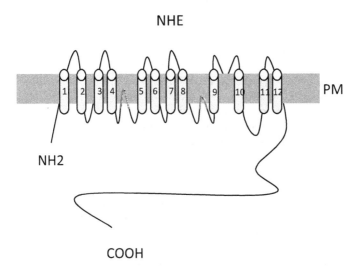

Fig. 2. Secondary structure of sodium/proton exchangers (NHE). All NHE isoforms have a membrane topology of 10-12 transmembrane segments. The carboxy-terminal region at the cytoplasmic site of NHE3 has consensus phosphorylation sites for PKA.

using isoform-specific antibodies have demonstrated that NHE3 is expressed on the brush border membrane of proximal tubule cells (Biemesderfer et al., 1993b; Biemesderfer et al., 1997). Second, the pattern of sensitivity to different inhibitors of Na^+/H^+ exchange in renal brush border vesicles and of bicarbonate absorption in microperfused proximal tubules is most consistent with the properties of NHE3 among the known NHE isoforms (Vallon et al., 2000). Third, micropuncture analysis on NHE3 knockout mice revealed a remarkable reduction of fluid and bicarbonate reabsorption in renal proximal tubule further supporting the concept that NHE3 is the isoform that accounts for most Na^+/H^+ exchange in this nephron segment (Lorenz et al., 1999; Wang et al., 1999). Indeed, mice lacking NHE3 display hypotension and had a mild hyperchloremic metabolic acidosis (Schultheis et al., 1998). Despite the reduced salt reabsortive capacity in the renal proximal tubule, the NHE3 deficient mice grows well when fed a normal sodium diet, mostly due to reduced glomerular filtration rate and increased sodium and bicarbonate reabsorption in the distal nephron. However, if these animals are subjected to dietary salt restriction the adaptative mechanisms are not sufficient to fully compensate for the large defect on proximal reabsorption and they may die from hypovolemic shock (Ledoussal et al., 2001).

Given the important role of NHE3 in mediating $NaHCO_3$ and NaCl reabsorption in the proximal tubule, this transporter is subject to acute and chronic regulation in response to a variety of conditions and humoral factors affecting acid-base or salt balance. In this chapter we will focus on the regulation of NHE3 by protein kinase A and the implications of this regulatory mechanism on renal function under physiological and pathophysiological conditions.

2. Regulation of NHE3 activity by hormones that activate cAMP-dependent protein kinase A (PKA) in renal proximal tubule

The signal transduction cascade mediating the acute effect of NHE3 agonists and antagonists involves multiple pathways. One of the best studied regulatory mechanisms affecting NHE3 activity is the inhibition resulting from protein kinase A (PKA) activation. Hormones activating cAMP-dependent PKA have been shown to reduce sodium and bicarbonate reabsorption in renal proximal tubule by inhibiting NHE3 transport activity. Table 1 presents a summary of hormones and molecular mechanisms associated with inhibition of NHE3 by PKA. These hormones act via G-protein coupled receptors (GPCR) expressed in the apical membrane of the renal proximal tubule (Felder et al., 1984; Muff et al., 1992; Marks et al., 2003; Schlatter et al., 2007; Crajoinas et al., 2011) . GPCR signal transduction occurs through coupling to heterotrimeric G proteins on the intracellular side of the membrane. Heterotrimeric G proteins contain three subunits referred as Gα, Gβ, and Gγ. Upon ligand binding, the GPCR undergoes a conformational change that promotes the exchange of bound guanosine diphosphate (GDP) from the Gα subunit for guanosine triphosphate (GTP). The G protein Gα subunit bound to GTP can then dissociate from the Gβγ dimer and initiate the intracellular signaling cascade that leads to cAMP dependent PKA activation that further elicits NHE3 inhibition (Fig. 3).

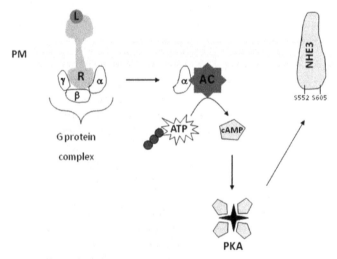

Fig. 3. Downstream effect of a receptor coupled to adenylyl cyclase. The ligand (L) binds to its receptor coupled to protein G (αs) and adenylyl cyclise (AC) is activated. The activated enzyme converts ATP to cAMP, which activates protein kinase A (PKA) which, in turn, is ready to phosphorylate NHE3 at the specific PKA consensus sites.

2.1 Parathyroid hormone

The kidney is a principal target organ for the action of the parathyroid hormone (PTH). PTH is primarily involved in modulation of serum calcium and phosphate homeostasis but also acts on the proximal tubule, the thick ascending limb, and the distal convoluted tubule to alter urinary electrolyte and fluid excretion. The inhibitory effect of PTH on renal proximal

Hormone/Condition	Associated Mechanism	References
Parathyroid hormone	$\downarrow$ of NHE3 affinity for protons due to phosphorylation of the exchanger $\downarrow$ of NHE3 maximum velocity due to a decrement of the number of NHE3 molecules at the plasma membrane $\downarrow$ of NHE3 promoter activity, NHE3 mRNA and protein abundance	(Fan et al., 1999; Collazo et al., 2000; Zhang et al., 1999; Girardi et al., 2000; Bezerra et al., 2008)
Dopamine	$\downarrow$ NHE3-mediated Na⁺/H⁺ exchange due to increased endocytosis caused by increased PKA-mediated NHE3 phosphorylation	(Gomes & Soares-da-Silva, 2004; Hu et al., 2001; Bacic et al., 2003)
Glucagon-like peptide 1	$\downarrow$ NHE3 activity due to increased NHE3 phosphorylation	(Crajoinas et al., 2011; Carraro-Lacroix et al., 2009)
Glucagon	Acutely - $\downarrow$ NHE via a PKA-dependent pathway Chronically - $\uparrow$ NHE3 mRNA and protein expression at the plasma membrane	(Amemiya et al., 2002)
Guanylins	$\downarrow$ NHE3-mediated Na⁺/H⁺ due to increased levels of NHE3 PKA-dependent phosphorylation and reduction of the exchanger at the plasma membrane	(Amorim et al., 2006; Lessa LM, Girardi AC, and Malnic G, unpublished observations)
High Salt Diet	$\downarrow$ NHE3-mediated Na⁺/H⁺ due to higher NHE3 phosphorylation on serine 552, redistribution from microvilli to the intermicrovillar region together with its regulatory partner dipeptidyl peptidase IV	(Yang et al., 2008)
Hypertension	$\downarrow$ NHE3-mediated Na⁺/H⁺ due to higher NHE3 phosphorylation on serine 552, redistribution of the transporter to the intermicrovillar region	(Magyar et al., 2000; Panico et al., 2009; Crajoinas et al., 2010)
Heart Failure	$\uparrow$ NHE3-mediated sodium reabsorption due to increased renal cortical NHE3 mRNA and protein levels and lower levels of NHE3 phosphorylation at serine 552	(Lutken et al., 2009; Inoue et al., 2012)

Table 1. Major factors and hormones that inhibit NHE3 activity via cAMP-dependent PKA activation.

tubule NHE3 activity has been consistently reported by several laboratories on both *in vivo* (Bank and Aynediian, 1976; Girardi et al., 2000) and *in vitro* studies (Pollock et al., 1986; Mrkic et al., 1992; Mrkic et al., 1993). Experiments performed *in vitro* and *ex vivo* provide evidence that the acute inhibition of NHE3 by PTH is mediated by molecular mechanisms the include reduction of the transporter's apparent affinity for protons in consequence of the direct phosphorylation of the exchanger followed by reduction of its maximum velocity due to a decrement of the number of NHE3 molecules expressed at the plasma membrane (Fan et al., 1999; Collazo et al., 2000). Consistent with a decrease of NHE3 surface expression in response to PTH, studies carried out by the McDonough laboratory have shown that reduction of NHE3 activity in response to acute treatment with this hormone is a consequence of NHE3 redistribution from the apical microvilli to the base of the intermicrovillar region of the proximal tubule brush border (Zhang et al., 1999).

The chronic effect of PTH on NHE3 regulation has also been evaluated (Girardi et al., 2000; Bezerra et al., 2008). Long term inhibition of NHE3 by PTH is associated with a reduction on NHE3 protein and mRNA levels. PTH also provokes a mild inhibitory effect on NHE3 promoter that seems to be PKA-dependent (Bezerra et al., 2008).

2.2 Dopamine

The intrarenal dopamine natriuretic system is critical for mammalian sodium homeostasis. Numerous studies have demonstrated that dopamine remarkably increases urinary sodium excretion mainly by inhibiting tubular sodium reabsorption. The inhibitory effect of dopamine on NHE3 transport activity is mediated mainly via the dopamine D1 receptor and stimulation of adenylyl cyclase/PKA system and phospholipase C/PKC (Gomes and Soares-da-Silva, 2004). The underlying molecular mechanisms by which dopamine decreases NHE3-mediated Na^+/H^+ exchange in renal proximal tubule involves increased endocytosis and is associated with increased PKA-mediated NHE3 phosphorylation (Hu et al., 2001; Bacic et al., 2003).

2.3 Glucagon-like peptide-1

Glucagon-like peptide-1 (GLP-1) is produced by posttranslational modification of the proglucagon gene product in the intestinal L-cells, predominantly localized in the colon and ileum (Holst, 1997; Drucker, 2005). This incretin hormone plays an important role on the maintenance of systemic glucose homeostasis by stimulating insulin secretion and improving insulin sensitivity (Drucker, 2005). Numerous reports in the literature have demonstrated that GLP-1 also exerts renoprotective actions. In this regard, continuous administration of GLP-1 induces diuresis and natriuresis in both humans (Gutzwiller et al., 2004; Gutzwiller et al., 2006) and experimental animal models (Moreno et al., 2002; Yu et al., 2003).

The molecular mechanisms underlying the renal actions of GLP-1 seems to involve increases of GFR and RPF and decrease of NHE3-mediated Na^+/H^+ exchange in the renal proximal tubule (Carraro-Lacroix et al., 2009). Recent studies by our group have demonstrated that binding of GLP-and/or the GLP-1R agonist exendin-4 to its receptor in the renal proximal tubule activates the cAMP/PKA signaling pathway, leading, in turn, to phosphorylation of the PKA consensus sites located at the C-terminal region of the exchanger (Carraro-Lacroix et al., 2009; Crajoinas et al., 2011). Increased NHE3 phosphorylation levels induced by GLP-1 was not accompanied by a decrease of NHE3 expression at the microvillar microdomain of the brush border, suggesting that the mechanism by which GLP-1 inhibits NHE3 activity does not involve subcellular redistribution of the exchanger between the subcompartiments of the renal proximal tubule brush border.

2.4 Glucagon

Glucagon is a 29-amino-acid pancreatic peptide produced by the α-cells present at the periphery of the islets of Langerhans (Baum et al., 1962) and its major function is the maintenance of plasma glucose homeostasis between meals and during fasting. Glucagon binding to its receptor primarily activates adenylyl cyclase and increases cAMP (Pohl et al., 1971; Rodbell et al., 1971). The tissue distribution of the glucagon receptor is broad, with higher levels of expression in liver and kidney (Svoboda et al., 1993; Dunphy et al., 1998).

In the kidney, glucagon affects renal glomerular filtration, renal blood flow, and decreases renal tubular sodium reabsorption (Pullman et al., 1967). Part of the acute natriuretic action of glucagon are mediated by inhibition of NHE3 in the renal proximal tubule via a cAMP/PKA-dependent pathway. Interestingly, in vitro studies using OKP cells have shown that glucagon acutely inihibits and chronically stimulates NHE3 activity (Amemiya et al., 2002).

2.5 Guanylins

The guanylin and uroguanylin are endogenous ligands of the Escherichia coli heat-stable enterotoxin (STa) receptor, guanylate cyclase C (Currie et al., 1992; Fonteles et al., 1998) and are known to be involved in a control system that regulates salt balance in response to oral salt intake. Both guanylin and uroguanylin are synthesized in the intestine and in the kidney and have already been identified in several animal species, including as mammals, fishes and birds (Forte, 2004).

The renal effects of uroguanylin are much more pronounced than the ones produced by guanylin and include natriuresis, kaliuresis, diuresis and increased excretion of cGMP (Forte et al., 1996; Greenberg et al., 1997; Fonteles et al., 1998). In the renal proximal tubule, uroguanylin significantly inhibits NHE3 transport function (Amorim et al., 2006). Ongoing studies by the Malnic laboratory have demonstrated that the mechanism by which uroguanyn inhibits NHE3 involves increased levels of NHE3 phosphorylation followed by retrieval of the exchanger from the plasma membrane (unpublished observations, Lessa LM, Girardi AC, and Malnic G)). The mechanism by which NHE3 is phosphorylated by PKA in response to coupling of uroguanylin to its receptor in the renal proximal tubule possibly involves a crosstalk mechanism between cGMP and cAMP pathways.

3. The Na⁺/H⁺ exchanger regulatory factor NHERF

Although a large number of hormones reported to affect NHE3 share the same signal pathways, the molecular mechanisms by which they regulate NHE3 may differ greatly among them. The identification of regulatory proteins that interact with NHE3 has unraveled some aspects of the molecular mechanisms underlying this transporter regulation.

The first NHE3 regulatory factor was isolated and characterized byWeinman and Shenolikar (Weinman et al., 1995; Weinman et al., 2000a; Weinman et al., 2000b). These investigators demonstrated that the presence of this cofactor was essential for PKA-mediated inhibition of NHE3. This protein was cloned and termed NHERF-1 (Na/H exchanger regulatory factor). Subsequently, in an attempt to identify proteins that interact with NHE3, Yun and coworkers used the C-terminus of NHE3 as a bait in a yeast two-hybrid screen and isolated E3KARP (exchanger-3 kinase A regulatory protein, or NHERF-2) . NHERF-1 and NHERF-2 are highly homologous proteins (52% sequence identity for the human orthologs) (Yun et al., 1997). Physical association of NHERF-1 and NHERF-2 with NHE3 has been demonstrated by binding assays using fusion proteins or by co-precipitation experiments using transfected cells overexpressing NHE3.

NHERF-1 and NHERF-2 are both members of a family of proteins that contain two tandem PDZ domains (that are conserved modules that mediate protein-protein interaction) and a C-terminal ezrin-radixin-moesin (ERM) binding domain which anchors the proteins to the actin cytoskeleton through ezrin. Lamprecht and Yun have proposed a model whereby the complex NHERF/ezrin acts as a functional AKAP (A kinase anchoring protein) for NHE3, serving as a structural link between not only NHE3 and the cytoskeleton but also between NHE3 and PKA, since ezrin is capable of binding the RII regulatory subunit of PKA (Dransfield et al., 1997; Lamprecht et al., 1998). The current model suggests that NHERF-1 is required for cAMP-dependent regulation of NHE3 and acts as an adapter to link NHE3 to ezrin, which then serves as a PKA anchoring protein. Upon activation of PKA by hormones and/or factors that increase intracellular cAMP, PKA phosphorylates serine residues in the C-terminal hydrophilic domain of NHE3 (Fig. 4). Biochemical experiments with brush border vesicles isolated from NHERF-1 knockout mice corroborate with this model (Weinman et al., 2003). These studies showed that NHERF-1 is crucial for PKA-mediated phosphorylation and inhibition of NHE3 transport activity. Moreover, NHE3 expression was not affected in these animals, showing that NHERF-1 plays an essential role on NHE3 modulation but it is not required for expression or apical targeting of the transporter in the proximal tubule (Weinman et al., 2003).

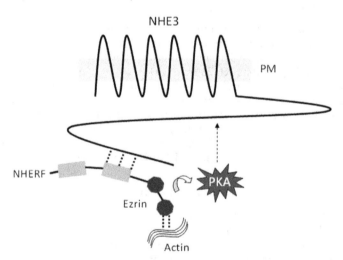

Fig. 4. Model of NHERF requirement for PKA-mediated NHE3 inhibition. NHERF facilitates cAMP-dependent regulation of NHE3 by interacting with the cytoskeleton to target PKA to phosphorylation sites within the cytoplasmic domain of NHE3.

To date, four members of the NHERF family of PDZ domain proteins have been described. These proteins bind to a variety of membrane transporters regulating their cell surface expression, protein interactions as well as the formation of signaling complexes.

4. NHE3 phosphorylation at the PKA consensus sites

Moe and coworkers were the first to demonstrate that the cytoplasmic domain of NHE3 is a substrate for PKA in vitro (Moe et al., 1995). There are multiple putative consensus motifs for PKA on the NHE3 C-terminal region. Three of these sites (S552, S605 and S634) are highly conserved among species. Based on that, transfection studies using truncation of the NHE3 C-terminal domain and site direct mutagenesis of the above mentioned serine residues were carried out to evaluate whether PKA directly phosphorylates one or more of these consensus sites in vivo (Cabado et al., 1996; Kurashima et al., 1997; Zhao et al., 1999). Kurashima and coworkers have shown that both serines 605 and 634 are important for PKA-mediated inhibition of NHE3, although only serine 605 is phosphorylated in vivo (Kurashima et al., 1997). The studies by Zhao and colleagues confirmed that phosphorylation of serine 605 is increased by cAMP/PKA. These investigators also found that PKA directly phosphorylates serine 552 and that both 552 and 605 residues appear to be critical for inhibition of NHE3 by PKA (Zhao et al., 1999).

Years later, the Aronson laboratory generated phosphospecific antibodies directed to the PKA consensus sites S552 and S605 of rat NHE3 (Kocinsky et al., 2005). These reagents are of great value, since they have enable investigators to evaluate the phosphorylation state of these two residues in endogenous NHE3 under basal conditions, under a variety of physiological and pharmacological maneuvers and in disease states. Indeed, increments in the phosphorylation status of NHE3 at serines 552 and 605 have been shown to occur in response to dopamine (Kocinsky et al., 2005), PTH (Kocinsky et al., 2007), and GLP-1 (Crajoinas et al., 2011).

Studies by Kocinsky and colleagues have also demonstrated that serine 552 is phosphorylated to a much greater extent than serine 605 in baseline in vivo. Moreover, these investigators found that when NHE3 is phosphorylated at serine 552, it mainly resides at the intermicrovillar domain of the brush border (Kocinsky et al., 2005). This observation is consistent with a decrease on NHE3-mediated Na⁺/H⁺ exchange in the renal proximal tubule, since within the intermicrovillar subcompartment of the brush border, this transporter must have very limited assess to the tubular fluid.

The precise mechanism by which NHE3 phosphorylation leads to NHE3 inhibition remains obscure. Although phosphorylation of NHE3 at the 552 and 605 residues are necessary for the PKA-dependent inhibitory effect, phosphorylation of NHE3 at these PKA consensus sites precedes transport inhibition (Kocinsky et al., 2007) indicating that phosphorylation per se is not sufficient to inhibit NHE3 activity. The current body of data suggests that PKA phosphorylation may ultimately result in inhibition of NHE3 by modulating NHE3 subcellular trafficking, interaction with regulatory proteins, or localization within the plasma membrane.

5. Pathophysiological Implications of NHE3 Phosphorylation by Protein Kinase A

As mentioned above, NHE3 is phosphorylated on serine 552 under basal conditions by the adenylyl cyclase/cAMP-activated-protein kinase A (PKA) and the endogenous levels of phosphorylation is often affected as part of acute NHE3 regulation. Interestingly, baseline levels of NHE3 phosphorylation at this residue may also be associated with chronic regulation of NHE3 activity.

5.1 High Salt Diet

During high salt diet the kidneys increase sodium and volume excretion to match intake. IYang and colleagues have demonstrated that three weeks of high salt diet (4%) doubled the levels of NHE3 phosphorylation at the serine 552 (Yang et al., 2008), possibly contributing to the natriuretic effect triggered by high sodium load.

5.2 Hypertension

Mice lacking Na^+/H^+ exchanger NHE3 are hypotensive and hypovolemic, underscoring the importance of the transporter on blood pressure control.

We have recently assessed in vivo NHE3 transport activity and defined the mechanisms underlying NHE3 regulation before and after development of hypertension in spontaneously hypertensive rats (SHR). By means of in vivo stationary microperfusion, we found that NHE3-mediated bicarbonate reabsorption is higher in the proximal tubule of 5-week-old pre-hypertensive spontaneously hypertensive rat (SHR) and lower in 14-week-old SHR compared to age-matched normotensive rats (Wistar Kyoto, WKY). Higher NHE3 activity in young pre-hypertensive SHR is associated with lower phosphorylation levels (serine 552) and increased NHE3 expression at the microvillar brush border. During the hypertensive stage, NHE3 was found to be mainly confined to the intermicrovillar region and the relative abundance of NHE3 phosphorylated on serine 552 is increased compared to normotensive animals (Fig. 5) (Crajoinas et al., 2010).

5.3 Heart failure

Heart failure (HF) is associated with sodium and water retention and extracellular volume expansion. A principal site of renal salt and water reabsorption is the proximal tubule. We therefore hypothesized that NHE3, the major apical transcellular pathway for sodium reabsorption in the proximal tubule, might be upregulated in heart failure. To test this hypothesis, we employed both stationary in vivo microperfusion and pH-dependent sodium uptake to verify whether NHE3 activity would be altered in the proximal tubule of an experimental model of heart failure (Antonio et al., 2009). Our data demonstrated that heart failure rats display enhanced NHE3-mediated sodium reabsorption in the proximal tubule which may contribute to extracellular volume expansion and edema. In addition to increased renal cortical NHE3 expression at both protein and mRNA levels, we have also observed that the levels of NHE3 phosphorylation at serine 552 in renal cortical membranes of heart failure rats are lower than in Sham animals. Thus, the molecular mechanisms mediating enhanced sodium reabsorption in the renal proximal tubule of heart failure rats also involves posttranslational covalent modification of NHE3 (Fig. 5) (Inoue et al., 2012).

6. Conclusion

PKA activation plays an important role in inhibiting the activity NHE3, the major apical transcellular pathway for sodium reabsorption in the proximal tubule. Similarly, PKA inhibition may be involved in NHE3 stimulation, as suggested by the studies carried out in young hypertensive and in heart failure animals. The elucidation of the mechanisms by which phosphorylation of NHE3 at the PKA consensus sites leads to inhibition of NHE3-mediated Na⁺/H⁺ exchange in the renal proximal tubule may also unravel important molecular underpinnings that lead to the development and/or progression of primary kidney diseases and other conditions that affect the kidneys.

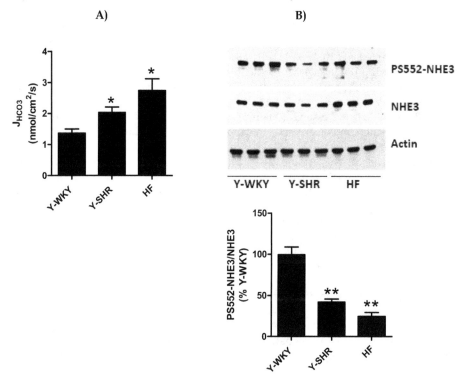

Fig. 5. NHE3 activity negatively correlates with the levels of endogenous phosphorylation of the PKA consensus site, serine 552, located at the C-terminus region of the exchanger Adapted from (Crajoinas et al., 2010; Inoue et al., 2012). (A) Stationary microperfusion was employed to measure NHE3-mediated bicarbonate reabsorption (J_{HCO3^-}) in the proximal tubules of 5-week-old Wistar Kyoto rats (Y-WKY, n = 5 rats, 12 tubules), 5-week old SHR (Y-SHR, n = 5 rats, 12 tubules, and heart failure rats (HF, n = 5 rats, 15 tubules). Data are expressed as means ± SE. *P < 0.05 *vs.* Y-WKY. (B) *Top*: Representative immunoblot of phosphorylated and total NHE3 expression in renal cortical membranes isolated from Y-WKY, Y-SHR or HF rats. Equivalent samples (15 μg of protein for NHE3 and 5 μg for PS552-NHE3 and actin) of renal cortical membranes were prepared for immunoblot analysis. The membranes were incubated with monoclonal antibodies against phosphorylated NHE3 at serine 552 (PS552-NHE3 (1:1,000)), total NHE3 (1:1,000) or anti-actin (1:50,000). *Bottom: Left -* Graphical representation of the phosphorylation ratio of NHE3 at Serine 552 to total NHE3 in renal cortical membranes (PS552-NHE3/NHE3). Values are means ± SE. n = 3/group, **P < 0.001 *vs.* Y-WKY.

7. Acknowledgments

This work was supported by Fundação de Amparo à Pesquisa do Estado de São Paulo, São Paulo, SP, Brazil (Grant 2007/52945-8).

8. References

Alpern RJ (1990) Cell mechanisms of proximal tubule acidification. *Physiol Rev* 70:79-114.

Amemiya M, Kusano E, Muto S, Tabei K, Ando Y, Alpern RJ and Asano Y (2002) Glucagon acutely inhibits but chronically activates Na(+)/H(+) antiporter 3 activity in OKP cells. *Exp Nephrol* 10:26-33.

Amemiya M, Loffing J, Lotscher M, Kaissling B, Alpern RJ and Moe OW (1995) Expression of NHE-3 in the apical membrane of rat renal proximal tubule and thick ascending limb. *Kidney International* 48:1206-1215.

Amorim JB, Musa-Aziz R, Lessa LM, Malnic G and Fonteles MC (2006) Effect of uroguanylin on potassium and bicarbonate transport in rat renal tubules. *Can J Physiol Pharmacol* 84:1003-1010.

Antonio EL, Dos Santos AA, Araujo SR, Bocalini DS, Dos Santos L, Fenelon G, Franco MF and Tucci PJ (2009) Left ventricle radio-frequency ablation in the rat: a new model of heart failure due to myocardial infarction homogeneous in size and low in mortality. *J Card Fail* 15:540-548.

Aronson PS (1985) Kinetic properties of the plasma membrane Na+-H+ exchanger. *Annual Review of Physiology* 47:545-560.

Aronson PS and Giebisch G (1997) Mechanisms of chloride transport in the proximal tubule. *Am J Physiol* 273:F179-192.

Bacic D, Kaissling B, McLeroy P, Zou L, Baum M and Moe OW (2003) Dopamine acutely decreases apical membrane Na/H exchanger NHE3 protein in mouse renal proximal tubule. *Kidney Int* 64:2133-2141.

Bank N and Aynediian HS (1976) A micropuncture study of the effect of parathyroid hormone on renal bicarbonate reabsorption. *J Clin Invest* 58:336-344.

Baum J, Simons BE, Jr., Unger RH and Madison LL (1962) Localization of glucagon in the alpha cells in the pancreatic islet by immunofluorescent technics. *Diabetes* 11:371-374.

Bezerra CN, Girardi AC, Carraro-Lacroix LR and Reboucas NA (2008) Mechanisms underlying the long-term regulation of NHE3 by parathyroid hormone. *Am J Physiol Renal Physiol* 294:F1232-1237.

Biemesderfer D, Pizzonia J, Abu-Alfa A, Exner M, Reilly R, Igarashi P and Aronson PS (1993a) NHE3: a Na+/H+ exchanger isoform of renal brush border. *American Journal of Physiology* 265:F736-742.

Biemesderfer D, Pizzonia J, Abu-Alfa A, Exner M, Reilly R, Igarashi P and Aronson PS (1993b) NHE3: a Na+/H+ exchanger isoform of renal brush border. *Am J Physiol* 265:F736-742.

Biemesderfer D, Reilly RF, Exner M, Igarashi P and Aronson PS (1992) Immunocytochemical characterization of Na(+)-H+ exchanger isoform NHE-1 in rabbit kidney. *American Journal of Physiology* 263:F833-840.

Biemesderfer D, Rutherford PA, Nagy T, Pizzonia JH, Abu-Alfa AK and Aronson PS (1997) Monoclonal antibodies for high-resolution localization of NHE3 in adult and neonatal rat kidney. *Am J Physiol* 273:F289-299.

Cabado AG, Yu FH, Kapus A, Lukacs G, Grinstein S and Orlowski J (1996) Distinct structural domains confer cAMP sensitivity and ATP dependence to the Na+/H+ exchanger NHE3 isoform. *J Biol Chem* 271:3590-3599.

Carraro-Lacroix LR, Malnic G and Girardi AC (2009) Regulation of Na+/H+ exchanger NHE3 by glucagon-like peptide 1 receptor agonist exendin-4 in renal proximal tubule cells. *Am J Physiol Renal Physiol* 297:F1647-1655.

Chambrey R, Achard JM, St John PL, Abrahamson DR and Warnock DG (1997) Evidence for an amiloride-insensitive Na+/H+ exchanger in rat renal cortical tubules. *Am J Physiol* 273:C1064-1074.

Chambrey R, Warnock DG, Podevin RA, Bruneval P, Mandet C, Belair MF, Bariety J and Paillard M (1998) Immunolocalization of the Na+/H+ exchanger isoform NHE2 in rat kidney. *American Journal of Physiology* 275:F379-386.

Collazo R, Fan L, Hu MC, Zhao H, Wiederkehr MR and Moe OW (2000) Acute regulation of Na+/H+ exchanger NHE3 by parathyroid hormone via NHE3 phosphorylation and dynamin-dependent endocytosis. *J Biol Chem* 275:31601-31608.

Crajoinas RO, Lessa LM, Carraro-Lacroix LR, Davel AP, Pacheco BP, Rossoni LV, Malnic G and Girardi AC (2010) Posttranslational mechanisms associated with reduced NHE3 activity in adult vs. young prehypertensive SHR. *Am J Physiol Renal Physiol* 299:F872-881.

Crajoinas RO, Oricchio FT, Pessoa TD, Pacheco BP, Lessa LM, Malnic G and Girardi AC (2011) Mechanisms mediating the diuretic and natriuretic actions of the incretin hormone glucagon-like peptide-1. *Am J Physiol Renal Physiol* 301:F355-363.

Currie MG, Fok KF, Kato J, Moore RJ, Hamra FK, Duffin KL and Smith CE (1992) Guanylin: an endogenous activator of intestinal guanylate cyclase. *Proc Natl Acad Sci U S A* 89:947-951.

Dransfield DT, Yeh JL, Bradford AJ and Goldenring JR (1997) Identification and characterization of a novel A-kinase-anchoring protein (AKAP120) from rabbit gastric parietal cells. *Biochem J* 322 (Pt 3):801-808.

Drucker DJ (2005) Biologic actions and therapeutic potential of the proglucagon-derived peptides. *Nat Clin Pract Endocrinol Metab* 1:22-31.

Dunphy JL, Taylor RG and Fuller PJ (1998) Tissue distribution of rat glucagon receptor and GLP-1 receptor gene expression. *Mol Cell Endocrinol* 141:179-186.

Fan L, Wiederkehr MR, Collazo R, Wang H, Crowder LA and Moe OW (1999) Dual mechanisms of regulation of Na/H exchanger NHE-3 by parathyroid hormone in rat kidney. *J Biol Chem* 274:11289-11295.

Felder RA, Blecher M, Calcagno PL and Jose PA (1984) Dopamine receptors in the proximal tubule of the rabbit. *Am J Physiol* 247:F499-505.

Fonteles MC, Greenberg RN, Monteiro HS, Currie MG and Forte LR (1998) Natriuretic and kaliuretic activities of guanylin and uroguanylin in the isolated perfused rat kidney. *Am J Physiol* 275:F191-197.

Forte LR, Fan X and Hamra FK (1996) Salt and water homeostasis: uroguanylin is a circulating peptide hormone with natriuretic activity. *Am J Kidney Dis* 28:296-304.

Forte LR, Jr. (2004) Uroguanylin and guanylin peptides: pharmacology and experimental therapeutics. *Pharmacol Ther* 104:137-162.

Girardi AC, Titan SM, Malnic G and Reboucas NA (2000) Chronic effect of parathyroid hormone on NHE3 expression in rat renal proximal tubules. *Kidney Int* 58:1623-1631.

Gomes P and Soares-da-Silva P (2004) Dopamine acutely decreases type 3 Na(+)/H(+) exchanger activity in renal OK cells through the activation of protein kinases A and C signalling cascades. *Eur J Pharmacol* 488:51-59.

Greenberg RN, Hill M, Crytzer J, Krause WJ, Eber SL, Hamra FK and Forte LR (1997) Comparison of effects of uroguanylin, guanylin, and Escherichia coli heat-stable enterotoxin STa in mouse intestine and kidney: evidence that uroguanylin is an intestinal natriuretic hormone. *J Investig Med* 45:276-282.

Holst JJ (1997) Enteroglucagon. *Annu Rev Physiol* 59:257-271.

Hu MC, Fan L, Crowder LA, Karim-Jimenez Z, Murer H and Moe OW (2001) Dopamine acutely stimulates Na+/H+ exchanger (NHE3) endocytosis via clathrin-coated vesicles: dependence on protein kinase A-mediated NHE3 phosphorylation. *J Biol Chem* 276:26906-26915.

Inoue BH, Dos Santos L, Pessoa TD, Antonio EL, Pacheco BP, Savignano FA, Carraro-Lacroix LR, Tucci PJ, Malnic G and Girardi AC (2012) Increased NHE3 abundance and transport activity in renal proximal tubule of rats with heart failure. *Am J Physiol Regul Integr Comp Physiol* 302:R166-174.

Kocinsky HS, Dynia DW, Wang T and Aronson PS (2007) NHE3 phosphorylation at serines 552 and 605 does not directly affect NHE3 activity. *Am J Physiol Renal Physiol* 293:F212-218.

Kocinsky HS, Girardi AC, Biemesderfer D, Nguyen T, Mentone S, Orlowski J and Aronson PS (2005) Use of phospho-specific antibodies to determine the phosphorylation of endogenous Na+/H+ exchanger NHE3 at PKA consensus sites. *Am J Physiol Renal Physiol* 289:F249-258.

Kurashima K, Yu FH, Cabado AG, Szabo EZ, Grinstein S and Orlowski J (1997) Identification of sites required for down-regulation of Na+/H+ exchanger NHE3 activity by cAMP-dependent protein kinase. phosphorylation-dependent and - independent mechanisms. *J Biol Chem* 272:28672-28679.

Lamprecht G, Weinman EJ and Yun CH (1998) The role of NHERF and E3KARP in the cAMP-mediated inhibition of NHE3. *J Biol Chem* 273:29972-29978.

Ledoussal C, Lorenz JN, Nieman ML, Soleimani M, Schultheis PJ and Shull GE (2001) Renal salt wasting in mice lacking NHE3 Na+/H+ exchanger but not in mice lacking NHE2. *Am J Physiol Renal Physiol* 281:F718-727.

Lorenz JN, Schultheis PJ, Traynor T, Shull GE and Schnermann J (1999) Micropuncture analysis of single-nephron function in NHE3-deficient mice. *Am J Physiol* 277:F447-453.

Lutken SC, Kim SW, Jonassen T, Marples D, Knepper MA, Kwon TH, Frokiaer J and Nielsen S (2009) Changes of renal AQP2, ENaC, and NHE3 in experimentally induced heart failure: response to angiotensin II AT1 receptor blockade. *Am J Physiol Renal Physiol* 297:F1678-1688.

Magyar CE, Zhang Y, Holstein-Rathlou NH and McDonough AA (2000) Proximal tubule Na transporter responses are the same during acute and chronic hypertension. *Am J Physiol Renal Physiol* 279:F358-369.

Marks J, Debnam ES, Dashwood MR, Srai SK and Unwin RJ (2003) Detection of glucagon receptor mRNA in the rat proximal tubule: potential role for glucagon in the control of renal glucose transport. *Clin Sci (Lond)* 104:253-258.

Moe OW, Amemiya M and Yamaji Y (1995) Activation of protein kinase A acutely inhibits and phosphorylates Na/H exchanger NHE-3. *J Clin Invest* 96:2187-2194.

Moreno C, Mistry M and Roman RJ (2002) Renal effects of glucagon-like peptide in rats. *Eur J Pharmacol* 434:163-167.

Mrkic B, Forgo J, Murer H and Helmle-Kolb C (1992) Apical and basolateral Na/H exchange in cultured murine proximal tubule cells (MCT): effect of parathyroid hormone (PTH). *J Membr Biol* 130:205-217.

Mrkic B, Tse CM, Forgo J, Helmle-Kolb C, Donowitz M and Murer H (1993) Identification of PTH-responsive Na/H-exchanger isoforms in a rabbit proximal tubule cell line (RKPC-2). *Pflugers Arch* 424:377-384.

Muff R, Fischer JA, Biber J and Murer H (1992) Parathyroid hormone receptors in control of proximal tubule function. *Annu Rev Physiol* 54:67-79.

Panico C, Luo Z, Damiano S, Artigiano F, Gill P and Welch WJ (2009) Renal proximal tubular reabsorption is reduced in adult spontaneously hypertensive rats: roles of superoxide and Na+/H+ exchanger 3. *Hypertension* 54:1291-1297.

Pohl SL, Birnbaumer L and Rodbell M (1971) The glucagon-sensitive adenyl cyclase system in plasma membranes of rat liver. I. Properties. *J Biol Chem* 246:1849-1856.

Pollock AS, Warnock DG and Strewler GJ (1986) Parathyroid hormone inhibition of Na+-H+ antiporter activity in a cultured renal cell line. *Am J Physiol* 250:F217-225.

Pouyssegur J (1994) Molecular biology and hormonal regulation of vertebrate Na+/H+ exchanger isoforms. *Renal Physiology & Biochemistry* 17:190-193.

Pullman TN, Lavender AR and Aho I (1967) Direct effects of glucagon on renal hemodynamics and excretion of inorganic ions. *Metabolism* 16:358-373.

Rodbell M, Birnbaumer L, Pohl SL and Krans HM (1971) The glucagon-sensitive adenyl cyclase system in plasma membranes of rat liver. V. An obligatory role of guanylnucleotides in glucagon action. *J Biol Chem* 246:1877-1882.

Schlatter P, Beglinger C, Drewe J and Gutmann H (2007) Glucagon-like peptide 1 receptor expression in primary porcine proximal tubular cells. *Regul Pept* 141:120-128.

Schultheis PJ, Clarke LL, Meneton P, Miller ML, Soleimani M, Gawenis LR, Riddle TM, Duffy JJ, Doetschman T, Wang T, Giebisch G, Aronson PS, Lorenz JN and Shull GE (1998) Renal and intestinal absorptive defects in mice lacking the NHE3 Na+/H+ exchanger. *Nat Genet* 19:282-285.

Svoboda M, Ciccarelli E, Tastenoy M, Robberecht P and Christophe J (1993) A cDNA construct allowing the expression of rat hepatic glucagon receptors. *Biochem Biophys Res Commun* 192:135-142.

Vallon V, Schwark JR, Richter K and Hropot M (2000) Role of Na(+)/H(+) exchanger NHE3 in nephron function: micropuncture studies with S3226, an inhibitor of NHE3. *Am J Physiol Renal Physiol* 278:F375-379.

Wang T, Yang CL, Abbiati T, Schultheis PJ, Shull GE, Giebisch G and Aronson PS (1999) Mechanism of proximal tubule bicarbonate absorption in NHE3 null mice. *Am J Physiol* 277:F298-302.

Warnock DG and Yee VJ (1981) Chloride uptake by brush border membrane vesicles isolated from rabbit renal cortex. Coupling to proton gradients and K+ diffusion potentials. *J Clin Invest* 67:103-115.

Weinman EJ, Minkoff C and Shenolikar S (2000a) Signal complex regulation of renal transport proteins: NHERF and regulation of NHE3 by PKA. *Am J Physiol Renal Physiol* 279:F393-399.

Weinman EJ, Steplock D, Donowitz M and Shenolikar S (2000b) NHERF associations with sodium-hydrogen exchanger isoform 3 (NHE3) and ezrin are essential for cAMP-mediated phosphorylation and inhibition of NHE3. *Biochemistry* 39:6123-6129.

Weinman EJ, Steplock D and Shenolikar S (2003) NHERF-1 uniquely transduces the cAMP signals that inhibit sodium-hydrogen exchange in mouse renal apical membranes. *FEBS Lett* 536:141-144.

Weinman EJ, Steplock D, Wang Y and Shenolikar S (1995) Characterization of a protein cofactor that mediates protein kinase A regulation of the renal brush border membrane Na(+)-H+ exchanger. *J Clin Invest* 95:2143-2149.

Yang LE, Sandberg MB, Can AD, Pihakaski-Maunsbach K and McDonough AA (2008) Effects of dietary salt on renal Na+ transporter subcellular distribution, abundance, and phosphorylation status. *Am J Physiol Renal Physiol* 295:F1003-1016.

Yu M, Moreno C, Hoagland KM, Dahly A, Ditter K, Mistry M and Roman RJ (2003) Antihypertensive effect of glucagon-like peptide 1 in Dahl salt-sensitive rats. *J Hypertens* 21:1125-1135.

Yun CH, Oh S, Zizak M, Steplock D, Tsao S, Tse CM, Weinman EJ and Donowitz M (1997) cAMP-mediated inhibition of the epithelial brush border Na+/H+ exchanger, NHE3, requires an associated regulatory protein. *Proc Natl Acad Sci U S A* 94:3010-3015.

Zhang Y, Norian JM, Magyar CE, Holstein-Rathlou NH, Mircheff AK and McDonough AA (1999) In vivo PTH provokes apical NHE3 and NaPi2 redistribution and Na-K-ATPase inhibition. *Am J Physiol* 276:F711-719.

Zhao H, Wiederkehr MR, Fan L, Collazo RL, Crowder LA and Moe OW (1999) Acute inhibition of Na/H exchanger NHE-3 by cAMP. Role of protein kinase a and NHE-3 phosphoserines 552 and 605. *J Biol Chem* 274:3978-3987.

6

Protein Kinases and Pancreatic Islet Function

Gabriela Da Silva Xavier
Imperial College London, Section of Cell Biology,
Division of Diabetes, Endocrinology and Metabolism, London
UK

1. Introduction

Intense research on pancreatic islet function is fuelled by its link to the disease diabetes mellitus. Diabetes is a chronic disease that is characterised by inappropriate regulation of blood glucose levels. This dysregulation of blood glucose homeostasis occurs when the pancreas produces insufficient amounts of the hormone, insulin, or when insulin sensitive organs lose sensitivity to insulin (insulin resistance). Increasingly there is evidence to support the idea that abnormal release of another pancreatic hormone, glucagon, may be involved in the dysregulation of blood glucose homeostasis.

There are two types of diabetes. Type 1 diabetes is characterised by the failure of the pancreas to produce insulin. In most cases, type 1 diabetes is caused by autoimmune destruction of the pancreatic β cells that produce insulin and release the hormone in response to changes in blood glucose levels. Type 2 diabetes is characterised by relative insulin insufficiency and insulin resistance. This is the more common form of diabetes, comprising of 90 % of people with diabetes worldwide. The prominence of this form of diabetes is associated with lifestyle choices: the obesity epidemic has led to an increase in the incidence of diseases associated with metabolic imbalance such as diabetes. Energy homeostasis and obesity are intimately linked. Between 45-80 % of our energy intake is in the form of carbohydrates (1;2), which are converted to glucose and transported in the blood stream (3). Thus, there has been much interest in the mechanisms that regulate glucose homeostasis, particularly those involving the endocrine pancreas. The pancreatic hormones glucagon and insulin, produced and released in the pancreatice α and β cells, repectively, are involved in maintaining blood glucose homeostasis.

Uncontrolled diabetes leads to hyperglycaemia and can lead to a number of diabetes related complications over time, such as increased risk of cardiovascular disease, diabetic retinopathy, kidney failure, and diabetic neuropathy. Current figures published by the World Health Organisation (WHO) indicate that 346 million people worldwide have diabetes. It is estimated that 5 % of all deaths worldwide each year are as a consequence of diabetes and its associated complications, with more than 80 % of these deaths occuring in low- and middle-income countries (WHO figures). It is projected that the number of diabetes related deaths will double between 2005 and 2030 (WHO figures), making diabetes a major burden on society and the health system. There is currently no cure for diabetes.

2. The pancreatic endocrine compartment

Blood glucose homeostasis is regulated by the pancreatic hormones, glucagon and insulin, which are secreted from the pancreatic islets of Langerhans. There are approximately one million islets in the adult human pancreas; this equates to about 2 % of total pancreatic mass. Pancreatic islets of Langerhans consist of four different cell types: glucagon producing α cells, insulin producing β cells, somatostatin producing δ cells and pancreatic polypeptide producing PP cells. The hormones insulin and glucagon are the principal islet hormones regulating blood glucose levels. Insulin is characteristic of the fed state and is released in response to hyperglycaemia. Glucagon is characteristic of the fasted state and is released in response to hypoglycaemia. Multiple other factors, including neural factors, regulate hormone release. The regulation of hormone release occurs at the level of the single hormone producing cell, the islet of Langerhans, the pancreas and the whole organism (Fig. 1).

Each islet of Langerhans is composed of about 2000 cells; typically, 60 % of these cells are insulin producing β cells. Insulin ensures that glucose is taken up and stored by peripheral tissues (4). Insulin secretion from islet β cells is regulated by nutrient availability,

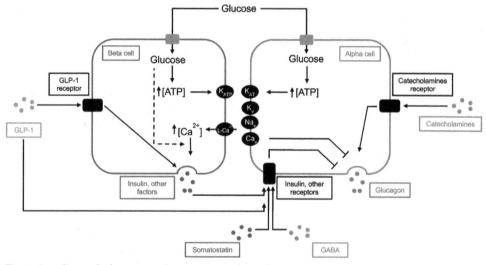

Fig. 1. **Insulin and glucagon release are regulated by nutrient availability, and a range of other factors.** Glucose entry in pancreatic α and β cells leads to increased ATP production, which in turn leads to closure of K-ATP (K_{ATP}) channels and membrane depolarisation. In the β cell this leads to opening of L-type voltage-gated calcium channels (L-Ca²⁺) and subsequent release of insulin via the triggering pathway. Glucose metabolism augment the calcium signal to enhance insulin secretion through the amplification pathways (dashed line). The incretin, glucagon-like peptide 1 (GLP-1), which is secreted from L-cells in the gut, can enhance insulin secretion from pancreatic β cells. In the α cells the presence of voltage gated sodium (Na_v) and potassium (K_v) channels keep voltage-gated calcium (Ca_v) channels closed, and leads to inhibition of glucagon release when extracellular glucose concentrations are high. Insulin and other factors secreted by the pancreatic β cell, e.g. zinc ions and gamma aminobutyric acid (GABA), somatostatin (secreted by pancreatic δ cells), and catecholamines all play a part in the regulation of glucagon release (42;220-224).

neurotransmitters, and hormones. β cells are electrically excitable and transduce variations in circulating glucose concentrations into secretory signals through changes in their own metabolic state. β cells are equipped with a high capacity glucose transporter (5-7) and the high K_m type IV hexokinase (or glucokinase) (8;9). Thus, glucose phosphorylation is a rate limiting step in glucose metabolism in the β cell and glucokinase is generally known as the β cell glucose sensor (10). However, other steps in the lower glycolytic pathway are also important in regulating glucose metabolism (11-13; reviewed in 14). Ultimately, the resultant increase in cellular ATP leads to changes in the activity of the K_{ATP} and L-type voltage gated Ca^{2+} channels, influx of calcium into the cell and, subsequently, insulin secretion (15-19; Fig. 1).

2.1 Insulin release from the islet β cell

Glucose stimulated insulin secretion from islets of Langerhans is biphasic (20), with a rapid first phase and more sustained second phase. The first phase is activated by the triggering pathway- K_{ATP} channel closure following increased glucose influx into pancreatic β cells, as described above. The second phase involves the activation of amplification pathways, also called K_{ATP} channel independent mechanisms (18;21), whereby the increase in intracellular calcium concentrations following K_{ATP} channel closure leads to changes in the sensitivity of the secretory machinery (18;22). Protein kinase A and C (21), AMP activated protein kinase (AMPK; 23-27), and insulin sensitive protein kinases such as protein kinase B and p70S6 kinase (28-33), are involved in regulating the amplifying pathways. The phosphoinositide 3 kinases (PI3Ks) are also thought to be important in the regulation of insulin secretion and synthesis (34-40). The roles of some of these protein kinases will be discussed in section 3.

2.2 Glucagon release from the islet α cell

Glucagon is released from pancreatic α cells in response to low blood glucose concentration, amongst other stimuli, to maintain blood glucose levels in the fasted state (Fig. 1). Elevated glucose concentrations (> 3.5 mM) normally suppress the release of glucagon from pancreatic α cells and dysregulation of this process is a feature of both types 1 and 2 diabetes (41;42). The lack of a counter regulatory response leads to potential danger from episodes of hypoglycaemia and is a limiting factor for good glycaemic management in diabetes (43).

Many of the proteins that are involved in glucose sensing in the β cell, such as glucokinase (44), are also present in the α cell and glucose is able to raise intracellular ATP concentrations in the α cells (45). Both intrinsic (45;46) and extrinsic (42;47;48) mechanisms for the regulation of glucagon secretion have been proposed. Recent evidence suggest that two fuel sensitive protein kinase, PAS-domain containing protein kinase (PASK) and AMPK, may be involved in the regulation of glucagon release and may have a role in the pathophysiology of type 2 diabetes (49; see section 3.1 and 3.2).

3. Glucose sensing and hormone production/release

A number of protein kinases are known to play crucial roles in the regulation of islet α and β cell function. These protein kinases represent potential drug targets for the treatment of type 2 diabetes, as glucose sensitivity and secretory capacity of the pancreatic islets may be improved (or restored) through the manipulation of the action of these proteins.

3.1 AMP activated protein kinase (AMPK)

AMPK is an evolutionarily conserved fuel-sensitive protein kinase that plays a role in glucose homeostasis (23-25;50-53). It is a target of the glucose-lowering drugs, metformin and thiazolidinediones, which act to improve insulin sensitivity in insulin-sensitive tissues such as muscle and liver (54). However, the long term effects of these drugs on β cell survival and function are less clear (53).

AMPK is a heterotrimeric protein consisting of an α catalytic subunit (two isoforms, α1 and 2), a β (two isoforms, β1 and 2) scaffold subunit, and a gamma (three isoforms, gamma 1, 2, 3) regulatory subunit (55;56). It is activated by increased intracellular AMP concentrations, i.e. at times of fuel deprivation. Salt and colleagues (23) showed that both AMPK catalytic subunits are present in the clonal rat pancreatic β cell line, INS-1 (57), and that AMPK activity is regulated by extracellular glucose concentrations in this cell type (23). Moreover, they demonstrated that insulin secretion and AMPK activity were inversely related (23). AICA riboside (5-aminoimidazole-4-carboxamide riboside; AICAR) (23;25), an activator of AMPK, and metformin (58), inhibited glucose stimulated insulin secretion in both clonal β cell lines and primary rat islets. Similarly, overexpression of a constitutively active form of AMPK in clonal β cells (25) and primary islets (59) led to impaired β cell function, due in part to the inhibition of secretory granule transport to the cell surface (27). In contrast, overexpression of a dominant negative form of AMPK led to increased insulin secretion at non-premissive glucose concentrations (25) without affecting release at elevated glucose concentrations (60).

In support of the above findings, it was recently demonstrated that transgenic mice over-expressing constitutively active AMPK specifically in pancreatic β cells were glucose intolerant and displayed defective insulin secretion (26). Mice in which expression of both AMPK catalytic subunits was ablated selectively in pancreatic β cells displayed defective glucose homeostasis due to defective insulin secretion in response to hyperglycaemia *in vivo* (26). Ablation of the expression of AMPK catalytic subunits in the β cell was protective against the deleterious effects of exposure to a high fat diet on insulin secretion (26).

Aside from insulin secretion, data from work in clonal cell lines indicate that AMPK may also be involved in the regulation of insulin gene expression in response to changing glucose concentrations (24;25). Thus, inhibition of the AMPK α2 catalytic subunit (24), which displays substantial nuclear localisation (24;57), led to increased insulin gene expression, while inhibition of the AMPK α1 catalytic subunit, which is predominantly cytosolic (57), had no effect on insulin gene expression (24).

Recently, Leclerc and colleagues showed that AMPK activity is modulated by glucose in a mouse clonal α cell line, αTC1-9 cells (61). Overexpression of constitutively active AMPK in αTC1-9 cells and specifically in pancreatic α cells in primary mouse islets of Langerhans led to activation of glucagon release at inhibitory glucose concentrations (61). Activation of AMPK with metformin, phenformin (62) and a selective activator of AMPK, A-769662 (63-65), in αTC1-9 cells led to increased AMPK activity and stimulated glucagon secretion at both permissive and non-permissive glucose concentrations (61). In contrast, overexpression of dominant negative AMPK and/or the AMPK inhibitor, compound C (66), inhibited glucagon release (61).

Thus, AMPK is involved in the regulation of both the release and biosynthesis of insulin in pancreatic β cells, and glucagon release from pancreatic α cells, in response to glucose challenge. Drugs that specifically inhibit AMPK activity in the pancreatic islet are likely to be useful in the treatment of diabetes by performing the dual role of increasing insulin secretion and inhibiting glucagon secretion.

3.2 PAS domain containing protein kinase (PASK)

Whilst homologous genes are common in prokaryotes, there is only one known mammalian PASK (67;68). The enzyme has one well-defined ligand-binding domain with potential for drug-targeting (67). We (49;69) and others (70;71) have shown that PASK is important for energy-sensing and maintenance of normal cellular energy balance in mammalian systems. Thus, PASK/Pask is expressed in human and rodent pancreatic islets of Langerhans and its expression is regulated by glucose (49;69;71). PASK is expressed in both α and β cells in human pancreatic islets (49). Importantly, PASK expression is lower in the β cells of patients with type 2 diabetes in comparison to β cells of non-diabetic individuals (49). Very recently, an activating mutation in PASK was identified which is associated with non-autoimmune early-onset diabetes in humans (72). However, this mutant did not fully co-segregate with the disease and appears to serve as a modifier for a separate disease-causing mutation in another gene. Thus, the mutated PASK (G1117E) has ~25 % higher activity than wild-type PASK and overexpression of this kinase variant led to a left-shift in the glucose response in mouse pancreatic islets. As a result, glucose-stimulated insulin secretion and insulin gene expression were increased at normally non-permissive (3 mM) glucose concentrations (72).

Pask activity is regulated by glucose and the enzyme is involved in the regulation of glucose-induced preproinsulin and pancreatic duodenum homeobox-1 (PDX-1) gene expression in the mouse pancreatic β cell line, MIN6 (69;71). Recently, PASK was implicated in the regulation of lipogenic gene expression (70) and might, therefore, influence glucose signaling through lipid intermediates as proposed for glucose-induced insulin secretion (73). Pask null (Pask-/-) mice (74) have normal glucose tolerance (49;70;74) and lower plasma insulin content than control littermates (49;70), but increased insulin sensitivity in peripheral tissues (70).

Glucose-stimulated insulin secretion from Pask-/- islets was variously shown to be not different (74) or lower (70) vs control islets. However, the lack of corresponding total islet insulin measurements in these studies made it difficult to evaluate islet secretory capacity (70). In our hands, Pask-/- islets were able to release insulin in response to glucose (49) but islet insulin content (49) and, thus, the amount of insulin released, was lower, in agreement with (70). Our data also indicate that the total pancreatic insulin content in Pask-/- is lower than control mice (49).

Pask-/- mice, when maintained on a high fat diet (HFD), develop glucose intolerance (70). Furthermore, inhibition of insulin gene expression by palmitate was reversed by PASK over-expression in MIN6 β cells (71), suggesting a role for PASK in protection against lipotoxicity. Thus, PASK appears to exert a protective effect in mature β cells and aberrant PASK expression and function may play a role in the development of diabetes.

Our recent data (49) indicate that PASK may regulate glucagon secretion in rodent and human pancreatic α cells. Eight-week old Pask-/- mice displayed higher plasma glucose after

16 h of fasting than wild-type littermate controls, but normal glucose tolerance after intraperitoneal glucose injection. After fasting, plasma glucagon was higher in Pask-/- mice than littermate controls. This increased glucagon concentration may account for the increased levels of blood glucose after the 16 h fast. This is physiologically relevant as long-term elevated plasma glucagon and glucose is a feature of type 2 diabetes (75).

The regulation of glucagon secretion by glucose was also impaired in Pask-/- islets (49). Interestingly, we observed a slight inhibitory effect of glucose on glucagon secretion, suggesting either a cell autonomous role for PASK in the α cell and/or reflecting altered secretion of regulatory factors from neighbouring β cells, e.g. insulin (49). Forced changes in PASK content affected the regulation of glucagon secretion by glucose in αTC1-9 cells (a mouse clonal α cell line) and human islets (49). RNAi-mediated silencing of Pask expression in αTC1-9 cells led to constitutive release of glucagon (49), while over-expression of PASK in αTC1-9 cells or in human islets led to inhibition of glucagon secretion, suggesting that PASK may be involved in glucose-sensing in pancreatic α cells (49). Inhibition of glucagon secretion by insulin was not affected in Pask silenced αTC1-9 cells indicating that the insulin signalling pathway was intact (49). Interestingly, there was still an apparent effect of glucose on glucagon secretion in human islets over-expressing PASK, possibly due to insulin release from β cells. Thus, the dysregulation of glucagon secretion in Pask-/- islets may be due, in part, to the decrease in insulin secretion.

To identify the mechanism(s) through which PASK may regulate glucagon secretion, we measured the expression of a number of potential target genes in Pask-/- islets (49) and αTC1-9 cells (49). The expression of both the insulin and Pdx-1 genes was impaired in Pask-/- islets (49;69;71). Thus, the decrease in insulin release may be due to decreased insulin synthesis and/or islet number in Pask-/- pancreata (49). Preproglucagon gene expression was increased, although total glucagon protein content was unaltered (49), consistent with the relatively slow turnover of mature glucagon. AMPKα2, but not AMPKα1, gene expression was increased by loss of Pask expression (49). This is an interesting observation since AMPK activity is glucose-responsive in pancreatic α and β cells (24;61) and regulates insulin (25;27) and glucagon (61) release.

Interestingly, we also observed that E13.5 rat pancreatic epithelial explants (76) in which PASK gene expression was silenced also display similar gene expression changes following culture to allow the development of endocrine cells (49) indicating that loss of PASK gene expression may have effects on pancreatic development that lead to the dysregulation of glucagon release. These data, and those from pancreatic β cells (69;71;72), suggest that changes in PASK activity may be important for appropriate glucose signalling in both α and β cells.

3.3 Protein kinase A (PKA)

Glucose administered via the gastointestinal tract leads to a greater induction of glucose-stimulated insulin secretion than a comparable intravenous adminstration of glucose (77;78) due to stimulation of the release of the incretin hormones, glucagon-like peptide 1 (GLP-1) and gastric inhibitory peptide (GIP), from intestinal L- and K-cells respectively (79;80). Potentiation of glucose stimulated insulin secretion only occurs at permissive glucose concentrations (81), making the use of incretins attractive in the treatment of diabetes (80). Both hormones potentiate glucose stimulated insulin secretion (82) via binding to their cognate G-protein coupled receptors on pancreatic β cells, activating adenylyl cyclase and thereby increasing cytosolic cyclic AMP (cAMP) levels (83-85). cAMP regulates insulin secretion, in part, by inducing the phosphorylation of proteins involved in the secretory process by PKA (86-88). For example, there is evidence that PKA activation is involved in controlling vesicle exocytosis by regulating Munc 13-1 function (89;90).

3.4 Protein Kinase C (PKC)

There is evidence both for and against the involvement of PKC in stimulus secretion coupling in the β cell (reviewed in (91;92). Early studies using pharmacological modulation of PKC activity in β cells gave contradictory results (91;92). Conventional PKCs are recruited to the plasma membrane by calcium-dependent binding of the C2 domain to the phospholipids, which is potentiated by the binding of diacyglycerol to the C1 domains (93). Conventional, novel and atypical PKCs are found in pancreatic islet cells (94-97), and PKC activity is present in primary pancreatic islets (98) and clonal β cell lines (99;100). It was proposed that the increase in cytosolic calcium concentrations, following exposure of pancreatic β cells to high glucose concentrations, leads to increased diacyglycerol (DAG) production which leads to the activation of PKC, and the translocation of PKC to the plasma membrane leading to potentiation of insulin release (92;101). DAG is produced by activated phospholipase C (PLC) from phosphatidyl-4,5-bisphosphate, and PLC activity in pancreatic β cells has been shown to occur in a dose-dependent manner in parallel to physiologically relevant increases in glucose that lead to insulin secretion (102-104).

The use of PKC isoform and green fluorescent protein chimeras, coupled with the use of total internal reflection fluorescence (TIRF) microscopy, demonstrated that elevated glucose concentrations led to complex oscillatory translocation of the conventional PKC, PKCβ2, to the plasma membrane, in primary β cells and clonal β cell lines, in response to the formation of calcium microdomains following transient depolarisation of the plasma membrane (101). Moreover, PKCβ2 was shown to migrate to the surface of secretory vesicles, suggesting that the process of vesicle fusion may be regulated locally (101) by this kinase.

3.5 Phosphoinositide 3 Kinases (PI3Ks)

The autocrine feedback action of insulin on pancreatic β cell function is a subject of much debate. β cells secrete insulin at basal glucose concentrations and increase the level of secretion in response to a glucose challenge. Thus, much of the discussion has been on whether the insulin signalling pathway in β cells is desensitised as these cells are constantly exposed to the hormone. Recent studies, using approaches that circumvent some of the confounding factors in earlier studies, have provided evidence that insulin acts as a positive regulator of its own secretion and synthesis in pancreatic β cells, and of β cell mass and survival (30;32;34-36;38;39;105-130). Activation of the insulin and insulin-like growth factor 1 (IGF-1) receptor (IGF-1R), through insulin and/or IGF-1 binding to these receptors, leads to activation of downstream signalling cascades including, those involving the PI3Ks (35;36;110;111;116;126;130), resulting in the moderation of β cell function. There are three classes of mammalian PI3Ks, class I-III (131). Class Ia and II PI3Ks have been reported to be activated by insulin and will be reviewed in the following subsections. The class I PI3Ks generate phosphatidyl 3,4,5-trisphosphate (132), while the class II PI3Ks generate phosphatidyl 3-phosphate (PI3P) *in vivo* (133;134), both of which interact with distinct molecules. Therefore, the activation of class I and II PI3Ks result in the activation of distinct signaling cascades.

3.5.1 PI3Ks and insulin gene expression

There is evidence in the literature suggesting that insulin regulates the expression of its own gene via activation of Class Ia PI3K (PI3K-1a) (35;36). Of a number of transcription factors that regulate the expression of the insulin gene, the action of the transcription factors pancreatic duodenal homeobox-1 (PDX-1) (110;135-141) and FoxO-1 (142) are thought to be regulated by insulin in a PI3K-1a dependent manner.

Insulin gene expression is upregulated in response to increasing glucose concentrations in a PI3K-1a-dependent manner (30;35;36;110). The increase in insulin gene expression in response to elevated glucose concentrations is due, at least in part, to the the the activation of PI3K-1a by secreted insulin (35;36;107;143;144). Thus, PDX-1 translocates from the cytosol to the nucleus in response to an increase in circulating glucose concentrations (110;144;145) and binds to a region upstream of the insulin gene called the A3 box (139). This binding is upregulated as glucose concentrations are increased in the near physiological range (140;141), and in response to insulin (146), to activate insulin gene expression. The expression of the PDX-1 gene itself is regulated by insulin (147), further implicating insulin and PI3K-1a in the feed-forward regulation of insulin gene expression.

Activation of PI3K-1a by insulin also leads to enhanced expressionof glucose/insulin-responsive genes through the removal of transcriptional repressors. For example, the transcription factor, FoxO1, is phosphorylated at Ser-256 in response to activation of PI3Ks and PKB by insulin (142;148), translocates from the nucleus to the cytosol (149;150) and is degraded (151). The degradation of FoxO1 leads to enhanced expression of glucose/insulin-responsive genes (152) such as the L-type pyruvate kinase (36), insulin 2 (*Ins2*) and *Pdx-1* genes in pancreatic β cells (142).

3.5.2 PI3Ks and insulin secretion

Data is available in the published literature suggesting that insulin has an inhibitory, activatory or no role in the regulation of its own secretion (153-160). Whilst most studies have focussed on the action of the class I PI3Ks on insulin secretion, recent data implicated the class II PI3K, PI3K-C2α, in the regulation of insulin secretion (126;130).

PI3K-C2α gene expression was shown to be down-regulated in islets of Langerhans of patients with type 2 diabetes (130). Leibiger and colleagues showed that PI3K-C2α is able to enhance glucose-stimulated insulin release in a feed-forward (126) mechanism, in response to stimuli from insulin itself, via the activation of PKB by PI3P. Subsequently, it was shown that PI3K-C2α may have a role in the late stages of insulin granule release through its action on the synaptosomal-associated protein of 25 kDA (SNAP25), which is independent of nutrient control (130). Insulin granules dock at the plasma mebrane via the interaction of SNAP25 within a protein complex (161-163), with granule fusion and insulin release occuring post proteolysis of SNAP25 (164). PI3K-C2α was shown to regulate the degradation rate of SNAP2, thereby controlling insulin granule fusion with the plasma membrane (130).

3.6 Homeodomain interacting protein kinase 2 (HIPK2)

Homeodomain interacting protein kinase 2 (HIPK2) belongs to the family of homeodomain interacting protein kinases, which was originally identified as binding partners of the homeodomain protein neurokinin-3 (NK-3) (165). Studies have shown that members of the HIPK family interact with, phosphorylate and modulate the function of other homeodomain containing proteins and transcription factors (166-170), indicating that the HIPKs may have an important role in the control of transcription. Recently it was shown that HIPK2 is expressed in the developing pancreatic epithelium from E12 to E15 and that its expression is confined preferentially to pancreatic endocrine cells later in development (171). Phosphorylation of the transcription factor, PDX-1, in the C-terminus by HIPK2, possibly at Ser-214 (172), was reported to increase the stability and transcriptional activity of PDX-1 (171). Our own data indicate that HIPK2 posphorylates PDX-1 at Ser-269 in the C-terminal portion of PDX-1 in pancreatic β cells *in vivo* and that phosphorylation at this site leads to nuclear exclusion of PDX-1 and a decrease in PDX-1 target gene expression (172). As PDX-1 has been shown to directly bind to and regulate the promoter activity of various β cell genes, e.g. insulin (110;135;140;144-146), glucose transporter 2 (GLUT2;173), glucokinase (GCK)(174;175), and islet amyloid polypeptide (176;177), HIPK2 represents an important regulator of β cell gene expression during development and in adult β cells.

4. β cell survival, growth and proliferation

Maintenance of an adequate functional β cell mass is a potential therapeutic target for diabetes. In this section we will look at the evidence that indicate protein kinases such as AMPK, Serine/threonine protein kinase 11 (STK11/LKB1), mammalian target of rapamycin (mTOR), and protein kinases in the Wnt signalling pathway may be involved in the regulation of β cell proliferation.

4.1 AMPK and LKB1

In section 3.1, we discussed the role of AMPK in the regulation of pancreatic β cell function. Recent data indicate that AMPK may also have a role in the maintenance of adequate β cell mass. Activation of AMPK has been shown to lead to decreased β cell viability (178;179), potentially through its action on the cell cycle regulator, p53 (180). Mice in which the expression of both AMPK catalytic subunits was ablated selectively in pancreatic β cells have normal β cell mass but smaller β cells and pancreatic islets (26). Islets of Langerhans in which the expression of both AMPK catalytic subunits was ablated did not display apparent differences in the ratio of α to β cells or in islet architecture. β cell proliferation was enhanced in islets from mice in which the expression of both AMPK catalytic subunits was ablated selectively in pancreatic β cells (26). In contrast, islets of Langerhans from mice in which the expression of an AMPK upstream kinase liver kinase B1 (LKB1) was ablated, had increased islet and β cell size (181) and altered islet architecture (181-184). Thus, the number of large islets, which may account for as much as 50 % of the total pancreatic β cell mass in normal pancreata (185), was increased in pancreata from mice in which LKB1 expression was selectively ablated in the β cell (181). These data indicate that AMPK and LKB1 play distinct roles in the control of islet development and cell proliferation and may impact on the development of pharmacological reagents targetting these pathways for the treatment of diabetes. In particular, inhibition of LKB1, or its downstream targets, may be a means by which β cell mass may be increased for the treatment of diabetes.

4.2 mTOR

Activation of the PI3K-PKB pathway by growth factors such as insulin and IGF-1, and subsequent activation of mammalian target of rapamycin (mTOR), is involved in β cell compensation in animals with genetic or high-fat diet induced insulin resistance (122;186). mTOR is an important nutrient sensor that plays a central role in the regulation of cellular metabolism, growth, proliferation and apoptosis (187-190). Signalling by (mTOR) to eukaryotic initiation factor 4-binding protein-1 (4E-BP1) and ribosomal S6 kinase (S6K) was enhanced in islets of Langerhans from mice in which LKB1 expression was specifically ablated in the β cell (181). Likewise, there is evidence for crosstalk between mTOR and AMPK as a regulatory pathway that couples cellular fuel availability to β cell apoptosis (58;59;191). S6K1 knockout mice are glucose intolerant, despite increased insulin sensitivity, which is associated with depletion of pancreatic insulin content, hypoinsulinaemia and reduced β cell mass (192;193), indicating that S6K1 is required for β cell growth and function. In addition, crosstalk between the mTOR and JNK pathways (194) is thought to regulate β cell survival through action on FoxO1 (195;196), whereby activation of FoxO1 in β cells protects the cells from oxidative stress by reducing cellular metabolic activity and energy-consuming processes, e.g. proliferation, and cell-specific function, e.g. insulin secretion (195;196).

4.3 Wnt signalling

The Wnt proteins are a family of cysteine-rich glycoproteins involved in intracellular signalling during vertebrate development. Activation of Wnt signalling leads to the expression of genes that are involved in promoting stem cell fate and inhibiting cell differentiation (197). One of the prominent biological phenomena controlled by Wnt signaling is the expansion of cells with predefined fates (198;199). It has previously been shown to be involved in the regulation of pancreatic development at all stages during organogenesis from specification to maintenance of normal function (200-206;206-211). Thus, tight temporal regulation of the Wnt signalling pathway is required for normal development.

Activation of the Wnt signalling pathway was shown to upregulate β cell proliferation in mouse islets (208;212;213) with upregulation of cell cycle genes that have been shown to regulate β cell proliferation (214) such as cyclin D1 and D2, and cyclin-dependent kinase 4 (CDK4) (208;213). The Wnt signalling pathway was also shown to be involved in the neogenesis of human β cells *in vitro* (215).

It was recently demonstrated that the incretin, GLP-1, induced β cell proliferation via activation of the Wnt signaling pathway (216). GLP-1 has previously been shown to increase β cell proliferation and survival (217-219). Liu and colleagues showed that Wnt signaling can be activated by downstream events from GLP-1 receptor activation in a Protein kinase B (PKB) and PKA-dependent manner (216).

5. Prospects for the development of treatment

Type 2 diabetes is fast becoming a major global problem and understanding the signalling molecules that regulate the maintenance of glucose homeostasis, particularly those that regulate pancreatic islet function, could lead to better therapeutic intervention for the disease. Increasing functional β cell mass is a particularly promising strategy: although islet transplantation is an effective means to restore glucose homeostasis, the lack of transplantable material makes this a non-viable treatment module for the masses. Thus, strategies that can lead to the generation and proliferation of β cells *in vivo* and/or *in vitro* may be important for the treatment of the disease.

6. Acknowledgements

Work in the author's laboratory is funded by the European Foundation for the Study of Diabetes, Diabetes U.K. and the Juvenile Diabetes Research Foundation.

7. References

[1] Henderson L, Gregory J, Irving K, and Swan G. The national diet and nutrition survey: adults aged 19 to 64 years. Volume 2. Energy, protein, carbohydrate , Fat and alcohol intake. HMSO: Norwich, 2003.

[2] Wright JD, Wang CY, Kennedy-Stephenson J, Ervin RB (2003) Dietary intake of ten key nutrients for public health, United States: 1999-2000. Adv.Data 1-4

[3] Burelle Y, Lamoureux MC, Peronnet F, Massicotte D, Lavoie C (2006) Comparison of exogenous glucose, fructose and galactose oxidation during exercise using 13C-labelling. Br.J.Nutr. 96: 56-61

[4] Ashcroft FM, Ashcroft SJM (1992) Insulin, Molecular Biology to Pathology. Oxford University Press, Oxford,

[5] Bell GI, Kayano T, Buse JB, et al (1990) Molecular biology of mammalian glucose transporters. Diabetes Care 13: 198-208

[6] Thorens B, Sarkar HK, Kaback HR, Lodish HF (1988) Cloning and functional expression in bacteria of a novel glucose transporter present in liver, intestine, kidney, and beta-pancreatic islet cells. Cell 55: 281-290

[7] Schuit FC (1997) Is GLUT2 required for glucose sensing? Diabetologia 40: 104-111

[8] Matschinsky FM, Meglasson M, Ghosh A, et al (1986) Biochemical design features of the pancreatic islet cell glucose-sensory system. Adv.Exp.Med.Biol. 211: 459-469

[9] Iynedjian PB (1993) Mammalian glucokinase and its gene. Biochem.J. 293 (Pt 1): 1-13

[10] Meglasson MD, Matschinsky FM (1986) Pancreatic islet glucose metabolism and regulation of insulin secretion. Diabetes Metab Rev. 2: 163-214

[11] Sekine N, Cirulli V, Regazzi R, et al (1994) Low lactate dehydrogenase and high mitochondrial glycerol phosphate dehydrogenase in pancreatic beta-cells. Potential role in nutrient sensing. J.Biol.Chem. 269: 4895-4902

[12] Schuit F, De VA, Farfari S, et al (1997) Metabolic fate of glucose in purified islet cells. Glucose-regulated anaplerosis in beta cells. J.Biol.Chem. 272: 18572-18579

[13] Berman HK, Newgard CB (1998) Fundamental metabolic differences between hepatocytes and islet beta-cells revealed by glucokinase overexpression. Biochemistry 37: 4543-4552

[14] Rutter GA (2001) Nutrient-secretion coupling in the pancreatic islet beta-cell: recent advances. Mol.Aspects Med. 22: 247-284

[15] Rorsman P (1997) The pancreatic beta-cell as a fuel sensor: an electrophysiologist's viewpoint. Diabetologia 40: 487-495

[16] Aguilar-Bryan L, Bryan J (1999) Molecular biology of adenosine triphosphate-sensitive potassium channels. Endocr.Rev. 20: 101-135

[17] Seino S, Iwanaga T, Nagashima K, Miki T (2000) Diverse roles of K(ATP) channels learned from Kir6.2 genetically engineered mice. Diabetes 49: 311-318

[18] Henquin JC (2000) Triggering and amplifying pathways of regulation of insulin secretion by glucose. Diabetes 49: 1751-1760

[19] Ashcroft FM (2007) The Walter B. Cannon Physiology in Perspective Lecture, 2007. ATP-sensitive K+ channels and disease: from molecule to malady. Am.J.Physiol Endocrinol.Metab 293: E880-E889

[20] Curry DL, Bennett LL, Grodsky GM (1968) Dynamics of insulin secretion by the perfused rat pancreas. Endocrinology 83: 572-584

[21] Aizawa T, Komatsu M, Asanuma N, Sato Y, Sharp GW (1998) Glucose action 'beyond ionic events' in the pancreatic beta cell. Trends Pharmacol.Sci. 19: 496-499

[22] Jonas JC, Gilon P, Henquin JC (1998) Temporal and quantitative correlations between insulin secretion and stably elevated or oscillatory cytoplasmic Ca2+ in mouse pancreatic beta-cells. Diabetes 47: 1266-1273

[23] Salt IP, Johnson G, Ashcroft SJ, Hardie DG (1998) AMP-activated protein kinase is activated by low glucose in cell lines derived from pancreatic beta cells, and may regulate insulin release. Biochem.J. 335 (Pt 3): 533-539

[24] da Silva Xavier G, Leclerc I, Salt IP, et al (2000) Role of AMP-activated protein kinase in the regulation by glucose of islet beta cell gene expression. Proc.Natl.Acad.Sci.U.S.A 97: 4023-4028

[25] da Silva Xavier G, Leclerc I, Varadi A, Tsuboi T, Moule SK, Rutter GA (2003) Role for AMP-activated protein kinase in glucose-stimulated insulin secretion and preproinsulin gene expression. Biochem.J. 371: 761-774

[26] Sun G, Tarasov AI, McGinty J, et al (2010) Ablation of AMP-activated protein kinase alpha1 and alpha2 from mouse pancreatic beta cells and RIP2.Cre neurons suppresses insulin release in vivo. Diabetologia 53: 924-936

[27] Tsuboi T, da Silva Xavier G, Leclerc I, Rutter GA (2003) 5'-AMP-activated protein kinase controls insulin-containing secretory vesicle dynamics. J.Biol.Chem. 278: 52042-52051

[28] Rutter GA (1999) Insulin secretion: feed-forward control of insulin biosynthesis? Curr.Biol. 9: R443-R445

[29] Leibiger B, Leibiger IB, Moede T, et al (2001) Selective insulin signaling through A and B insulin receptors regulates transcription of insulin and glucokinase genes in pancreatic beta cells. Mol.Cell 7: 559-570

[30] Leibiger B, Wahlander K, Berggren PO, Leibiger IB (2000) Glucose-stimulated insulin biosynthesis depends on insulin-stimulated insulin gene transcription. J.Biol.Chem. 275: 30153-30156

[31] Khan FA, Goforth PB, Zhang M, Satin LS (2001) Insulin activates ATP-sensitive K(+) channels in pancreatic beta-cells through a phosphatidylinositol 3-kinase-dependent pathway. Diabetes 50: 2192-2198

[32] Aspinwall CA, Lakey JR, Kennedy RT (1999) Insulin-stimulated insulin secretion in single pancreatic beta cells. J.Biol.Chem. 274: 6360-6365

[33] Kulkarni RN, Bruning JC, Winnay JN, Postic C, Magnuson MA, Kahn CR (1999) Tissue-specific knockout of the insulin receptor in pancreatic beta cells creates an insulin secretory defect similar to that in type 2 diabetes. Cell 96: 329-339

[34] Kulkarni RN, Bruning JC, Winnay JN, Postic C, Magnuson MA, Kahn CR (1999) Tissue-specific knockout of the insulin receptor in pancreatic beta cells creates an insulin secretory defect similar to that in type 2 diabetes. Cell 96: 329-339

[35] Leibiger IB, Leibiger B, Moede T, Berggren PO (1998) Exocytosis of insulin promotes insulin gene transcription via the insulin receptor/PI-3 kinase/p70 s6 kinase and CaM kinase pathways. Mol.Cell 1: 933-938

[36] da Silva Xavier G, Varadi A, Ainscow EK, Rutter GA (2000) Regulation of gene expression by glucose in pancreatic beta -cells (MIN6) via insulin secretion and activation of phosphatidylinositol 3'-kinase. J.Biol.Chem. 275: 36269-36277

[37] Andreolas C, da Silva Xavier G, Diraison F, et al (2002) Stimulation of acetyl-CoA carboxylase gene expression by glucose requires insulin release and sterol regulatory element binding protein 1c in pancreatic MIN6 beta-cells. Diabetes 51: 2536-2545

[38] da Silva Xavier G, Qian Q, Cullen PJ, Rutter GA (2004) Distinct roles for insulin and insulin-like growth factor-1 receptors in pancreatic beta-cell glucose sensing revealed by RNA silencing. Biochem.J. 377: 149-158

[39] Kulkarni RN (2002) Receptors for insulin and insulin-like growth factor-1 and insulin receptor substrate-1 mediate pathways that regulate islet function. Biochem.Soc.Trans. 30: 317-322

[40] Burks DJ, White MF (2001) IRS proteins and beta-cell function. Diabetes 50 Suppl 1: S140-S145

[41] Unger RH (1985) Glucagon physiology and pathophysiology in the light of new advances. Diabetologia 28: 574-578

[42] Gromada J, Franklin I, Wollheim CB (2007) Alpha-cells of the endocrine pancreas: 35 years of research but the enigma remains. Endocr.Rev. 28: 84-116

[43] Cryer PE (2002) Hypoglycaemia: the limiting factor in the glycaemic management of Type I and Type II diabetes. Diabetologia 45: 937-948

[44] Heimberg H, De VA, Moens K, et al (1996) The glucose sensor protein glucokinase is expressed in glucagon-producing alpha-cells. Proc.Natl.Acad.Sci.U.S.A 93: 7036-7041

[45] Ravier MA, Rutter GA (2005) Glucose or insulin, but not zinc ions, inhibit glucagon secretion from mouse pancreatic alpha-cells. Diabetes 54: 1789-1797

[46] Rorsman P, Salehi SA, Abdulkader F, Braun M, MacDonald PE (2008) K(ATP)-channels and glucose-regulated glucagon secretion. Trends Endocrinol.Metab 19: 277-284

[47] Ishihara H, Maechler P, Gjinovci A, Herrera PL, Wollheim CB (2003) Islet beta-cell secretion determines glucagon release from neighbouring alpha-cells. Nat.Cell Biol. 5: 330-335

[48] Zhou H, Zhang T, Harmon JS, Bryan J, Robertson RP (2007) Zinc, not insulin, regulates the rat alpha-cell response to hypoglycemia in vivo. Diabetes 56: 1107-1112

[49] da Silva Xavier G, Farhan H, Kim H, et al (2011) Per-arnt-sim (PAS) domain-containing protein kinase is downregulated in human islets in type 2 diabetes and regulates glucagon secretion. Diabetologia 54: 819-827

[50] Shaw RJ, Lamia KA, Vasquez D, et al (2005) The kinase LKB1 mediates glucose homeostasis in liver and therapeutic effects of metformin. Science 310: 1642-1646

[51] Towler MC, Hardie DG (2007) AMP-activated protein kinase in metabolic control and insulin signaling. Circ.Res. 100: 328-341

[52] Kurth-Kraczek EJ, Hirshman MF, Goodyear LJ, Winder WW (1999) 5' AMP-activated protein kinase activation causes GLUT4 translocation in skeletal muscle. Diabetes 48: 1667-1671

[53] Rutter GA, Leclerc I (2009) The AMP-regulated kinase family: enigmatic targets for diabetes therapy. Mol.Cell Endocrinol. 297: 41-49

[54] Long YC, Zierath JR (2006) AMP-activated protein kinase signaling in metabolic regulation. J.Clin.Invest 116: 1776-1783

[55] Stapleton D, Woollatt E, Mitchelhill KI, et al (1997) AMP-activated protein kinase isoenzyme family: subunit structure and chromosomal location. FEBS Lett. 409: 452-456

[56] Hardie DG, Carling D (1997) The AMP-activated protein kinase--fuel gauge of the mammalian cell? Eur.J.Biochem. 246: 259-273

[57] Salt I, Celler JW, Hawley SA, et al (1998) AMP-activated protein kinase: greater AMP dependence, and preferential nuclear localization, of complexes containing the alpha2 isoform. Biochem.J. 334 (Pt 1): 177-187

[58] Leclerc I, Woltersdorf WW, da Silva Xavier G, et al (2004) Metformin, but not leptin, regulates AMP-activated protein kinase in pancreatic islets: impact on glucose-stimulated insulin secretion. Am.J.Physiol Endocrinol.Metab 286: E1023-E1031

[59] Richards SK, Parton LE, Leclerc I, Rutter GA, Smith RM (2005) Over-expression of AMP-activated protein kinase impairs pancreatic {beta}-cell function in vivo. J.Endocrinol. 187: 225-235

[60] Gleason CE, Lu D, Witters LA, Newgard CB, Birnbaum MJ (2007) The role of AMPK and mTOR in nutrient sensing in pancreatic beta-cells. J.Biol.Chem. 282: 10341-10351

[61] Leclerc I, Sun G, Morris C, Fernandez-Millan E, Nyirenda M, Rutter GA (2011) AMP-activated protein kinase regulates glucagon secretion from mouse pancreatic alpha cells. Diabetologia 54: 125-134

[62] Hardie DG (2006) Neither LKB1 nor AMPK are the direct targets of metformin. Gastroenterology 131: 973-975

[63] Hawley SA, Boudeau J, Reid JL, et al (2003) Complexes between the LKB1 tumor suppressor, STRAD alpha/beta and MO25 alpha/beta are upstream kinases in the AMP-activated protein kinase cascade. J.Biol. 2: 28

[64] Scott JW, van Denderen BJ, Jorgensen SB, et al (2008) Thienopyridone drugs are selective activators of AMP-activated protein kinase beta1-containing complexes. Chem.Biol. 15: 1220-1230

[65] Cool B, Zinker B, Chiou W, et al (2006) Identification and characterization of a small molecule AMPK activator that treats key components of type 2 diabetes and the metabolic syndrome. Cell Metab 3: 403-416

[66] Zhou G, Myers R, Li Y, et al (2001) Role of AMP-activated protein kinase in mechanism of metformin action. J.Clin.Invest 108: 1167-1174

[67] Rutter J, Michnoff CH, Harper SM, Gardner KH, McKnight SL (2001) PAS kinase: an evolutionarily conserved PAS domain-regulated serine/threonine kinase. Proc.Natl.Acad.Sci.U.S.A 98: 8991-8996

[68] Hofer T, Spielmann P, Stengel P, et al (2001) Mammalian PASKIN, a PAS-serine/threonine kinase related to bacterial oxygen sensors. Biochem.Biophys.Res.Commun. 288: 757-764

[69] da Silva Xavier G, Rutter J, Rutter GA (2004) Involvement of Per-Arnt-Sim (PAS) kinase in the stimulation of preproinsulin and pancreatic duodenum homeobox 1 gene expression by glucose. Proc.Natl.Acad.Sci.U.S.A 101: 8319-8324

[70] Hao HX, Cardon CM, Swiatek W, et al (2007) PAS kinase is required for normal cellular energy balance. Proc.Natl.Acad.Sci.U.S.A 104: 15466-15471

[71] Fontes G, Semache M, Hagman DK, et al (2009) Involvement of Per-Arnt-Sim Kinase and extracellular-regulated kinases-1/2 in palmitate inhibition of insulin gene expression in pancreatic beta-cells. Diabetes 58: 2048-2058

[72] Semplici F, Vaxillaire M, Fogarty S, et al (2011) A human mutation within the per-ARNT-sim (PAS) domain-containing protein kinase (PASK) causes basal insulin hypersecretion. J.Biol.Chem.

[73] Corkey BE, Deeney JT, Yaney GC, Tornheim K, Prentki M (2000) The role of long-chain fatty acyl-CoA esters in beta-cell signal transduction. J.Nutr. 130: 299S-304S

[74] Borter E, Niessen M, Zuellig R, et al (2007) Glucose-stimulated insulin production in mice deficient for the PAS kinase PASKIN. Diabetes 56: 113-117

[75] Li XC, Liao TD, Zhuo JL (2008) Long-term hyperglucagonaemia induces early metabolic and renal phenotypes of Type 2 diabetes in mice. Clin.Sci.(Lond) 114: 591-601

[76] Miralles F, Serup P, Cluzeaud F, Vandewalle A, Czernichow P, Scharfmann R (1999) Characterization of beta cells developed in vitro from rat embryonic pancreatic epithelium. Dev.Dyn. 214: 116-126

[77] Elrick H, Stimmler L, Hlad CJ, Rai Y (1964) Plasma insulin response to oral and intravenous glucose administration. J.Clin.Endocrinol.Metab 24: 1076-1082

[78] Perley MJ, Kipnis DM (1967) Plasma insulin responses to oral and intravenous glucose: studies in normal and diabetic sujbjects. J.Clin.Invest 46: 1954-1962

[79] Holst JJ, Gromada J (2004) Role of incretin hormones in the regulation of insulin secretion in diabetic and nondiabetic humans. Am.J.Physiol Endocrinol.Metab 287: E199-E206

[80] Baggio LL, Drucker DJ (2007) Biology of incretins: GLP-1 and GIP. Gastroenterology 132: 2131-2157

[81] Weir GC, Mojsov S, Hendrick GK, Habener JF (1989) Glucagonlike peptide I (7-37) actions on endocrine pancreas. Diabetes 38: 338-342

[82] Dupre J, Ross SA, Watson D, Brown JC (1973) Stimulation of insulin secretion by gastric inhibitory polypeptide in man. J.Clin.Endocrinol.Metab 37: 826-828

[83] Thorens B (1992) Expression cloning of the pancreatic beta cell receptor for the gluco-incretin hormone glucagon-like peptide 1. Proc.Natl.Acad.Sci.U.S.A 89: 8641-8645

[84] Usdin TB, Mezey E, Button DC, Brownstein MJ, Bonner TI (1993) Gastric inhibitory polypeptide receptor, a member of the secretin-vasoactive intestinal peptide receptor family, is widely distributed in peripheral organs and the brain. Endocrinology 133: 2861-2870

[85] Yasuda K, Inagaki N, Yamada Y, Kubota A, Seino S, Seino Y (1994) Hamster gastric inhibitory polypeptide receptor expressed in pancreatic islets and clonal insulin-secreting cells: its structure and functional properties. Biochem.Biophys.Res.Commun. 205: 1556-1562

[86] Seino S, Shibasaki T (2005) PKA-dependent and PKA-independent pathways for cAMP-regulated exocytosis. Physiol Rev. 85: 1303-1342

[87] Malaisse WJ, Malaisse-Lagae F (1984) The role of cyclic AMP in insulin release. Experientia 40: 1068-1074

[88] Prentki M, Matschinsky FM (1987) Ca2+, cAMP, and phospholipid-derived messengers in coupling mechanisms of insulin secretion. Physiol Rev. 67: 1185-1248

[89] Kwan EP, Gao X, Leung YM, Gaisano HY (2007) Activation of exchange protein directly activated by cyclic adenosine monophosphate and protein kinase A regulate common and distinct steps in promoting plasma membrane exocytic and granule-to-granule fusions in rat islet beta cells. Pancreas 35: e45-e54

[90] Kwan EP, Xie L, Sheu L, Ohtsuka T, Gaisano HY (2007) Interaction between Munc13-1 and RIM is critical for glucagon-like peptide-1 mediated rescue of exocytotic defects in Munc13-1 deficient pancreatic beta-cells. Diabetes 56: 2579-2588

[91] Jones PM, Persaud SJ, Howell SL (1991) Protein kinase C and the regulation of insulin secretion from pancreatic B cells. J.Mol.Endocrinol. 6: 121-127

[92] Wollheim CB, Regazzi R (1990) Protein kinase C in insulin releasing cells. Putative role in stimulus secretion coupling. FEBS Lett. 268: 376-380

[93] Nishizuka Y (1988) The molecular heterogeneity of protein kinase C and its implications for cellular regulation. Nature 334: 661-665

[94] Tian YM, Urquidi V, Ashcroft SJ (1996) Protein kinase C in beta-cells: expression of multiple isoforms and involvement in cholinergic stimulation of insulin secretion. Mol.Cell Endocrinol. 119: 185-193

[95] Onoda K, Hagiwara M, Hachiya T, Usuda N, Nagata T, Hidaka H (1990) Different expression of protein kinase C isozymes in pancreatic islet cells. Endocrinology 126: 1235-1240

[96] Arkhammar P, Juntti-Berggren L, Larsson O, et al (1994) Protein kinase C modulates the insulin secretory process by maintaining a proper function of the beta-cell voltage-activated Ca2+ channels. J.Biol.Chem. 269: 2743-2749

[97] Kaneto H, Suzuma K, Sharma A, Bonner-Weir S, King GL, Weir GC (2002) Involvement of protein kinase C beta 2 in c-myc induction by high glucose in pancreatic beta-cells. J.Biol.Chem. 277: 3680-3685

[98] Tanigawa K, Kuzuya H, Imura H, et al (1982) Calcium-activated, phospholipid-dependent protein kinase in rat pancreas islets of langerhans. Its possible role in glucose-induced insulin release. FEBS Lett. 138: 183-186

[99] Lord JM, Ashcroft SJ (1984) Identification and characterization of Ca2+-phospholipid-dependent protein kinase in rat islets and hamster beta-cells. Biochem.J. 219: 547-551

[100] Zawalich W, Brown C, Rasmussen H (1983) Insulin secretion: combined effects of phorbol ester and A23187. Biochem.Biophys.Res.Commun. 117: 448-455

[101] Pinton P, Tsuboi T, Ainscow EK, Pozzan T, Rizzuto R, Rutter GA (2002) Dynamics of glucose-induced membrane recruitment of protein kinase C beta II in living pancreatic islet beta-cells. J.Biol.Chem. 277: 37702-37710

[102] Best L, Malaisse WJ (1984) Nutrient and hormone-neurotransmitter stimuli induce hydrolysis of polyphosphoinositides in rat pancreatic islets. Endocrinology 115: 1814-1820

[103] Zawalich W (1990) Multiple effects of increases in phosphoinositide hydrolysis on islets and their relationship to changing patterns of insulin secretion. Diabetes Res. 13: 101-111

[104] Zawalich WS, Zawalich KC (1996) Regulation of insulin secretion by phospholipase C. Am.J.Physiol 271: E409-E416

[105] Rothenberg PL, Willison LD, Simon J, Wolf BA (1995) Glucose-induced insulin receptor tyrosine phosphorylation in insulin-secreting beta-cells. Diabetes 44: 802-809

[106] Harbeck MC, Louie DC, Howland J, Wolf BA, Rothenberg PL (1996) Expression of insulin receptor mRNA and insulin receptor substrate 1 in pancreatic islet beta-cells. Diabetes 45: 711-717

[107] Xu GG, Rothenberg PL (1998) Insulin receptor signaling in the beta-cell influences insulin gene expression and insulin content: evidence for autocrine beta-cell regulation. Diabetes 47: 1243-1252

[108] Porzio O, Federici M, Hribal ML, et al (1999) The Gly972-->Arg amino acid polymorphism in IRS-1 impairs insulin secretion in pancreatic beta cells. J.Clin.Invest 104: 357-364

[109] Xu GG, Gao ZY, Borge PD, Jr., Jegier PA, Young RA, Wolf BA (2000) Insulin regulation of beta-cell function involves a feedback loop on SERCA gene expression, Ca(2+) homeostasis, and insulin expression and secretion. Biochemistry 39: 14912-14919

[110] Rafiq I, da Silva Xavier G, Hooper S, Rutter GA (2000) Glucose-stimulated preproinsulin gene expression and nuclear trans-location of pancreatic duodenum homeobox-1 require activation of phosphatidylinositol 3-kinase but not p38 MAPK/SAPK2. J.Biol.Chem. 275: 15977-15984

[111] Aspinwall CA, Qian WJ, Roper MG, Kulkarni RN, Kahn CR, Kennedy RT (2000) Roles of insulin receptor substrate-1, phosphatidylinositol 3-kinase, and release of intracellular Ca2+ stores in insulin-stimulated insulin secretion in beta -cells. J.Biol.Chem. 275: 22331-22338

[112] Assmann A, Ueki K, Winnay JN, Kadowaki T, Kulkarni RN (2009) Glucose effects on beta-cell growth and survival require activation of insulin receptors and insulin receptor substrate 2. Mol.Cell Biol. 29: 3219-3228

[113] Hennige AM, Ozcan U, Okada T, et al (2005) Alterations in growth and apoptosis of insulin receptor substrate-1-deficient beta-cells. Am.J.Physiol Endocrinol.Metab 289: E337-E346

[114] Hennige AM, Burks DJ, Ozcan U, et al (2003) Upregulation of insulin receptor substrate-2 in pancreatic beta cells prevents diabetes. J.Clin.Invest 112: 1521-1532

[115] Hisanaga E, Nagasawa M, Ueki K, Kulkarni RN, Mori M, Kojima I (2009) Regulation of calcium-permeable TRPV2 channel by insulin in pancreatic beta-cells. Diabetes 58: 174-184

[116] Kaneko K, Ueki K, Takahashi N, et al (2010) Class IA phosphatidylinositol 3-kinase in pancreatic beta cells controls insulin secretion by multiple mechanisms. Cell Metab 12: 619-632

[117] Kulkarni RN, Almind K, Goren HJ, et al (2003) Impact of genetic background on development of hyperinsulinemia and diabetes in insulin receptor/insulin receptor substrate-1 double heterozygous mice. Diabetes 52: 1528-1534

[118] Kulkarni RN, Holzenberger M, Shih DQ, et al (2002) beta-cell-specific deletion of the Igf1 receptor leads to hyperinsulinemia and glucose intolerance but does not alter beta-cell mass. Nat.Genet. 31: 111-115

[119] Kulkarni RN, Winnay JN, Daniels M, et al (1999) Altered function of insulin receptor substrate-1-deficient mouse islets and cultured beta-cell lines. J.Clin.Invest 104: R69-R75

[120] Leibiger B, Leibiger IB, Moede T, et al (2001) Selective insulin signaling through A and B insulin receptors regulates transcription of insulin and glucokinase genes in pancreatic beta cells. Mol.Cell 7: 559-570

[121] Liu S, Okada T, Assmann A, et al (2009) Insulin signaling regulates mitochondrial function in pancreatic beta-cells. PLoS.One. 4: e7983

[122] Okada T, Liew CW, Hu J, et al (2007) Insulin receptors in beta-cells are critical for islet compensatory growth response to insulin resistance. Proc.Natl.Acad.Sci.U.S.A 104: 8977-8982

[123] Otani K, Kulkarni RN, Baldwin AC, et al (2004) Reduced beta-cell mass and altered glucose sensing impair insulin-secretory function in betaIRKO mice. Am.J.Physiol Endocrinol.Metab 286: E41-E49

[124] Roper MG, Qian WJ, Zhang BB, Kulkarni RN, Kahn CR, Kennedy RT (2002) Effect of the insulin mimetic L-783,281 on intracellular Ca2+ and insulin secretion from pancreatic beta-cells. Diabetes 51 Suppl 1: S43-S49

[125] Ueki K, Okada T, Hu J, et al (2006) Total insulin and IGF-I resistance in pancreatic beta cells causes overt diabetes. Nat.Genet. 38: 583-588

[126] Leibiger B, Moede T, Uhles S, et al (2010) Insulin-feedback via PI3K-C2alpha activated PKBalpha/Akt1 is required for glucose-stimulated insulin secretion. FASEB J. 24: 1824-1837

[127] Leibiger B, Moede T, Uhles S, Berggren PO, Leibiger IB (2002) Short-term regulation of insulin gene transcription. Biochem.Soc.Trans. 30: 312-317

[128] Leibiger IB, Leibiger B, Berggren PO (2008) Insulin signaling in the pancreatic beta-cell. Annu.Rev.Nutr. 28: 233-251

[129] Leibiger IB, Leibiger B, Berggren PO (2002) Insulin feedback action on pancreatic beta-cell function. FEBS Lett. 532: 1-6

[130] Dominguez V, Raimondi C, Somanath S, et al (2011) Class II phosphoinositide 3-kinase regulates exocytosis of insulin granules in pancreatic beta cells. J.Biol.Chem. 286: 4216-4225

[131] Fruman DA, Meyers RE, Cantley LC (1998) Phosphoinositide kinases. Annu. Rev. Biochem. 67: 481-507

[132] Cantley LC (2002) The phosphoinositide 3-kinase pathway. Science 296: 1655-1657

[133] Maffucci T, Cooke FT, Foster FM, Traer CJ, Fry MJ, Falasca M (2005) Class II phosphoinositide 3-kinase defines a novel signaling pathway in cell migration. J.Cell Biol. 169: 789-799

[134] Falasca M, Hughes WE, Dominguez V, et al (2007) The role of phosphoinositide 3-kinase C2alpha in insulin signaling. J.Biol.Chem. 282: 28226-28236

[135] Ohlsson H, Karlsson K, Edlund T (1993) IPF1, a homeodomain-containing transactivator of the insulin gene. EMBO J. 12: 4251-4259

[136] Leonard J, Peers B, Johnson T, Ferreri K, Lee S, Montminy MR (1993) Characterization of somatostatin transactivating factor-1, a novel homeobox factor that stimulates somatostatin expression in pancreatic islet cells. Mol.Endocrinol. 7: 1275-1283

[137] Miller CP, McGehee RE, Jr., Habener JF (1994) IDX-1: a new homeodomain transcription factor expressed in rat pancreatic islets and duodenum that transactivates the somatostatin gene. EMBO J. 13: 1145-1156

[138] Boam DS, Clark AR, Docherty K (1990) Positive and negative regulation of the human insulin gene by multiple trans-acting factors. J.Biol.Chem. 265: 8285-8296

[139] Petersen HV, Serup P, Leonard J, Michelsen BK, Madsen OD (1994) Transcriptional regulation of the human insulin gene is dependent on the homeodomain protein STF1/IPF1 acting through the CT boxes. Proc.Natl.Acad.Sci.U.S.A 91: 10465-10469

[140] MacFarlane WM, Read ML, Gilligan M, Bujalska I, Docherty K (1994) Glucose modulates the binding activity of the beta-cell transcription factor IUF1 in a phosphorylation-dependent manner. Biochem.J. 303 (Pt 2): 625-631

[141] Marshak S, Totary H, Cerasi E, Melloul D (1996) Purification of the beta-cell glucose-sensitive factor that transactivates the insulin gene differentially in normal and transformed islet cells. Proc.Natl.Acad.Sci.U.S.A 93: 15057-15062

[142] Meur G, Qian Q, da Silva Xavier G, et al (2011) Nucleo-cytosolic shuttling of FoxO1 directly regulates mouse Ins2 but not Ins1 gene expression in pancreatic beta cells (MIN6). J.Biol.Chem. 286: 13647-13656

[143] Rutter GA (1999) Insulin secretion: feed-forward control of insulin biosynthesis? Curr.Biol. 9: R443-R445

[144] MacFarlane WM, McKinnon CM, Felton-Edkins ZA, Cragg H, James RF, Docherty K (1999) Glucose stimulates translocation of the homeodomain transcription factor PDX1 from the cytoplasm to the nucleus in pancreatic beta-cells. J.Biol.Chem. 274: 1011-1016

[145] Rafiq I, Kennedy HJ, Rutter GA (1998) Glucose-dependent translocation of insulin promoter factor-1 (IPF-1) between the nuclear periphery and the nucleoplasm of single MIN6 beta-cells. J.Biol.Chem. 273: 23241-23247

[146] Wu H, MacFarlane WM, Tadayyon M, Arch JR, James RF, Docherty K (1999) Insulin stimulates pancreatic-duodenal homoeobox factor-1 (PDX1) DNA-binding activity and insulin promoter activity in pancreatic beta cells. Biochem.J. 344 Pt 3: 813-818

[147] Campbell SC, MacFarlane WM (2002) Regulation of the pdx1 gene promoter in pancreatic beta-cells. Biochem.Biophys.Res.Commun. 299: 277-284

[148] Cahill CM, Tzivion G, Nasrin N, et al (2001) Phosphatidylinositol 3-kinase signaling inhibits DAF-16 DNA binding and function via 14-3-3-dependent and 14-3-3-independent pathways. J.Biol.Chem. 276: 13402-13410

[149] Trumper A, Trumper K, Trusheim H, Arnold R, Goke B, Horsch D (2001) Glucose-dependent insulinotropic polypeptide is a growth factor for beta (INS-1) cells by pleiotropic signaling. Mol.Endocrinol. 15: 1559-1570

[150] Rena G, Guo S, Cichy SC, Unterman TG, Cohen P (1999) Phosphorylation of the transcription factor forkhead family member FKHR by protein kinase B. J.Biol.Chem. 274: 17179-17183

[151] Matsuzaki H, Daitoku H, Hatta M, Tanaka K, Fukamizu A (2003) Insulin-induced phosphorylation of FKHR (Foxo1) targets to proteasomal degradation. Proc.Natl.Acad.Sci.U.S.A 100: 11285-11290

[152] Brunet A, Bonni A, Zigmond MJ, et al (1999) Akt promotes cell survival by phosphorylating and inhibiting a Forkhead transcription factor. Cell 96: 857-868

[153] Zawalich WS, Zawalich KC (2000) Glucose-induced insulin secretion from islets of fasted rats: modulation by alternate fuel and neurohumoral agonists. J.Endocrinol. 166: 111-120

[154] Zawalich WS, Tesz GJ, Zawalich KC (2002) Inhibitors of phosphatidylinositol 3-kinase amplify insulin release from islets of lean but not obese mice. J.Endocrinol. 174: 247-258

[155] Hagiwara S, Sakurai T, Tashiro F, et al (1995) An inhibitory role for phosphatidylinositol 3-kinase in insulin secretion from pancreatic B cell line MIN6. Biochem.Biophys.Res.Commun. 214: 51-59

[156] Turner MD, Arvan P (2000) Protein traffic from the secretory pathway to the endosomal system in pancreatic beta-cells. J.Biol.Chem. 275: 14025-14030

[157] Li LX, MacDonald PE, Ahn DS, Oudit GY, Backx PH, Brubaker PL (2006) Role of phosphatidylinositol 3-kinasegamma in the beta-cell: interactions with glucagon-like peptide-1. Endocrinology 147: 3318-3325

[158] MacDonald PE, Joseph JW, Yau D, et al (2004) Impaired glucose-stimulated insulin secretion, enhanced intraperitoneal insulin tolerance, and increased beta-cell mass in mice lacking the p110gamma isoform of phosphoinositide 3-kinase. Endocrinology 145: 4078-4083

[159] Pigeau GM, Kolic J, Ball BJ, et al (2009) Insulin granule recruitment and exocytosis is dependent on p110gamma in insulinoma and human beta-cells. Diabetes 58: 2084-2092

[160] Aoyagi K, Ohara-Imaizumi M, Nishiwaki C, Nakamichi Y, Nagamatsu S (2010) Insulin/phosphoinositide 3-kinase pathway accelerates the glucose-induced first-phase insulin secretion through TrpV2 recruitment in pancreatic beta-cells. Biochem.J. 432: 375-386

[161] Rorsman P, Renstrom E (2003) Insulin granule dynamics in pancreatic beta cells. Diabetologia 46: 1029-1045

[162] Iezzi M, Kouri G, Fukuda M, Wollheim CB (2004) Synaptotagmin V and IX isoforms control Ca2+ -dependent insulin exocytosis. J.Cell Sci. 117: 3119-3127

[163] Gauthier BR, Duhamel DL, Iezzi M, et al (2008) Synaptotagmin VII splice variants alpha, beta, and delta are expressed in pancreatic beta-cells and regulate insulin exocytosis. FASEB J. 22: 194-206

[164] Marshall C, Hitman GA, Partridge CJ, et al (2005) Evidence that an isoform of calpain-10 is a regulator of exocytosis in pancreatic beta-cells. Mol.Endocrinol. 19: 213-224

[165] Kim YH, Choi CY, Lee SJ, Conti MA, Kim Y (1998) Homeodomain-interacting protein kinases, a novel family of co-repressors for homeodomain transcription factors. J.Biol.Chem. 273: 25875-25879

[166] Kim EA, Noh YT, Ryu MJ, et al (2006) Phosphorylation and transactivation of Pax6 by homeodomain-interacting protein kinase 2. J.Biol.Chem. 281: 7489-7497

[167] D'Orazi G, Cecchinelli B, Bruno T, et al (2002) Homeodomain-interacting protein kinase-2 phosphorylates p53 at Ser 46 and mediates apoptosis. Nat.Cell Biol. 4: 11-19

[168] Hofmann TG, Moller A, Sirma H, et al (2002) Regulation of p53 activity by its interaction with homeodomain-interacting protein kinase-2. Nat.Cell Biol. 4: 1-10

[169] Zhang Q, Yoshimatsu Y, Hildebrand J, Frisch SM, Goodman RH (2003) Homeodomain interacting protein kinase 2 promotes apoptosis by downregulating the transcriptional corepressor CtBP. Cell 115: 177-186

[170] Kanei-Ishii C, Ninomiya-Tsuji J, Tanikawa J, et al (2004) Wnt-1 signal induces phosphorylation and degradation of c-Myb protein via TAK1, HIPK2, and NLK. Genes Dev. 18: 816-829

[171] Boucher MJ, Simoneau M, Edlund H (2009) The homeodomain-interacting protein kinase 2 regulates insulin promoter factor-1/pancreatic duodenal homeobox-1 transcriptional activity. Endocrinology 150: 87-97

[172] An R, da Silva Xavier G, Semplici F, et al (2010) Pancreatic and duodenal homeobox 1 (PDX1) phosphorylation at serine-269 is HIPK2-dependent and affects PDX1 subnuclear localization. Biochem.Biophys.Res.Commun. 399: 155-161

[173] Waeber G, Thompson N, Nicod P, Bonny C (1996) Transcriptional activation of the GLUT2 gene by the IPF-1/STF-1/IDX-1 homeobox factor. Mol.Endocrinol. 10: 1327-1334

[174] Watada H, Kajimoto Y, Miyagawa J, et al (1996) PDX-1 induces insulin and glucokinase gene expressions in alphaTC1 clone 6 cells in the presence of betacellulin. Diabetes 45: 1826-1831

[175] Watada H, Kajimoto Y, Umayahara Y, et al (1996) The human glucokinase gene beta-cell-type promoter: an essential role of insulin promoter factor 1/PDX-1 in its activation in HIT-T15 cells. Diabetes 45: 1478-1488

[176] Bretherton-Watt D, Gore N, Boam DS (1996) Insulin upstream factor 1 and a novel ubiquitous factor bind to the human islet amyloid polypeptide/amylin gene promoter. Biochem.J. 313 (Pt 2): 495-502

[177] Carty MD, Lillquist JS, Peshavaria M, Stein R, Soeller WC (1997) Identification of cis- and trans-active factors regulating human islet amyloid polypeptide gene expression in pancreatic beta-cells. J.Biol.Chem. 272: 11986-11993

[178] Kefas BA, Heimberg H, Vaulont S, et al (2003) AICA-riboside induces apoptosis of pancreatic beta cells through stimulation of AMP-activated protein kinase. Diabetologia 46: 250-254

[179] Riboulet-Chavey A, Diraison F, Siew LK, Wong FS, Rutter GA (2008) Inhibition of AMP-activated protein kinase protects pancreatic beta-cells from cytokine-mediated apoptosis and CD8+ T-cell-induced cytotoxicity. Diabetes 57: 415-423

[180] Jones RG, Plas DR, Kubek S, et al (2005) AMP-activated protein kinase induces a p53-dependent metabolic checkpoint. Mol.Cell 18: 283-293

[181] Sun G, Tarasov AI, McGinty JA, et al (2010) LKB1 deletion with the RIP2.Cre transgene modifies pancreatic beta-cell morphology and enhances insulin secretion in vivo. Am.J.Physiol Endocrinol.Metab 298: E1261-E1273

[182] Williams T, Brenman JE (2008) LKB1 and AMPK in cell polarity and division. Trends Cell Biol. 18: 193-198

[183] Fu A, Ng AC, Depatie C, et al (2009) Loss of Lkb1 in adult beta cells increases beta cell mass and enhances glucose tolerance in mice. Cell Metab 10: 285-295

[184] Granot Z, Swisa A, Magenheim J, et al (2009) LKB1 regulates pancreatic beta cell size, polarity, and function. Cell Metab 10: 296-308

[185] Alanentalo T, Asayesh A, Morrison H, et al (2007) Tomographic molecular imaging and 3D quantification within adult mouse organs. Nat.Methods 4: 31-33

[186] Jetton TL, Lausier J, LaRock K, et al (2005) Mechanisms of compensatory beta-cell growth in insulin-resistant rats: roles of Akt kinase. Diabetes 54: 2294-2304

[187] Montagne J, Stewart MJ, Stocker H, Hafen E, Kozma SC, Thomas G (1999) Drosophila S6 kinase: a regulator of cell size. Science 285: 2126-2129

[188] Shima H, Pende M, Chen Y, Fumagalli S, Thomas G, Kozma SC (1998) Disruption of the p70(s6k)/p85(s6k) gene reveals a small mouse phenotype and a new functional S6 kinase. EMBO J. 17: 6649-6659

[189] Kwon G, Marshall CA, Liu H, Pappan KL, Remedi MS, McDaniel ML (2006) Glucose-stimulated DNA synthesis through mammalian target of rapamycin (mTOR) is regulated by KATP channels: effects on cell cycle progression in rodent islets. J.Biol.Chem. 281: 3261-3267

[190] Rajasekhar VK, Viale A, Socci ND, Wiedmann M, Hu X, Holland EC (2003) Oncogenic Ras and Akt signaling contribute to glioblastoma formation by differential recruitment of existing mRNAs to polysomes. Mol.Cell 12: 889-901

[191] Eto K, Yamashita T, Matsui J, Terauchi Y, Noda M, Kadowaki T (2002) Genetic manipulations of fatty acid metabolism in beta-cells are associated with dysregulated insulin secretion. Diabetes 51 Suppl 3: S414-S420

[192] Ruvinsky I, Sharon N, Lerer T, et al (2005) Ribosomal protein S6 phosphorylation is a determinant of cell size and glucose homeostasis. Genes Dev. 19: 2199-2211

[193] Pende M, Kozma SC, Jaquet M, et al (2000) Hypoinsulinaemia, glucose intolerance and diminished beta-cell size in S6K1-deficient mice. Nature 408: 994-997

[194] Fraenkel M, Ketzinel-Gilad M, Ariav Y, et al (2008) mTOR inhibition by rapamycin prevents beta-cell adaptation to hyperglycemia and exacerbates the metabolic state in type 2 diabetes. Diabetes 57: 945-957

[195] Kawamori D, Kaneto H, Nakatani Y, et al (2006) The forkhead transcription factor Foxo1 bridges the JNK pathway and the transcription factor PDX-1 through its intracellular translocation. J.Biol.Chem. 281: 1091-1098

[196] Buteau J, Shlien A, Foisy S, Accili D (2007) Metabolic diapause in pancreatic beta-cells expressing a gain-of-function mutant of the forkhead protein Foxo1. J.Biol.Chem. 282: 287-293

[197] De BJ, Wang HJ, Van BC (2004) Effects of Wnt signaling on proliferation and differentiation of human mesenchymal stem cells. Tissue Eng 10: 393-401

[198] van de WM, Sancho E, Verweij C, et al (2002) The beta-catenin/TCF-4 complex imposes a crypt progenitor phenotype on colorectal cancer cells. Cell 111: 241-250

[199] Korinek V, Barker N, Moerer P, et al (1998) Depletion of epithelial stem-cell compartments in the small intestine of mice lacking Tcf-4. Nat.Genet. 19: 379-383

[200] Heller RS, Dichmann DS, Jensen J, et al (2002) Expression patterns of Wnts, Frizzleds, sFRPs, and misexpression in transgenic mice suggesting a role for Wnts in pancreas and foregut pattern formation. Dev.Dyn. 225: 260-270

[201] Papadopoulou S, Edlund H (2005) Attenuated Wnt signaling perturbs pancreatic growth but not pancreatic function. Diabetes 54: 2844-2851

[202] Heiser PW, Lau J, Taketo MM, Herrera PL, Hebrok M (2006) Stabilization of beta-catenin impacts pancreas growth. Development 133: 2023-2032

[203] McLin VA, Rankin SA, Zorn AM (2007) Repression of Wnt/beta-catenin signaling in the anterior endoderm is essential for liver and pancreas development. Development 134: 2207-2217

[204] Dessimoz J, Bonnard C, Huelsken J, Grapin-Botton A (2005) Pancreas-specific deletion of beta-catenin reveals Wnt-dependent and Wnt-independent functions during development. Curr.Biol. 15: 1677-1683

[205] Wells JM, Esni F, Boivin GP, et al (2007) Wnt/beta-catenin signaling is required for development of the exocrine pancreas. BMC.Dev.Biol. 7: 4

[206] Pedersen AH, Heller RS (2005) A possible role for the canonical Wnt pathway in endocrine cell development in chicks. Biochem.Biophys.Res.Commun. 333: 961-968

[207] Fujino T, Asaba H, Kang MJ, et al (2003) Low-density lipoprotein receptor-related protein 5 (LRP5) is essential for normal cholesterol metabolism and glucose-induced insulin secretion. Proc.Natl.Acad.Sci.U.S.A 100: 229-234

[208] Rulifson IC, Karnik SK, Heiser PW, et al (2007) Wnt signaling regulates pancreatic beta cell proliferation. Proc.Natl.Acad.Sci.U.S.A 104: 6247-6252

[209] Grant SF, Thorleifsson G, Reynisdottir I, et al (2006) Variant of transcription factor 7-like 2 (TCF7L2) gene confers risk of type 2 diabetes. Nat.Genet. 38: 320-323

[210] Florez JC (2007) The new type 2 diabetes gene TCF7L2. Curr.Opin.Clin.Nutr.Metab Care 10: 391-396

[211] Murtaugh LC, Law AC, Dor Y, Melton DA (2005) Beta-catenin is essential for pancreatic acinar but not islet development. Development 132: 4663-4674

[212] Welters HJ, Kulkarni RN (2008) Wnt signaling: relevance to beta-cell biology and diabetes. Trends Endocrinol.Metab 19: 349-355

[213] Schinner S, Ulgen F, Papewalis C, et al (2008) Regulation of insulin secretion, glucokinase gene transcription and beta cell proliferation by adipocyte-derived Wnt signalling molecules. Diabetologia 51: 147-154

[214] Cozar-Castellano I, Fiaschi-Taesch N, Bigatel TA, et al (2006) Molecular control of cell cycle progression in the pancreatic beta-cell. Endocr.Rev. 27: 356-370

[215] Dodge R, Loomans C, Sharma A, Bonner-Weir S (2009) Developmental pathways during in vitro progression of human islet neogenesis. Differentiation 77: 135-147

[216] Liu Z, Habener JF (2008) Glucagon-like peptide-1 activation of TCF7L2-dependent Wnt signaling enhances pancreatic beta cell proliferation. J.Biol.Chem. 283: 8723-8735

[217] Egan JM, Bulotta A, Hui H, Perfetti R (2003) GLP-1 receptor agonists are growth and differentiation factors for pancreatic islet beta cells. Diabetes Metab Res.Rev. 19: 115-123

[218] Xu G, Stoffers DA, Habener JF, Bonner-Weir S (1999) Exendin-4 stimulates both beta-cell replication and neogenesis, resulting in increased beta-cell mass and improved glucose tolerance in diabetic rats. Diabetes 48: 2270-2276

[219] List JF, Habener JF (2004) Glucagon-like peptide 1 agonists and the development and growth of pancreatic beta-cells. Am.J.Physiol Endocrinol.Metab 286: E875-E881

[220] Ashcroft FM (2006) K(ATP) channels and insulin secretion: a key role in health and disease. Biochem.Soc.Trans.34: 243-246.

[221] Braun M, Rorsman P (2006) The glucagon-producing alpha cell: an electrophysiologically exceptional cell. Diabetologia 53:1827-1830.

[222] Kawamori D, Kurpad AJ, Hu J, Liew CW, Shih JL, Ford EL, Herrera PL, Polonsky KS, McGuinness OP, Kulkarni RN (2009) Cell Metab 9:350-361.

[223] Henquin J.C, Nenquin M, Ravier M.A, Szollosi A (2009) Shortcomings of current models of glucose-induced insulin secretion. Diabetes Obes.Metab. 11 (Suppl 4): 168-179.

7

The ERK MAPK Pathway in Bone and Cartilage Formation

Takehiko Matsushita and Shunichi Murakami
Department of Orthopaedics, Case Western Reserve University,
USA

1. Introduction

The Extracellular signal-related kinase (ERK) Mitogen-activated protein kinase (MAPK) pathway is a signaling cascade that is activated by various extracellular stimuli including fibroblast growth factors. The ERK MAPK pathway has recently been shown to play critical roles in skeletal development. A number of human skeletal syndromes have been shown to result from mutations in this pathway. These include Noonan, Costello, and cardio-facio-cutaneous syndromes (Aoki et al., 2005; Pandit et al., 2007; Rodriguez-Viciana et al., 2006). In addition, activating mutations in FGFR2 cause craniosynostosis syndromes such as Apert and Crouzon syndromes, while activating mutations in fibroblast growth factor receptor 3 (FGFR3) are responsible for the most common forms of human dwarfism, achondroplasia, thanatophoric dysplasia, and hypochondroplasia (Bellus et al., 1995; Jabs et al., 1994; Rousseau et al., 1994, 1995; Rutland et al., 1995; Shiang et al., 1994; Wilcox et al., 1998; Wilkie et al., 1995).

Although a number of in vitro experiments indicated profound effects of the ERK MAPK pathway on chondrocyte and osteoblast phenotype, sometimes conflicting results were reported presumably due to variable culture conditions (Bobick & Kulyk, 2008; Schindeler & Little, 2006), and the roles of the ERK MAPK pathway in vivo remained elusive. Therefore, to examine the role of ERK MAPK in skeletal development, we used both gain-of-function and loss-of-function approaches to activate or inactivate the ERK MAPK pathway in skeletal tissues of genetically engineered mice. We used the Cre-*loxP* system to inactivate *ERK1* and *ERK2* in skeletal tissues (Logan et al., 2002; Matsushita et al., 2009a). By using the *Prx1-Cre* transgene, mice lacking *ERK1* and *ERK2* in the limb and head mesenchyme were created. We also generated a loss of function model in chondrocytes by using the *Col2a1-Cre* transgene (Matsushita et al., 2009a; Ovchinnikov et al., 2000). To induce postnatal inactivation of *ERK1* and *ERK2* in chondrocytes, we used the *Col2a1-CreER* transgene to express a tamoxifen-inducible form of Cre recombinase (Nakamura et al., 2006; Sebastian et al., 2011). For gain-of-function experiments, we generated *Prx1-MEK1* transgenic mice that express a constitutively active mutant of MEK1 in undifferentiated mesenchymal cells under the control of a *Prx1* promoter (Matsushita et al., 2009a). We also generated *Col2a1-MEK1* transgenic mice that express a constitutively active mutant of MEK1 in chondrocytes under the control of the regulatory sequences of *Col2a1* (Murakami et al., 2004). In this review, we will summarize the roles of the ERK MAPK pathway in skeletal development based on our recent studies using genetically engineered mouse models.

2. Skeletal development and ossification processes

Bone formation takes place in two major ossification processes, endochondral ossification and intramembranous ossification (Colnot, 2005; Hunziker, 1994; Opperman, 2000; Shapiro et al., 2005). Both chondrocytes and osteoblasts arise from common undifferentiated mesenchymal progenitor cells. In endochondral ossification, the skeletal element is formed as a cartilaginous template that is subsequently replaced by bone. Undifferentiated mesenchymal cells first aggregate to form mesenchymal condensation and differentiate into chondrocytes. Chondrocytes proliferate in columnar stacks to form the growth plate, then exit the cell cycle, and differentiate into hypertrophic chondrocytes. The cartilaginous matrix of hypertrophic chondrocytes is calcified and subsequently invaded by blood vessels. Hypertrophic chondrocytes are removed by apoptotic cell death, and the cartilaginous matrix is resorbed by chondroclasts/osteoclasts and replaced by trabecular bone. Chondroclast/osteoclast formation is supported by receptor activator of nuclear factor-kappa B ligand (RANKL) secreted from osteoblasts and bone marrow stromal cells (Kim et al., 2000; Yasuda et al., 1998). In intramembranous ossification, mesenchymal cells directly differentiate into bone-forming osteoblasts; cortical bone is formed by osteoblasts that arise from the osteochondro progenitor cells in the perichondrium. The entire process of endochondral ossification and intramembranous ossification is under the control of various hormones and growth factors. These include systemic factors such as growth hormone, estrogen, and glucocorticoids; and local factors such as Indian hedgehog (Ihh), parathyroid hormone-related peptide (PTHrP), fibroblast growth factors (FGF), transforming growth factor-β (TGF-β), and bone morphogenetic proteins (BMP) (DeLise et al., 2000; van der Eerden et al., 2003).

3. The ERK MAPK pathway and human syndromes

3.1 The ERK MAPK pathway

The ERK MAPK pathway (Fig. 1), which is activated by various stimuli in eukaryotic cells, transduces extracellular signals into cells and coordinates cellular responses. The MAPK pathways are generally organized into three kinase modules: MAPKK kinase (MAPKKK), MAPK kinase (MAPKK), and MAPK. MAPKKK phosphorylates and activates MAPKK, which in turn phosphorylates and activates MAPK. A diverse array of growth factors and cytokines transduce their signals through the activation of the small G protein Ras, which leads to the activation of the Raf members of MAPKKK, and then to the activation of MAPKK, MEK1 and MEK2. MEK1 and MEK2 then phosphorylate and activate MAPK, ERK1 and ERK2. ERK1 and ERK2 then phosphorylate various cytoplasmic and nuclear target proteins, ranging from cytoplasmic adaptor proteins and transcription factors to kinases including RSK (Cargnello & Roux, 2011; Roux & Blenis, 2004). The ERK MAPK pathway has been shown to mediate the intracellular signaling induced by a variety of growth factors such as FGFs, BMPs, and TGFs (Jun et al., 2010; Mu et al., 2011; Murakami et al., 2000; Osyczka & Leboy, 2005; Qureshi et al., 2005; Tuli et al., 2003).

3.2 Human syndromes caused by mutations in the MAPK pathway

Recently, a number of human mutations have been identified in the molecules in the MAPK cascade (Fig. 1). Missense activating mutations in KRAS, BRAF, MEK1, and MEK2 have been identified in Cardio-facio-cutaneous syndrome (Rodriguez-Viciana et al., 2006). KRAS mutations have been also identified in Noonan syndrome (Schubbert et al., 2006). HRAS mutations cause Costello syndrome (Aoki et al., 2005), and loss-of-function mutations in RSK2 cause Coffin-Lowry syndrome (Trivier et al., 1996). In addition, haploinsufficient expression of ERK2 has been associated with DiGeorge syndrome (Newbern et al., 2008). All of these syndromes present with various skeletal manifestations, including short stature and craniofacial and limb abnormalities, underscoring the importance of the MAPK pathway in human skeletal development (Hanauer & Young, 2002; Hennekam, 2003; Noonan, 2006; Reynolds et al., 1986; van der Burgt, 2007).

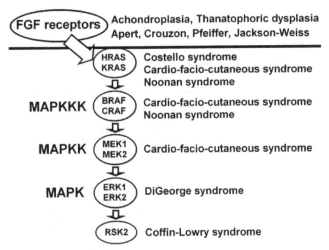

Fig. 1. The ERK MAPK pathway and human skeletal syndromes

4. Genetic manipulation of molecules in the ERK MAPK pathway in mice

To examine the role of the MAPK pathway, various mutant mice have been generated. Inactivation of Rsk2, a downstream kinase of the MAPK pathway, caused a widening of cranial sutures at birth, similar to the delayed closure of fontanelles in patients with Coffin-Lowry syndrome (David et al., 2005; Yang et al., 2004). These observations indicate that Rsk2 plays a critical role in osteoblast differentiation. In contrast to *Rsk2*-null mice, the inactivation of ERK, MEK, and Raf family members has provided little information regarding skeletal development. *ERK1*-null mice are viable and fertile and develop normally without obvious skeletal abnormalities, suggesting that ERK1 is dispensable for skeletal development (Pages et al., 1999; Selcher et al., 2001). In remarkable contrast, *ERK2*-null mice show early embryonic lethality at E6.5, precluding the analysis of skeletal development (Saba-El-Leil et al., 2003; Yao et al., 2003). *Mek1*-null embryos die at E10.5 due to placental defects, while *Mek2*-null mice develop normally without any obvious abnormalities (Belanger et al., 2003; Giroux et al., 1999). *Araf*-null mice show neurological and gastrointestinal defects, but do not show an obvious skeletal phenotype (Pritchard et al., 1996). *Braf*-deficient embryos die at midgestation due to vascular defects, precluding the analysis of skeletal development (Wojnowski et al., 1997). *Craf*-null mice show placental defects and die at around E10.5-12.5 on the C57BL/6 and 129 backgrounds. On the outbred

CD1 background, two-thirds of embryos reach term and die soon after birth. These surviving embryos show a mild delay in ossification; however, it is not clear whether the observed skeletal phenotype is primarily caused by *Craf* deficiency in the skeletal tissues (Wojnowski et al., 1998). Furthermore, mice with chondrocytes deficient in both *Araf* and *Braf* showed normal endochondral bone development (Provot et al., 2008). These observations suggest that members of the ERK, MEK, and Raf family are functionally redundant, while some of the tissue-specific functions are not fully compensated by other family members. To circumvent early embryonic lethality caused by the systemic inactivation of the target gene, tissue-specific inactivation would be essential. Furthermore, the inactivation of multiple family members may be necessary to uncover the roles of ERK, MEK, and Raf family members in skeletal development.

5. Inactivation of ERK1 and ERK2 in undifferentiated mesenchymal cells disrupts bone formation and induces ectopic cartilage formation

Since early embryonic lethality hampered researchers from analyzing the role of the ERK MAPK pathway in skeletal development in vivo as described above, we used the Cre-*loxP* system to inactivate *ERK1* and *ERK2* in skeletal tissues. We used *Prx1-Cre* transgenic mice that express Cre recombinase under the control of the 2.4 kb *Prx1* promoter (Logan et al., 2002) to inactivate *ERK1* and *ERK2* in the limb and head mesenchyme (Matsushita et al., 2009a). The *Prx1* promoter has been shown to direct transgene expression in undifferentiated mesenchyme in the developing limb buds and head mesenchyme. The transgene expression is detectable as early as E10.5, and the transgene expression is confined to the periosteum of the long bones and tendons of the limbs at E15.5 (Logan et al., 2002). We analyzed *ERK1*-null mice and *ERK2*flox/flox; *Prx1-Cre* mice, and these mice did not show obvious skeletal abnormalities. Therefore, we further inactivated *ERK2* in the *ERK1*-null background to totally inactivate *ERK1* and *ERK2* in the head and limb mesenchyme of mouse embryos using the *Prx1-Cre* transgene (Matsushita et al., 2009a).

5.1 Inactivation of ERK1 and ERK2 in mesenchymal cells disrupts osteoblast differentiation

Skeletal preparation of *ERK1* -/-; *ERK2* flox/flox; *Prx1-Cre* mutants revealed severe limb deformities as well as calvaria defects characterized by delayed closure of the cranial sutures (Fig. 2). Histological analysis of the long bones showed disruption of bone formation. These findings indicate that ERK1 and ERK2 play an essential role in bone formation. In situ hybridization analysis indicated that master osteogenic transcription factors *Runx2, Osterix,* and *Atf4* were expressed at normal levels, while expression of *Osteocalcin*, a marker of mature osteoblasts, was strongly decreased in *ERK1* -/-; *ERK2* flox/flox; *Prx1-Cre* mice. These observations suggest that osteoblast differentiation was blocked after *Runx2, Osterix,* and *Atf4* expression and before *Osteocalcin* expression. The impaired bone formation was associated with decreased beta-catenin protein levels in the periosteum, suggesting decreased Wnt signaling. We also found that other transcriptional regulators such as *Krox20, Fra1, Fra2, cFos,* and *Cbfb* were downregulated in *ERK1* -/-; *ERK2* flox/flox; *Prx1-Cre* embryos. While the regulatory mechanisms of osteoblast differentiation require further investigation, ERK1 and ERK2 are likely to control skeletal development and osteoblast differentiation through multiple downstream molecules. Consistent with our observation, Ge et al. reported that transgenic mice that express dominant negative MEK1 under an *Osteocalcin* promoter showed delayed bone formation and reduced mineralization of calvaria, while mice that express a constitutively active MEK1 under the *Osteocalcin* promoter showed accelerated bone formation (Ge et al., 2007). These observations also indicated a critical role of ERK1 and ERK2 in osteoblast differentiation. In vitro studies have also suggested that the ERK MAPK pathway regulates osteoblast differentiation through phosphorylation and acetylation of Runx2 (Ge et al., 2007; Park et al., 2010; Xiao et al., 2000, 2002). The importance of the ERK MAPK pathway in osteoblast differentiation in head mesenchyme was also demonstrated by Shukla et al., who showed that craniosynostosis in a mouse model of Apert syndrome that carries a mutation in *Fgfr2* was prevented by the treatment of MEK1/2 inhibitor (Shukla et al., 2007). These observations further link the activation of the ERK MAPK pathway to the pathogenesis of craniosynostosis syndromes caused by activating mutations in FGFR2.

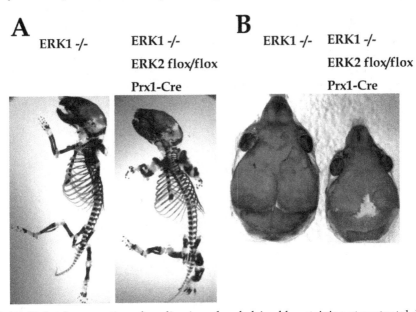

Fig. 2. (A) Skeletal preparation after alizarin red and alcian blue staining at postnatal day 1. (B) Skeletal preparation after alizarin red and alcian blue staining at postnatal day 5.

5.2 Inactivation of ERK1 and ERK2 disrupts osteocyte differentiation

Osteoblasts undergo sequential steps of differentiation and subsequently become embedded in bone matrix as osteocytes. Osteocytes function as a mechanosensor in the bone and secrete dentin matrix protein 1 (Dmp1) and FGF23 to regulate phosphate homeostasis (Dallas & Bonewald, 2010; Feng et al., 2006; Tatsumi et al., 2007; Xiao and Quarles, 2010). Although *ERK1-/-; ERK2flox/flox; Prx1-Cre* mice showed a remarkable impairment of bone formation, bone-like architecture was observed in the diaphyses of long bones, and osteocyte-like cells were found within the bone-like matrix. Our real-time PCR and immunohistochemical analysis indicated a strong decrease in Dmp1 expression in the skeletal elements of *ERK1-/-; ERK2 flox/flox; Prx1-Cre* mice. Furthermore, scanning electron microscopic analysis revealed that osteocytes in *ERK1-/-*; *ERK2flox/flox*; *Prx1-Cre* mice lack dendritic processes, indicating that *ERK1* and *ERK2* inactivation disrupts the formation of osteocyte-lacunar-canalicular system (Kyono et al., 2011). These observations indicate that ERK signaling is essential for *Dmp1* expression and osteocyte differentiation.

5.3 Inactivation of ERK1 and ERK2 in mesenchyme causes ectopic cartilage formation

While inactivation of *ERK1* and *ERK2* inhibited osteoblast differentiation and bone formation, we found ectopic cartilage formation in the perichondrium of *ERK1 -/-*; *ERK2 flox/flox*; *Prx1-Cre* mice (Fig. 3). The ectopic cartilage expressed *Sox9*, a transcription factor for chondrocyte differentiation, and *Col2a1*, the gene for type II collagen (Matsushita et al., 2009a). These findings suggest that inactivation of *ERK1* and *ERK2* in mesenchyme inhibited osteoblast differentiation and promoted chondrocyte differentiation. Ectopic cartilage formation has been also reported in the perichondrium of *Osterix*-null mice (Nakashima et al., 2002) and in the perichondrium of mice in which beta-catenin was disrupted in mesenchymal cells (Day et al., 2005; Hill et al., 2005). While normal *Osterix* expression was observed in *ERK1 -/-*; *ERK2 flox/flox*; *Prx1-Cre* mice, beta-catenin protein levels were decreased in the perichondrium of *ERK1 -/-*; *ERK2 flox/flox*; *Prx1-Cre* mice. These observations suggest a role for decreased beta-catenin in the ectopic cartilage formation in the perichondrium.

ERK1 -/-   **ERK1 -/-**

ERK2 flox/flox

Prx1-Cre

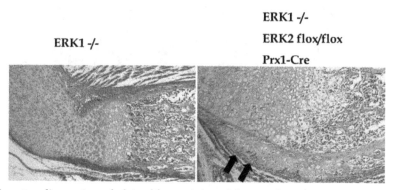

Fig. 3. Hematoxylin, eosin and alcian blue staining of femur of newborn mice. Arrows indicate ectopic cartilage formation in the perichondrium of *ERK1-/-;ERK2flox/flox;Prx1-Cre* mice.

6. Constitutive activation of MEK1 in undifferentiated mesenchymal cells leads to increased bone formation and inhibition of cartilage formation

As a complementary experiment, we generated *Prx1-MEK1* transgenic mice that express a constitutively active mutant of MEK1 in undifferentiated limb and cranial mesenchyme (Matsushita et al., 2009a). *Prx1-MEK1* mice showed a marked increase in cortical bone formation, fusion of long bones as well as carpal and tarsal bones, and an accelerated closure of cranial sutures, mimicking the phenotype of human craniosynostosis syndromes caused by activating mutations in FGFR2. The increase in bone formation was associated with increased expression of osteoblast markers, such as *Runx2, Osterix, Bsp*, and *Osteocalcin*. In contrast, cartilage formation was inhibited in *Prx1-MEK1* transgenic mice. There was a clear delay in the formation of cartilage anlagen as well as a decrease in anlagen size. In

Effects on the ERK MAPK pathway	Mouse models	Phenotypes
	Mesenchymal cells	
Loss of function	ERK1 $^{-/-}$; ERK2 $^{flox/flox}$; Prx1-Cre (Matsushita 2009a)	Limb deformity Delayed closure of the cranial suture Calvaria defect Inhibition of bone formation Ectopic cartilage formation in perichondrium Increase in terminally differentiated hypertrophic chondrocytes Absence of Osteocalcin expression Decreased Dmp1 expression Absence of osteocyte-lacunar-canalicur system
Gain of function	CA MEK1/Prx1 (Matsushita 2009a)	Increase in cortical bone formation Fusion of long bones, carpal and tarsal bones Accelerated closure of cranial suture Delayed and decreased formation of cartilage anlagen Increased Runx2, Osterix, Bsp, Osteocalcin expression
	Chondrocytes	
Loss of function	ERK1 $^{-/-}$; ERK2 $^{flox/flox}$; Col2a1-Cre (Matsushita 2009a)	Die immediately after birth Deformed rib cage Kyphotic deformity of the spine Disorganization of epiphyseal cartilage Widening of zone of hypertrophic chondrocytes
	ERK1 $^{-/-}$; ERK2 $^{flox/flox}$; Col2a1-CreER (Sebastian 2011)	Delayed synchondrosis closure of the vertebrae Increased vertebral foramen cross-sectional area
Gain of function	CA MEK1/Col2a1 (Murakami 2004)	Dwarfism Premature synchondrosis closure Narrower zone of hypertrophic chondrocytes Smaller hypertrophic chondrocytes Reduced rate of chondrocyte hypertrophy
	Osteoblasts	
Loss of function	DN MEK1/Osteocalcin (Ge 2007)	Delayed calvarial mineralization Delayed formation of primary ossification centers in the long bones
Gain of function	CA MEK1/Osteocalcin (Ge 2007)	Increased calvarial mineralization Accelerated trabecular bone formation

Table 1. Genetically engineered mouse models with increased or decreased ERK MAPK signaling in skeletal cells. CA MEK1/Prx1, Transgenic mice that express a constitutively active mutant of MEK1 under the control of a *Prx1* promoter. CA MEK1/Col2a1, Transgenic mice that express a constitutively active mutant of MEK1 under the control of a *Col2a1* promoter. DN MEK1/Osteocalcin, Transgenic mice that express a dominant-negative mutant of MEK1 under the control of an *Osteocalcin* promoter. CA MEK1/Osteocalcin, Transgenic mice that express a constitutively active mutant of MEK1 under the control of an *Osteocalcin* promoter.

addition, expression of *Col2a1* was reduced in the cartilage primordia. These observations are consistent with the phenotypes of *ERK1* -/-; *ERK2* *flox/flox*; *Prx1-Cre* mice, in which the lack of *ERK1* and *ERK2* disrupts bone formation and induces ectopic cartilage formation. These observations are also consistent with the in vitro studies showing inhibitory effects of the ERK MAPK pathway on chondrogenesis. The ERK MAPK inhibitors U0126 and PD98059 increased the expression of *Col2a1* and *aggrecan* in embryonic limb mesenchyme, and transfection of limb mesenchyme with constitutively active mutant of MEK decreased the activity of a Sox9-responsive *Col2a1* enhancer reporter gene (Bobick & Kulyk, 2004). Collectively, these observations indicate that ERK MAPK signaling plays an important role in the lineage specification of mesenchymal cells.

7. Inactivation of ERK1 and ERK2 in chondrocytes causes severe chondrodysplasia and enhances bone growth

Type II collagen is the most abundant collagen in cartilage. *Col2a1*, the gene encoding the proalpha1(II) collagen chain, is a principal marker of chondrocyte differentiation. We created a loss of function model of ERK1 and ERK2 in chondrocytes by using the *Col2a1-Cre* transgenic mice that express Cre recombinase under the regulatory sequences of *Col2a1* (Matsushita et al., 2009a; Ovchinnikov et al., 2000). *ERK1* -/-; *ERK2* *flox/flox*; *Col2a1-Cre* mutant mice died immediately after birth, likely secondary to respiratory insufficiency caused by rib cage deformity (Matsushita et al., 2009a). To circumvent the perinatal lethality, we also used the *Col2a1-CreER* transgene to express a tamoxifen-inducible form of Cre recombinase and examined the role of *ERK1* and *ERK2* in chondrocytes during postnatal growth (Nakamura et al., 2006; Sebastian et al., 2011). For the gain-of-function experiments, we generated *Col2a1-MEK1* transgenic mice that express a constitutively active mutant of MEK1 in chondrocytes under the control of the regulatory sequences of *Col2a1* (Murakami et al., 2004).

7.1 ERK1 and ERK2 are essential for proper organization of epiphyseal cartilage

A strong skeletal phenotype was observed in *ERK1* -/-; *ERK2* *flox/flox*; *Col2a1-Cre* embryos. We observed severe kyphotic deformities of the spine. Histological analysis of *ERK1* -/-; *ERK2* *flox/flox*; *Col2a1-Cre* embryos at embryonic day 16.5 showed an absence of primary ossification centers in the axial skeleton and a widening of the zone of hypertrophic chondrocytes in the long bones (Fig. 4). In addition, disorganization of the epiphyseal cartilage with lack of columnar growth structures was observed in *ERK1* -/-; *ERK2* *flox/flox*; *Col2a1-Cre* embryos at embryonic day 18.5. These findings indicate that ERK1 and ERK2 are essential for the proper organization of the epiphyseal cartilage.

ERK1 -/-

ERK1 -/-
ERK2 flox/flox
Col2a1-Cre

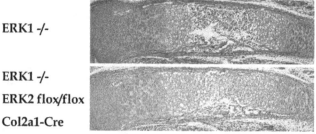

Fig. 4. Hematoxylin, eosin and alcian blue staining of the tibia showed delayed formation of primary ossification center in *ERK1* -/-; *ERK2* *flox/flox*; *Col2a1-Cre* embryos at embryonic day 16.5.

7.2 ERK1 and ERK2 inhibit hypertrophic chondrocyte differentiation

Both *ERK1 -/-; ERK2 flox/flox; Prx1-Cre* and *ERK1 -/-; ERK2 flox/flox; Col2a1-Cre* mice showed a remarkable expansion of the zone of hypertrophic chondrocytes. In both animal models, chondrocytes that were closer to the articular surface expressed hypertrophic chondrocyte marker *Col10a1*, suggesting premature chondrocyte hypertrophy. These observations are consistent with the growth plate phenotype of *Col2a1-MEK1* transgenic mice that express a constitutively active MEK1 mutant in chondrocytes (Murakami et al., 2004). The growth plate of *Col2a1-MEK1* transgenic mice was characterized by smaller than normal hypertrophic chondrocytes and narrower zone of hypertrophic chondrocytes. BrdU labeling of proliferating chondrocytes and subsequent identification of BrdU-labeled hypertrophic chondrocytes indicated reduced rate of chondrocyte hypertrophy in *Col2a1-MEK1* mice. Collectively, these observations indicate that ERK MAPK signaling inhibits hypertrophic chondrocyte differentiation.

The pronounced expansion of the zone of hypertrophic chondrocytes in *ERK1 -/-; ERK2 flox/flox; Prx1-Cre* mice was also characterized by an increase in terminally differentiated hypertrophic chondrocytes expressing *Vegf*, *Mmp13*, and *Osteopontin*, suggesting impaired removal of terminally differentiated hypertrophic chondrocytes. We also observed a decrease in TRAP-positive osteoclasts in association with reduced expression of *receptor activator of nuclear factor-kappa B ligand (RANKL)*. Therefore, decreased osteoclastogenesis may also account for the expansion of the zone of hypertrophic chondrocytes.

7.3 ERK1 and ERK2 inhibit growth of cartilaginous skeletal element

A number of genetic studies have indicated the growth inhibitory role of FGFR3 signaling. Mice with activating mutations in Fgfr3 show a dwarf phenotype similar to the human syndromes of achondroplasia and thanatophoric dysplasia (Chen et al., 1999; Iwata et al., 2000, 2001; Li et al., 1999; Naski et al., 1998; Wang et al., 1999). In contrast, *Fgfr3*-null mice show a skeletal overgrowth (Colvin et al., 1996; Deng et al., 1996). Our observations in genetically engineered mouse models have provided evidence indicating that the ERK MAPK pathway is a critical downstream effector of Fgfr3 signaling. We have shown that *Col2a1-MEK1* transgenic mice that express a constitutively active MEK1 mutant in chondrocytes show an achondroplasia-like dwarf phenotype (Murakami et al., 2004). We have also shown that *ERK1* and *ERK2* inactivation in chondrocytes promotes bone growth. We found increased length of the proximal long bones, specifically the humerus and femur, of *ERK1 -/-; ERK2 flox/flox; Col2a1-Cre* embryos (Sebastian et al., 2011). These embryos also showed an increase in the width of epiphyses in the humerus and femur. Histological analysis of the vertebrae also showed an overgrowth of cartilage in the vertebral body. These observations indicate that ERK1 and ERK2 negatively regulate the growth of cartilaginous skeletal elements.

8. Postnatal ERK1 and ERK2 inactivation delays synchondrosis closure and enlarges the spinal canal

Our studies have indicated that FGFR3 and the MAPK pathway are important regulators of synchondrosis closure. In bones such as the vertebrae, sternum, pelvis, and bones in the cranial base, a synchondrosis—a growth plate-like cartilaginous structure—connects the ossification centers and contributes to the bone growth. During postnatal skeletal development, the width of synchondroses reduces with age. Ossification centers eventually unite when synchondroses close. Histologically, a synchondrosis consists of two opposed growth plates with a common zone of resting chondrocytes (Fig. 5A). We have found premature synchondrosis closure in the vertebrae and cranial base of human samples of achondroplasia and thanatophoric dysplasia (Matsushita et al., 2009b). In addition, we have also observed premature synchondrosis closure in a mouse model of achondroplasia and *Col2a1-MEK1* transgenic mice that express a constitutively active MEK1 mutant in chondrocytes (Fig. 5B). Because growth at the synchondrosis determines the final dimension and shape of the endochondral skeletons, premature synchondrosis closure should play a critical role in the development of spinal canal stenosis that is frequently seen in patients with achondroplasia.

Since increased Fgfr3 and MEK1 signaling accelerates synchondrosis closure, we hypothesized that *ERK1* and *ERK2* inactivation delays synchondrosis closure and enlarges the spinal canal. To test this hypothesis, we inactivated *ERK2* in chondrocytes of *ERK1*-null

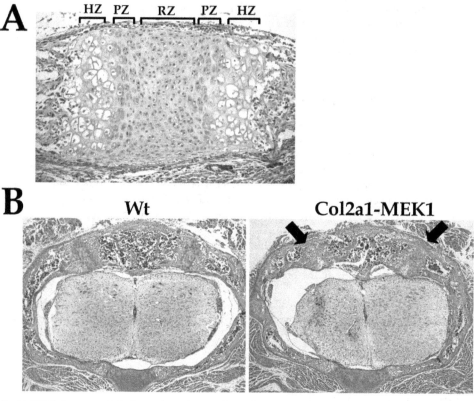

Fig. 5. (A) Spheno-occipital synchondrosis of a 4-day-old wild type mouse. HZ; hypertrophic zone. PZ; proliferation zone. RZ; resting zone. (B) Thoracic spine of wild type and *Col2a1-MEK1* transgenic mice at postnatal day 4. Arrows indicate prematurely closing synchondroses. Wt; wild type.

mice using the *Col2a1-CreER* transgene (Sebastian et al., 2011). Tamoxifen injection into *ERK1 -/-*; *ERK2 flox/flox*; *Col2a1-CreER* mice resulted in 60% inhibition of *ERK2* expression in the epiphyseal cartilage. Although these mice did not show an increased growth of the long bones presumably due to incomplete *ERK2* inactivation, we observed a significant delay in synchondrosis closure of the vertebrae and an increase in the cross-sectional area of vertebral foramen. The delayed synchondrosis closure was associated with a decreased expression of the endothelial marker CD31 surrounding the synchondroses, suggesting that *ERK1* and *ERK2* inactivation in chondrocytes causes reduced vascular invasion. These observations indicate the potential of ERK1 and ERK2 as therapeutic targets for spinal canal stenosis in achondroplasia.

9. Conclusion

By creating and analyzing gain-of-function and loss-of-function mouse models, we have identified multiple roles of the ERK MAPK pathway at successive steps of skeletal development. In undifferentiated mesenchymal cells, *ERK1* and *ERK2* inactivation causes a block in osteoblast differentiation and induces ectopic cartilage formation. In contrast, increased MEK1 signaling promotes bone formation and inhibits cartilage formation. These observations indicate that ERK MAPK signaling plays a critical role in the lineage specification of mesenchymal cells (Fig. 6). ERK MAPK signaling also inhibits hypertrophic chondrocyte differentiation and bone growth. Furthermore, ERK MAPK signaling regulates the timing of growth plate and synchondrosis closure, the very last step of endochondral ossification. A better understanding of the roles of the ERK MAPK pathway in skeletal tissues will lead to new insights in skeletal development and the treatment of various skeletal disorders.

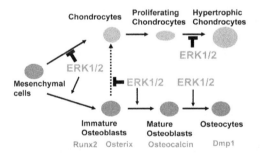

Fig. 6. Proposed model of the role of ERK1 and ERK2 in the regulation of osteoblast and chondrocyte differentiation.

10. Acknowledgments

We thank Ms. Valerie Schmedlen for editorial assistance. Work in the laboratory of the authors was supported by Arthritis Investigator Award of the Arthritis Foundation, March of Dimes Birth Defects Foundation, and NIH grants R21DE017406, R01AR055556, R03DE019814 to S.M.

11. References

Aoki, Y., Niihori, T., Kawame, H., Kurosawa, K., Ohashi, H., Tanaka, Y., Filocamo, M., Kato, K., Suzuki, Y., Kure, S., & Matsubara, Y. (2005). Germline mutations in HRAS proto-oncogene cause Costello syndrome. *Nat Genet*, Vol. 37, No. 10, pp. 1038-40, ISSN 1061-4036

Belanger, L.F., Roy, S., Tremblay, M., Brott, B., Steff, A.M., Mourad, W., Hugo, P., Erikson, R., & Charron, J. (2003). Mek2 is dispensable for mouse growth and development. *Mol Cell Biol*, Vol. 23, No. 14, pp. 4778-87. ISSN 0270-7306

Bellus, G.A., McIntosh, I., Smith, E.A., Aylsworth, A.S., Kaitila, I., Horton, W.A., Greenhaw, G.A., Hecht, J.T., & Francomano, C.A. (1995). A recurrent mutation in the tyrosine kinase domain of fibroblast growth factor receptor 3 causes hypochondroplasia. *Nat Genet*, Vol. 10, No. 3, pp. 357-9, ISSN 1061-4036

Bobick, B.E., & Kulyk, W.M. (2004). The MEK-ERK signaling pathway is a negative regulator of cartilage-specific gene expression in embryonic limb mesenchyme. *J Biol Chem*, Vol. 279, No. 6, pp. 4588-95, ISSN 0021-9258

Bobick, B.E., & Kulyk, W.M. (2008). Regulation of cartilage formation and maturation by mitogen-activated protein kinase signaling. *Birth Defects Res C Embryo Today*, Vol. 84, No. 2, pp. 131-54, ISSN 1542-975X

Cargnello, M., & Roux, P.P. (2011). Activation and function of the MAPKs and their substrates, the MAPK-activated protein kinases. *Microbiol Mol Biol Rev*, Vol. 75, No. 1, pp. 50-83, ISSN 1092-2171

Chen, L., Adar, R., Yang, X., Monsonego, E.O., Li, C., Hauschka, P.V., Yayon, A., & Deng, C.X. (1999). Gly369Cys mutation in mouse FGFR3 causes achondroplasia by affecting both chondrogenesis and osteogenesis. *J Clin Invest*, Vol. 104, No. 11, pp. 1517-25, ISSN 0021-9738

Colnot, C. (2005). Cellular and molecular interactions regulating skeletogenesis. *J Cell Biochem*, Vol.95, No. 4, pp. 688–97, ISSN 0730-2313

Colvin, J.S., Bohne, B.A., Harding, G.W., McEwen, D.G., & Ornitz, D.M. (1996). Skeletal overgrowth and deafness in mice lacking fibroblast growth factor receptor 3. *Nat Genet*, Vol. 12, No. 4, pp. 390-7, ISSN 1061-4036

Dallas, S.L., & Bonewald, L.F. (2010). Dynamics of the transition from osteoblast to osteocyte. *Ann N Y Acad Sci*, Vol. 1192, pp. 437-43, ISSN 0077-8923

David, J.P., Mehic, D., Bakiri, L., Schilling, A.F., Mandic, V., Priemel, M., Idarraga, M.H., Reschke, M.O., Hoffmann, O., Amling, M., & Wagner, E.F. (2005). Essential role of RSK2 in c-Fos-dependent osteosarcoma development. *J Clin Invest*, Vol. 115, No. 3, pp. 664-72, ISSN 0021-9738

Day, T.F., Guo, X., Garrett-Beal, L., & Yang, Y. (2005). Wnt/beta-catenin signaling in mesenchymal progenitors controls osteoblast and chondrocyte differentiation during vertebrate skeletogenesis. *Dev Cell*, Vol. 8, No. 5, pp. 739–50, ISSN 1534-5807

DeLise, A.M., Fischer, L., & Tuan, R.S. (2000). Cellular interactions and signaling in cartilage development. *Osteoarthritis Cartilage*, Vol. 8, No. 5, pp. 309-34, ISSN1063-4584

Deng, C., Wynshaw-Boris, A., Zhou, F., Kuo, A., & Leder, P. (1996). Fibroblast growth factor receptor 3 is a negative regulator of bone growth. *Cell*, Vol. 84, No. 6, pp. 911-21, ISSN 0092-8674

Feng, J.Q., Ward, L.M., Liu, S., Lu, Y., Xie, Y., Yuan, B., Yu, X., Rauch, F., Davis, S.I., Zhang, S., Rios, H., Drezner, M.K., Quarles, L.D., Bonewald, L.F., & White, K.E. (2006). Loss of DMP1 causes rickets and osteomalacia and identifies a role for osteocytes in mineral metabolism. *Nat Genet*, Vol. 38, No. 11, pp. 1310-5, ISSN 1061-4036

Ge, C., Xiao, G., Jiang, D., & Franceschi, R.T. (2007). Critical role of the extracellular signal-regulated kinase-MAPK pathway in osteoblast differentiation and skeletal development. *J Cell Biol*, Vol. 176, No. 5, pp. 709-18, ISSN 0021-9525

Giroux, S., Tremblay, M., Bernard, D., Cardin-Girard, J.F., Aubry, S., Larouche, L., Rousseau, S., Huot, J., Landry, J., Jeannotte, L., & Charron, J. (1999). Embryonic death of Mek1-deficient mice reveals a role for this kinase in angiogenesis in the labyrinthine region of the placenta. *Curr Biol*, Vol. 9, No. 7, pp. 369-72, ISSN 0960-9822

Hanauer, A., & Young, I.D. (2002). Coffin-Lowry syndrome: clinical and molecular features. *J Med Genet*, Vol. 39, No. 10, pp. 705-13, Review, ISSN 0022-2593

Hennekam, R.C. (2003). Costello syndrome: an overview. *Am J Med Genet C Semin Med Genet*, Vol. 117C, No. 1, pp. 42-8, Review, ISSN 1552-4868

Hill, T.P., Später, D., Taketo, M.M., Birchmeier, W., & Hartmann, C. (2005). Canonical Wnt/beta-catenin signaling prevents osteoblasts from differentiating into chondrocytes. *Dev Cell*, Vol. 8, No. 5, pp. 727–38, ISSN 1534-5807

Hunziker, E.B. (1994). Mechanism of longitudinal bone growth and its regulation by growth plate chondrocytes. *Microsc Res Tech*, Vol. 28, No. 6, pp. 505–19, ISSN 1059-910X

Iwata, T., Chen, L., Li, C., Ovchinnikov, D.A., Behringer, R.R., Francomano, C.A., & Deng, C.X. (2000). A neonatal lethal mutation in FGFR3 uncouples proliferation and differentiation of growth plate chondrocytes in embryos. *Hum Mol Genet*, Vol. 9, No. 11, pp. 1603-13, ISSN 0964-6906

Iwata, T., Li, C.L., Deng, C.X., & Francomano, C.A. (2001). Highly activated Fgfr3 with the K644M mutation causes prolonged survival in severe dwarf mice. *Hum Mol Genet*, Vol. 10, No. 12, pp. 1255-64. ISSN 0964-6906

Jabs, E.W., Li, X., Scott, A.F., Meyers, G., Chen, W., Eccles, M., Mao, J.I., Charnas, L.R., Jackson, C.E., & Jaye, M. (1994). Jackson-Weiss and Crouzon syndromes are allelic with mutations in fibroblast growth factor receptor 2. *Nat Genet*, Vol. 8, No. 3, pp. 275-9, ISSN 1061-4036

Jun, J.H., Yoon, W.J., Seo, S.B., Woo, K.M., Kim, G.S., Ryoo, H.M., & Baek, J.H. (2010). BMP2-activated Erk/MAP kinase stabilizes Runx2 by increasing p300 levels and histone acetyltransferase activity. *J Biol Chem*, Vol. 285, No. 47, pp. 36410-9, ISSN 0021-9258

Kim, N., Odgren, P.R., Kim, D.K., Marks, S.C.Jr., & Choi, Y. (2000). Diverse roles of the tumor necrosis factor family member TRANCE in skeletal physiology revealed by TRANCE deficiency and partial rescue by a lymphocyte-expressed TRANCE transgene. *Proc Natl Acad Sci U S A*, Vol. 97, No. 20, pp. 10905-10, ISSN 0027-8424

Kyono, A., Avishai, N., Ouyang, Z., Landreth, G.E., & Murakami, S. (2011). FGF and ERK signaling coordinately regulate mineralization-related genes and play essential roles in osteocyte differentiation. *J Bone Miner Metab*, (Jun 17). ISSN 0914-8779

Li, C., Chen, L., Iwata, T., Kitagawa, M., Fu, X.Y., & Deng, C.X. (1999). A Lys644Glu substitution in fibroblast growth factor receptor 3 (FGFR3) causes dwarfism in mice by activation of STATs and ink4 cell cycle inhibitors. *Hum Mol Genet*, Vol. 8, No. 1, pp. 35-44, ISSN 0964-6906

Logan, M., Martin, J.F., Nagy, A., Lobe, C., Olson, E.N., & Tabin, C.J. (2002). Expression of Cre Recombinase in the developing mouse limb bud driven by a Prxl enhancer. *Genesis*, Vol.33, No. 2, pp. 77-80, ISSN 1526-954X

Matsushita, T., Chan, Y., Kawanami, A., Balmes, G., Landreth, G.E., & Murakami, S. (2009a). Extracellular Signal-Regulated Kinase 1 (ERK1) and ERK2 play essential roles in osteoblast differentiation and in supporting osteoclastogenesis. *Mol Cell Biol*, Vol. 29, No. 21, pp. 5843-57, ISSN 0270-7306

Matsushita, T., Wilcox, W.R., Chan, Y.Y., Kawanami, A., Bükülmez, H., Balmes, G., Krejci, P., Mekikian, P.B., Otani, K., Yamaura, I., Warman, M.L., Givol, D., & Murakami, S. (2009b). FGFR3 promotes synchondrosis closure and fusion of ossification centers through the MAPK pathway. *Hum Mol Genet*, Vol. 18, No. 2, pp. 227-40, ISSN 0964-6906

Mu, Y., Gudey, S.K., & Landström, M. (2011). Non-Smad signaling pathways. *Cell Tissue Res*, (Jun 24).

Murakami, S., Kan, M., McKeehan, W.L., & de Crombrugghe, B. (2000). Up-regulation of the chondrogenic Sox9 gene by fibroblast growth factors is mediated by the mitogen-activated protein kinase pathway. *Proc Natl Acad Sci U S A*, Vol. 97, No. 3, pp. 1113-8, ISSN 0027-8424

Murakami, S., Balmes, G., McKinney, S., Zhang, Z., Givol, D., & de Crombrugghe, B. (2004). Constitutive activation of MEK1 in chondrocytes causes Stat1-independent achondroplasia-like dwarfism and rescues the Fgfr3-deficient mouse phenotype. *Genes Dev*, Vol. 18, No. 3, pp. 290-305, ISSN 0860-9369

Nakamura, E., Nguyen, M.T., & Mackem, S. (2006). Kinetics of tamoxifen-regulated Cre activity in mice using a cartilage-specific CreER(T) to assay temporal activity windows along the proximodistal limb skeleton. *Dev Dyn*, Vol. 235, No. 9, pp. 2603-12, ISSN 1058-8388

Nakashima, K., Zhou, X., Kunkel, G., Zhang, Z., Deng, J.M., Behringer, R.R., & de Crombrugghe, B. (2002). The novel zinc finger-containing transcription factor osterix is required for osteoblast differentiation and bone formation. *Cell*, Vol. 108, No. 1, pp. 17-29, ISSN 0092-8674

Naski, M.C., Colvin, J.S., Coffin, J.D., & Ornitz, D.M. (1998). Repression of hedgehog signaling and BMP4 expression in growth plate cartilage by fibroblast growth factor receptor 3. *Development*, Vol. 125, No. 24, pp. 4977-88, ISSN 1011-6370

Newbern, J., Zhong, J., Wickramasinghe, R.S., Li, X., Wu, Y., Samuels, I., Cherosky, N., Karlo, J.C., O'Loughlin, B., Wikenheiser, J., Gargesha, M., Doughman, Y.Q., Charron, J., Ginty, D.D., Watanabe, M., Saitta, S.C., Snider, W.D., & Landreth, G.E. (2008). Mouse and human phenotypes indicate a critical conserved role for ERK2 signaling in neural crest development. *Proc Natl Acad Sci U S A*, Vol. 105, No. 44, pp. 17115-20, ISSN 0027-8424

Noonan, J.A. (2006). Noonan syndrome and related disorders: Alterations in growth and puberty. *Rev Endocr Metab Disord*, Vol. 7, No. 4, pp. 251-5, ISSN 1389-9155

Opperman, L.A. (2000). Cranial sutures as intramembranous bone growth sites. *Dev Dyn*, Vol. 219, No. 4, pp. 472–85, ISSN 1058-8388

Osyczka, A.M., & Leboy, P.S. (2005). Bone morphogenetic protein regulation of early osteoblast genes in human marrow stromal cells is mediated by extracellular signal-regulated kinase and phosphatidylinositol 3-kinase signaling. *Endocrinology*, Vol. 146, No. 8, pp. 3428-37, ISSN 0013-7227

Ovchinnikov, D.A., Deng, J.M., Ogunrinu, G., & Behringer, R.R. (2000). Col2a1-directed expression of Cre recombinase in differentiating chondrocytes in transgenic mice. *Genesis*, Vol. 26, No. 2, pp. 145-6, ISSN 1526-954X

Pages, G., Guerin, S., Grall, D., Bonino, F., Smith, A., Anjuere, F., Auberger, P., & Pouyssegur, J. (1999). Defective thymocyte maturation in p44 MAP kinase (Erk 1) knockout mice. *Science*, Vol. 286, No. 5443, pp. 1374-7, ISSN 0036-8075

Pandit, B., Sarkozy, A., Pennacchio, L.A., Carta, C., Oishi, K., Martinelli, S., Pogna, E.A., Schackwitz, W., Ustaszewska, A., Landstrom, A., Bos, J.M., Ommen, S.R., Esposito, G., Lepri, F., Faul, C., Mundel, P., López Siguero, J.P., Tenconi, R., Selicorni, A., Rossi, C., Mazzanti, L., Torrente, I., Marino, B., Digilio, M.C., Zampino, G., Ackerman, M.J., Dallapiccola, B., Tartaglia, M., & Gelb, B.D. (2007). Gain-of-function RAF1 mutations cause Noonan and LEOPARD syndromes with hypertrophic cardiomyopathy. *Nat Genet*, Vol. 39, No. 8, pp. 1007-12, ISSN 1061-4036

Park, O.J., Kim, H.J., Woo, K.M., Baek, J.H., & Ryoo, H.M. (2010). FGF2-activated ERK mitogen-activated protein kinase enhances Runx2 acetylation and stabilization. *J Biol Chem*, Vol. 285, No. 6, pp. 3568-74, ISSN 0021-9258

Pritchard, C.A., Bolin, L., Slattery, R., Murray, R., & McMahon, M. (1996). Post-natal lethality and neurological and gastrointestinal defects in mice with targeted disruption of the A-Raf protein kinase gene. *Curr Biol*, Vol. 6, No. 5, pp. 614-7, ISSN 0960-9822

Provot, S., Nachtrab, G., Paruch, J., Chen, A.P., Silva, A., & Kronenberg, H.M. (2008). A-raf and B-raf are dispensable for normal endochondral bone development, and parathyroid hormone-related peptide suppresses extracellular signal-regulated kinase activation in hypertrophic chondrocytes. *Mol Cell Biol*, Vol. 28, No. 1, pp. 344-57, ISSN 0270-7306

Qureshi, H.Y., Sylvester, J., El Mabrouk, M., & Zafarullah, M. (2005). TGF-beta-induced expression of tissue inhibitor of metalloproteinases-3 gene in chondrocytes is mediated by extracellular signal-regulated kinase pathway and Sp1 transcription factor. *J Cell Physiol*, Vol. 203, No. 2, pp. 345-52, ISSN 0021-9541

Reynolds, J.F., Neri, G., Herrmann, J.P., Blumberg, B., Coldwell, J.G., Miles, P.V., & Opitz, J.M. (1986). New multiple congenital anomalies/mental retardation syndrome with cardio-facio-cutaneous involvement--the CFC syndrome. *Am J Med Genet*, Vol. 25, No. 3, pp. 413-27, ISSN 1552-4833

Rodriguez-Viciana, P., Tetsu, O., Tidyman, W.E., Estep, A.L., Conger, B.A., Santa Cruz, M., McCormick, F., & Rauen, K.A. (2006). Germline Mutations in Genes within the MAPK Pathway Cause Cardio-facio-cutaneous Syndrome. *Science*, Vol. 311, No. 5765, pp. 1287-90, ISSN 0036-8075

Rousseau, F., Bonaventure, J., Legeai-Mallet, L., Pelet, A., Rozet, J.M., Maroteaux, P., Le Merrer, M., & Munnich, A. (1994). Mutations in the gene encoding fibroblast growth factor receptor-3 in achondroplasia. *Nature,* Vol. 371, No. 6494, pp.252-4, ISSN 0028-0836

Rousseau, F., Saugier, P., Le Merrer, M., Munnich, A., Delezoide, A.L., Maroteaux, P., Bonaventure, J., Narcy, F., & Sanak, M. (1995). Stop codon FGFR3 mutations in thanatophoric dwarfism type 1. *Nat Genet,* Vol. 10, No. 1, pp. 11-2, ISSN 1061-4036

Roux, P.P., & Blenis, J. (2004). ERK and p38 MAPK-activated protein kinases: a family of protein kinases with diverse biological functions. *Microbiol Mol Biol Rev,* Vol. 68, No. 2, pp. 320-44, ISSN 1092-2171

Rutland, P., Pulleyn, L.J., Reardon, W., Baraitser, M., Hayward, R., Jones, B., Malcolm, S., Winter, R.M., Oldridge, M., Slaney, S.F., Poole, M.D., & Wilkie A.O.M. (1995). Identical mutations in the FGFR2 gene cause both Pfeiffer and Crouzon syndrome phenotypes. *Nat Genet,* Vol. 9, No. 2, 173-6, ISSN 1061-4036

Saba-El-Leil, M.K., Vella, F.D., Vernay, B., Voisin, L., Chen, L., Labrecque, N., Ang, S.L., & Meloche, S. (2003). An essential function of the mitogen-activated protein kinase Erk2 in mouse trophoblast development. *EMBO Rep,* Vol. 4, No. 2, pp. 964-8, ISSN 1469-221X

Schindeler, A., & Little, D.G. (2006). Ras-MAPK signaling in osteogenic differentiation: friend or foe? *J Bone Miner Res,* Vol. 21, No. 9, pp. 1331-8, ISSN 0914-8779

Schubbert, S., Zenker, M., Rowe, S.L., Boll, S., Klein, C., Bollag, G., van der Burgt, I., Musante, L., Kalscheuer, V., Wehner, L.E., Nguyen, H., West, B., Zhang, K.Y., Sistermans, E., Rauch, A., Niemeyer, C.M., Shannon, K., & Kratz, C.P. (2006). Germline KRAS mutations cause Noonan syndrome. *Nat Genet,* Vol. 38, No. 3, pp. 331-6, ISSN 1061-4036

Sebastian, A., Matsushita, T., Kawanami, A., Mackem, S., Landreth, G., & Murakami S. (2011). Genetic inactivation of ERK1 and ERK2 in chondrocytes promotes bone growth and enlarges the spinal canal. *J Orthop Res,* Vol. 29, No. 3, pp. 375-9, ISSN 0736-0266

Selcher, J.C., Nekrasova, T., Paylor, R., Landreth, G.E., & Sweatt, J.D. (2001). Mice lacking the ERK1 isoform of MAP kinase are unimpaired in emotional learning. *Learn Mem,* Vol. 8, No. 1, pp. 11-9, ISSN 1072-0502

Shapiro, I.M., Adams, C.S., Freeman, T. & Srinivas, V. (2005). Fate of the hypertrophic chondrocyte: microenvironmental perspectives on apoptosis and survival in the epiphyseal growth plate. *Birth Defects Res. C,* Vol. 75, No. 4, pp. 330-9, ISSN 1542-975X

Shiang, R., Thompson, L.M., Zhu, Y.Z., Church, D.M., Fielder, T.J., Bocian, M., Winokur, S.T., & Wasmuth, J.J. (1994). Mutations in the transmembrane domain of FGFR3 cause the most common genetic form of dwarfism, achondroplasia. *Cell,* Vol. 78, No. 2, pp. 335-42, ISSN 0092-8674

Shukla, V., Coumoul, X., Wang, R.H., Kim, H.S., & Deng, C.X. (2007). RNA interference and inhibition of MEK-ERK signaling prevent abnormal skeletal phenotypes in a mouse model of craniosynostosis. *Nat Genet,* Vol. 39, No. 9, pp. 1145-50, ISSN 1061-4036

Tatsumi, S., Ishii, K., Amizuka, N., Li, M., Kobayashi, T., Kohno, K., Ito, M., Takeshita, S., & Ikeda, K. (2007). Targeted ablation of osteocytes induces osteoporosis with defective mechanotransduction. *Cell Metab*, Vol. 5, No. 6, pp. 464-75, ISSN 1550-4131

Trivier, E., De Cesare, D., Jacquot, S., Pannetier, S., Zackai, E., Young, I., Mandel, J.L., Sassone-Corsi, P., & Hanauer, A. (1996). Mutations in the kinase Rsk-2 associated with Coffin-Lowry syndrome. *Nature*, Vol. 384, No. 1, pp. 567-70, ISSN 0028-0836

Tuli, R., Tuli, S., Nandi, S., Huang, X., Manner, P.A., Hozack, W.J., Danielson, K.G., Hall, D.J., & Tuan, R.S. (2003). Transforming growth factor-beta-mediated chondrogenesis of human mesenchymal progenitor cells involves N-cadherin and mitogen-activated protein kinase and Wnt signaling cross-talk. *J Biol Chem*, Vol. 278, No. 42, pp. 41227-36, ISSN 0021-9258

van der Burgt, I. (2007). Noonan syndrome. *Orphanet J Rare Dis*, Vol. 2, pp. 4, ISSN 1750-1172

van der Eerden, B.C., Karperien, M., & Wit, J.M. (2003). Systemic and local regulation of the growth plate. *Endocr Rev*, Vol. 24, No. 6, pp. 782-801, ISSN 0163-769X

Wang, Y., Spatz, M.K., Kannan, K., Hayk, H., Avivi, A., Gorivodsky, M., Pines, M., Yayon, A., Lonai, P., & Givol, D. (1999). A mouse model for achondroplasia produced by targeting fibroblast growth factor receptor 3. *Proc Natl Acad Sci U S A*, Vol. 96, No. 8, pp. 4455-60. ISSN 0027-8424

Wilcox, W.R., Tavormina, P.L., Krakow, D., Kitoh, H., Lachman, R.S., Wasmuth, J.J., Thompson, L.M., & Rimoin, D.L. (1998). Molecular, radiologic, and histopathologic correlations in thanatophoric dysplasia. *Am J Med Genet*, Vol. 78, No. 3, pp. 274-81, ISSN 1552-4833

Wilkie, A.O., Slaney, S.F., Oldridge, M., Poole, M.D., Ashworth, G.J., Hockley, A.D., Hayward, R.D., David, D.J., Pulleyn, L.J., Rutland, P., Malcolm, S., Winter, R.M., & Reardon, W. (1995). Apert syndrome results from localized mutations of FGFR2 and is allelic with Crouzon syndrome. *Nat Genet*, Vol. 9, No. 2, pp. 165-72, ISSN 1061-4036

Wojnowski, L., Zimmer, A.M., Beck, T.W., Hahn, H., Bernal, R., Rapp, U.R., & Zimmer, A. (1997). Endothelial apoptosis in Braf-deficient mice. *Nat Genet*, Vol. 16, No. 3, pp. 293-7, ISSN 1061-4036

Wojnowski, L., Stancato, L.F., Zimmer, A.M., Hahn, H., Beck, T.W., Larner, A.C., Rapp, U.R., & Zimmer, A. (1998). Craf-1 protein kinase is essential for mouse development. *Mech Dev*, Vol. 76, No. 1-2, pp. 141-9, ISSN 0925-4773

Xiao, G., Jiang, D., Gopalakrishnan, R., & Franceschi, R.T. (2002). Fibroblast growth factor 2 induction of the osteocalcin gene requires MAPK activity and phosphorylation of the osteoblast transcription factor, Cbfa1/Runx2. *J Biol Chem*, Vol. 277, No. 39, pp. 36181-7, ISSN 0021-9258

Xiao, G., Jiang, D., Thomas, P., Benson, M.D., Guan, K., Karsenty, G., & Franceschi, R.T. (2000). MAPK pathways activate and phosphorylate the osteoblast-specific transcription factor, Cbfa1. *J Biol Chem*, Vol. 275, No. 6, pp. 4453-9, ISSN 0021-9258

Xiao, Z.S., & Quarles, L.D. (2010). Role of the polycytin-primary cilia complex in bone development and mechanosensing. *Ann N Y Acad Sci*, Vol. 1192, pp. 410-21, ISSN 0077-8923

Yang, X., Matsuda, K., Bialek, P., Jacquot, S., Masuoka, H.C., Schinke, T., Li, L., Brancorsini, S., Sassone-Corsi, P., Townes, T.M., Hanauer, A., & Karsenty, G. (2004). ATF4 is a substrate of RSK2 and an essential regulator of osteoblast biology; implication for Coffin-Lowry Syndrome. *Cell,* Vol. 117, No. 3, pp. 387-98, ISSN 0092-8674

Yao, Y., Li, W., Wu, J., Germann, U.A., Su, M.S., Kuida, K., & Boucher, D.M. (2003). Extracellular signal-regulated kinase 2 is necessary for mesoderm differentiation. *Proc Natl Acad Sci U S A,* Vol. 100, No. 22, pp. 12759-64, ISSN 0027-8424

Yasuda, H., Shima, N., Nakagawa, N., Yamaguchi, K., Kinosaki, M., Mochizuki, S., Tomoyasu, A., Yano, K., Goto, M., Murakami, A., Tsuda, E., Morinaga, T., Higashio, K., Udagawa, N., Takahashi, N., & Suda, T. (1998). Osteoclast differentiation factor is a ligand for osteoprotegerin/osteoclastogenesis-inhibitory factor and is identical to TRANCE/RANKL. *Proc Natl Acad Sci U S A,* Vol. 95, No. 7, pp. 3597-602, ISSN 0027-8424

8

Raf Serine/Threonine Protein Kinases: Immunohistochemical Localization in the Mammalian Nervous System

András Mihály
University of Szeged, Faculty of Medicine,
Department of Anatomy, Histology and Embryology,
Hungary

1. Introduction

The Raf protein kinases are members of the mitogen-activated protein kinase (MAPK)/extracellular signal-regulated kinase (ERK) signaling pathway, which links the activation of cell surface receptors to intracellular cytoplasmic and nuclear molecular targets. Raf kinases relay signals from growth factor receptor-coupled GTPases of the Ras family to the MAPK/ERK kinase (MEK). This pathway is very important in developmental processes, cell survival and proliferation (Zebisch & Troppmair, 2006). Three different Raf isoenzymes exist in mammals: C-Raf, B-Raf and A-Raf. They originate from 3 independent genes (for review, see Matallanas et al., 2011). The Raf genes are protooncogenes; they are the cellular counterparts of the v-Raf oncogene, originally discovered in murine sarcoma cell lines (Rapp et al. 1988). The Raf genes are ubiquitously present in different tissues, showing a relatively segregated expression pattern: B-Raf gene is expressed mainly in the brain and testes, C-Raf gene is ubiquitous, although the degree of expression is uneven, A-Raf gene is expressed in gonads, kidney, spleen and bone mainly, as detected in mice (Storm et al., 1990). Accordingly, the Raf proteins are present in three isoforms: the A-Raf, B-Raf and C-Raf are protein kinase isozymes, showing similar molecular structure (Zebisch & Troppmair, 2006). The Raf proteins consist of three conserved regions (CR1, CR2, CR3), which display different functions: CR1 contains the Ras-binding region, which is responsible for the interaction with Ras and membrane phopholipids; CR2 contains phosphorylation site and CR3 is the kinase region of the molecule (Heidecker et al., 1991; Zebisch and Troppmair, 2006). The inactive Raf molecule is in a closed conformation: the N-terminal and C-terminal regions are above each other (Matallanas et al., 2011). When this molecule opens, activation and dimerization can happen (Heidecker et al., 1991; Matallanas et al., 2011; Zebisch & Troppmair, 2006). Activation-inactivation are regulated by phosphorylation-dephosphorylation of the molecule (Matallanas et al., 2011). Activation also needs the recruitment of Raf to the cell membrane (Leevers et al., 1994). Many details of the Raf kinase activation steps have been described in the last twenty years (for review, see Matallanas et al., 2011). The regulation of Raf signaling by intracellular proteins is complicated: (1) because of the membrane recruitment of Raf during activation, scaffolding proteins exert regulatory role on the Ras-Raf-MEK pathway (Matallanas et al., 2011). (2) Recently discovered Raf kinase inhibitor protein (RKIP) negatively modulates Raf kinase, being important mainly in cancer cells (Klysik et al., 2008; Matallanas et al., 2011). Over the years, B-Raf has been increasingly associated with cancer development (Niault & Baccarini, 2010). Mutations of B-Raf proved to be important in malignant melanoma, thyroid carcinomas and colorectal tumors (Niault & Baccarini, 2010). This underlines the utmost importance of B-Raf in mitogenic signal transduction.

Experimental data suggest, that Raf kinases are activated by growth factor- and cytokine receptors. Platelet-derived growth factor (PDGF; Morrison et al., 1989), epidermal growth factor (EGF; App et al., 1991), nerve growth factor (NGF; Oshima et al., 1991), insulin (Blackshear et al., 1990), interleukins (Carrol et al., 1990), neuronal angiotensin receptor (Yang et al., 1997) and vascular endothelial growth factor (VEGF; Lu et al., 2011) act through the Raf-MEK signaling pathway. Raf mediated signals are able to move in the cytoplasm; the Golgi-apparatus, the mitochondria and the cell nucleus are targets of Raf-mediated phosphorylation signals (Matallanas et al., 2011; Zebisch & Troppmair, 2006). The question arises, if these organelles contain Raf kinase, or Raf kinase translocates to them during the signalling process (Mor & Philips, 2006).

Immunohistochemical studies have revealed that B-raf and C-Raf are widely distributed in central nervous system (CNS) areas including the hippocampus, neocortex and spinal cord (Mihály et al. 1991, 1993; Mihály & Rapp 1993). They are present not only in neurons but also in astrocytes (Mihály & Rapp 1994). Although most CNS neurons express Raf immunoreactivity, the possibility remains that Raf expression varies according to the type of growth factor receptor that a neuron expresses (Mihály & Endrész, 2000). Our ultrastructural immunohistochemical studies proved that Raf protein-like immunoreactivity was localized primarily in postsynaptic densities, dendritic spines, dendrites and soma of the neurons (Mihály et al., 1991). On the basis of these observations the possibility arose, that Raf kinases participated in some receptor-mediated postsynaptic phenomena which might initiate long-term changes in postsynaptic neurons (Mihály et al., 1990). The present review collects the relevant localization data of Raf protein kinases in CNS structures. The localization of the Raf proteins will be discussed in light of some plasticity experiments (Mihály et al., 1990; 1991; 1996).

2. Detection of Raf protein kinases in the brain with Western blotting

The first literature data about the presence of Raf protein kinase in the mammalian brain are those of Mihály et al. (1991). In these studies we used the polyclonal v-Raf antiserum which was produced in the laboratory of Dr. Ulf Rapp. The serum was raised in rabbit, against a large (30 kDa) C-terminal Raf protein, expressed in *E. coli*, by inserting the v-Raf oncogene into the bacterial DNA (Kolch et al., 1988). Adult Wistar rats were deeply anesthetized with halothane, decapitated and the brains were homogenized, lysed and immunprecipitated. The precipitates were subjected to gel electrophoresis, then proteins were transferred to nitrocellulose membranes and incubated with the polyclonal v-Raf antiserum (Mihály et al., 1991). One conspicuous 95 kDa protein band was detected with this method (Mihály et al., 1991). The same antibody was used on isolated subcellular fractions of the rat brain, where the cytosolic- and microsome fractions displayed positive signals, again at 95 kDa (Mihály et al., 1996). Efforts aiming the detection of C-Raf and B-Raf separately, were also successful (Morice et al., 1999): Western blotting revealed the presence of C-Raf and B-Raf proteins in different areas of the rat brain (Morice et al., 1999). Immunoblotting of guinea pig brain homogenates also resulted in positive signals at 95 kDa with the aforementioned polyclonal anti-v-Raf serum (Mihály & Endrész, 2000). Several years later, using specific, polyclonal B-raf antibodies (Santa Cruz Biotechnology, Santa Cruz, CA, USA), we detected the 95 kDa protein again in rat brain homogenates (Mihály et al., 2007). Therefore, we conclude, that brain B-Raf kinase is ubiquitous and detectable through Western blotting in the rodent brain, as a 95 kDa protein (Fig. 1).

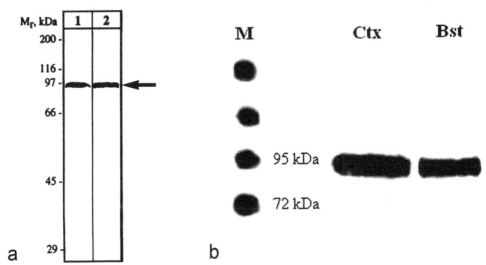

Fig. 1. Rat brain Western blots made with v-Raf-specific (a) and B-raf-specific (b) polyclonal antibodies. Molecular weight markers are on the left side (Mr, kDa; M). Fig. 1a: lane 1 represents the cytosolic fraction, lane 2 the microsome fraction (Mihály et al., 1996). Fig. 1b: „Ctx" represents the signal obtained from brain cortex homogenate; „Bst" represents the signal obtained from brain stem homogentes (Mihály et al., 2007). The two antibodies detect one 95 kDa protein, indicating the presence of B-raf kinase in the rat brain.

3. Light microscopic localization of Raf proteins in the central nervous system (CNS)

Light microscopic immunohistochemistry of brain areas from laboratory rats, laboratory guinea pigs and domestic cats have been performed with help of the polyclonal v-Raf antibodies (Mihály et al., 1991), polyclonal B-raf antibodies (Santa Cruz Biotechnology, Santa Cruz, CA, USA), monoclonal B-Raf antibodies (produced against the last 12 amino acids of the C-terminal region of B-Raf) and polyclonal C-Raf antibodies (anti-SP 63 serum). The polyclonal anti-v-Raf-, anti-SP 63- and monoclonal anti-B-Raf antibodies were prepared in the laboratories of Dr. Ulf Rapp (National Cancer Institute, Frederick Cancer Research and Development Center, Frederick, MD, USA; see descriptions of the antibodies in Mihály et al., 1993; 1996). Two human hippocampi were investigated, too. The human brain samples were obtained from autopsy in the Department of Pathology, Szeged University. The autopsy was performed 16 h after death. The hippocampal tissue was kept in fixative solution (4% paraformaldehyde in 0.1 M phosphate buffer) for 12 days, then transverse plane frozen sections were made and stained with the monoclonal B-Raf antibody. Rats, guinea pigs and cats were deeply anesthetized and perfused through the heart with fixative solution (4% paraformaldehyde in 0.1 M phosphate buffer, pH 7.4). The brains were sectioned with freezing microtome, cryostat and vibratome. The tissue sections were immunostained as glass-mounted- or free-floating sections. Immunofluorescence was performed on cryostat sections, mounted on glass slides. The slides were incubated with primary antibodies (anti-Raf serum), then with secondary antibodies conjugated with fluorescein isothiocyanate (FITC). In peroxidase-based techniques, free-floating sections were used, incubated with the primary antibody, then treated according to the avidin-biotin method, using the Vectastain ABC kits (Vector, Burlingame, CA, USA). Independent from the antibody and method used, the staining pattern of the different brain areas was consistent, showing neuronal and glial localization of Raf kinases. The differences between B-Raf and C-Raf localizations are indicated in the text. No differences were detected in the staining pattern comparing the different species (rat, cat, guinea pig, human). The descriptions below refer to the results obtained with polyclonal v-Raf antibodies, if not indicated otherwise. Staining obtained with v-Raf antibodies will be referred to as B-Raf-like staining, staining with anti-SP 63 will be described as C-Raf-like immunoreactivity. Controls of the immunohistochemical procedure included absorption controls performed with the recombinant v-Raf protein and the SP 63 peptide. One mg of the protein and 10 mg of the peptide were reconstituted in 1 ml of deionized water, mixed with the undiluted antibody (0.01 ml), and kept at room temperature overnight with slow agitation. The mixture was diluted to working concentration (1:500), centrifuged at 70,000xg for 20 min, and used for the immunohistochemical procedure (Mihály et al., 1993). No immunostaining was seen after the procedure (Mihály et al., 1993). Other controls included incubations without the primary anti-Raf sera: in these experiments the primary antibody was omitted, and the sections were incubated in normal horse serum, diluted to 1:100, and secondary antibodies as usual. In these sections, no specific immunostaining was detected (Fig. 6e).

3.1 Localization of B-Raf in the spinal cord and brain stem

Neuronal and glial localization of Raf-protein-like immunoreactivity (RPI) was encountered in the spinal cord of the laboratory rat and guinea pig (Mihály & Rapp, 1993; 1994). Neurons in laminae I, V, VI and IX displayed strong cytoplasmic staining. Glial cells in the white matter and grey substance were stained, too (Fig. 7). Axonal staining (e.g.: in the dorsal horn, where afferent axons terminate in large numbers) was not observed (Mihály et al., 1996). Different levels of the brain stem displayed similar staining pattern: cytoplasmic staining in large- and medium-sized nerve cells. The motor neurons of the cranial nerve nuclei were strongly stained. Scattered RPI was detected in sensory nuclei, such as the superior and medial vestibular nuclei and the ventral cochlear nucleus. The raphe nuclei of the pons and medulla also contained RPI (Mihály & Endrész, 2000; Mihály et al., 2007). Large neurons of the reticular formation and the inferior olive were containing strong RPI (Fig. 3). The neurons of the mesencephalic trigeminal nucleus were strongly stained (Mihály et al., 2007). On the other hand, the spinal trigeminal nucleus displayed only faint staining (except lamina I neurons, which were strongly stained; Mihály et al., 2007). The red nucleus displayed strong staining, whilst the substantia nigra cells contained less RPI (Fig. 3). The immunostaining was always localized in cell bodies, no axonal staining was observed in any of the brain stem areas. Detailed examination of the brain stem of the rat and the guinea pig did not reveal any species differences (Mihály & Endrész, 2000; Mihály et al., 2007).

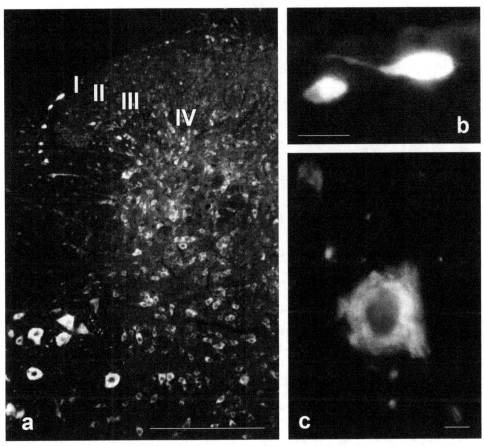

Fig. 2. B-Raf immunoreactivity in the spinal cord of the rat. The RPI is localized in cell bodies and proximal dendrites. No axonal staining is visible. Bar: 0.5 mm. Rexed laminae are indicated with roman numerals (a). Strong staing of lamina I neurons in the dorsal horn is conspicuous. Bar: 10 μm (b). Large motor neurons of lamina IX are labelled strongly, too. Bar: 10 μm (c). Immunolabeling was obtained with polyclonal anti-v-Raf serum.

3.2 Localization of B-Raf and C-Raf proteins in the cerebellum

The C-Raf and B-Raf proteins were detected in the vermis of the guinea pig cerebellum, using polyclonal anti-SP 63 (anti-C-Raf), and polyclonal anti-v-Raf (Mihály et al., 1993) antibodies. The B-Raf-like staining was very strong in neuronal cell bodies and glia-like cells. The staining was localized in neuronal cell bodies and dendrites. Dendritic staining was conspicuous in Purkinje cells, mainly in primary and secondary, large dendrites. Cell bodies of Purkinje cells were stained, too. Strong RPI was seen in the granular layer, where cell bodies of the granule cells and Golgi cells were immunoreactive. Small groups of glia cells stained in the white matter, and strong RPI was detected in the neurons of the fastigial nucleus (Fig. 4).

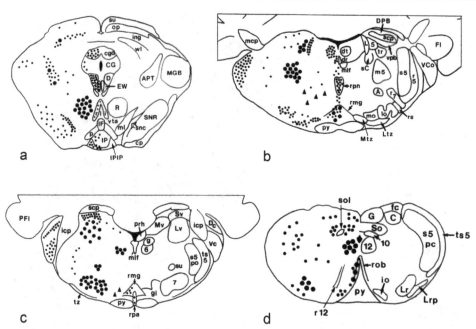

Fig. 3. Representative cross sections of the rat brain stem (Swanson, 1992), showing the cellular distrubution of Raf-protein-like immunoreactivity. Large dots represent strongly stained neurons, small dots represent medium or faint staining. Triangles represent large nerve cells with outstanding RPI. The upper mesencephalon (a), the upper pons (b), the lower pons (c) and the lower medulla (d) are represented (Mihály et al., 2007). Abbreviations: 5: mesencephalic nucleus of trigeminal nerve; m5: motor nucleus of trigeminal nerve; s5: principal sensory nucleus of trigeminal nerve; r5: radix of trigeminal nerve; s5po: spinal trigeminal nucleus pars oralis; s5pc: spinal trigeminal nucelus pars caudalis; ts5: spinal trigeminal tract; 6: abducens nucleus; 7: facial motor nucleus; g: genu of facial nerve; r7: radix of facial nerve; 10: dorsal nucleus of vagus nerve; 12: hypoglossal nucleus; r12: radix of hypoglossal nerve; D: Darkschewitsch nucleus; CG, cgd: periaqueductal gray; su, op, ing, wl: layers of the superior colliculus; R: red nucleus; SNR, snc: substantia nigra; EW: Edinger-Westphal nucleus; APT: anterior pretectal nucleus; MGB: medial geniculate body; IP, IPIP: interpeduncular nucleus; vta: ventral tegmental area; IF: interfascicular nucleus; p: paranigral nucleus; cp: cerebral peduncle; rli: raphe rostral linear nucleus; rmg: raphe magnus nucleus; rpn: raphe pontis nucleus; rpa: raphe pallidus nucleus; rob: raphe obscurus nucleus; dt: dorsal tegmental nucleus; mlf: medial longitudinal fasciculus; Lr, Lrp: lateral reticular nucleus; py: pyramidal tract; mo, lo: medial and lateral superior olivary nucleus; Mtz, Ltz: medial and lateral nucleus of the trapezoid body; Vco, Vc: ventral cochlear nucleus; Dc: dorsal cochlear nucleus; DPB, vpb: dorsal and ventral parabrachial nucleus; scp: superior cerebellar peduncle; mcp: middle peduncle; icp: inferior cerebellar peduncle; rs: rubrospinal tract; Mv, Lv, Sv: medial, lateral, superior vestibular nucleus; prh: prepositus hypoglossal nucleus; gi: gigantocellularis reticular nucleus; su: superior salivatory nucleus; sol: solitary tract; So: nucleus of the solitary tract; io: inferior olivary nucleus; Lr, Lrp: lateral reticular nucleus; G: gracile nucleus; C: cuneate nucleus; fc: fasciculus cuneatus; Fl: flocculus; PFl: paraflocculus.

The C-Raf staining was pale, displaying Purkinje cells and their large (primary and secondary) dendrites. Other neuronal elements (e.g.: granule cells, basket cells, Golgi cells) did not stain. C-Raf-like staining was seen in the nerve cells of the fastigial nucleus, and faint labeling was detected in some glia-like cells in white matter (Fig. 5).

3.3 Localization of B-Raf and C-Raf proteins in the cerebral cortex

RPI has been detected with polyclonal anti-v-Raf antibody and polyclonal anti-SP 63 antiserum. The latter is specific for C-Raf (Schultz et al., 1985; 1988; Morice et al., 1999). The RPI was detected in the neocortex (Fig. 6), cingular, pyriform, perirhinal, entorhinal areas and in the hippocampus (Mihály et al., 1993). RPI was localized in neurons and glia-like cells. No attempt was made to identify the cell types with double immunostainings. However, on the basis of the shape of the RPI-containing cells, we state that pyramidal and non-pyramidal cells of the Ammon's horn, granule cells of the dentate fascia and multipolar cells of the hilum of the dentate fascia were strongly stained (Mihály et al., 1993). Similar localizations were detected with the two antibodies, although C-Raf-like staining was faint compared to the B-Raf-like immunoreactivity. The intensity differences of the staining obtained with serial dilutions of the antibodies indicated that the cytoplasmic concentration of B-Raf kinase was much higher than that of the C-Raf in every region of the cerebral cortex. Detailed analysis of C-Raf-like staining and localization was not performed (Mihály et al., 1993). The brains of three domestic cats were also studied, using vibratome sections of the motor neocortex (sigmoid gyrus). Similarly to rats and guinea pigs, RPI has been detected in neuronal and glial cells bodies. Layer V pyramidal cells were stained in their cell bodies and proximal, large dendrites (Fig. 6). Scattered, small, glia-like cells were labeled, too. The human hippocampus displayed faint immunostaining with the B-Raf antiserum. Neuronal cell bodies were stained, which were similar to pyramidal cells, located in the stratum pyramidale of CA1 (Fig. 6).

3.4 Localization of Raf proteins in glia cells

B-Raf immunostaining was observed in small, glia-like cells, measuring 8-12 µm, in the cerebral cortex (gray and white matter), cerebellum (mentioned above), brain stem and spinal cord. Glial cells in the grey substance and in the whitte matter were regularly seen (Fig. 4; Fig. 7). Marginal astrocytes contained RPI in the spinal cord. Radial glia processes were stained in the cerebellar cortex (Fig. 4). No double immunolabelling was attempted in order to identify the cells. RPI was localized in the cytoplasm of the cell body and the processes (Fig. 7). No systematic study was made for the exploration of RPI in other glial cell types (oligodendroglia, microglia).

4. Electron microscopic localization of RPI in neurons

Ultrastructural immunohistochemistry was performed on vibratome sections of rat and cat cerebral cortex, using peroxidase-based preembedding methods (Mihály et al., 1991; Mihály & Rapp, 1994). These reports were the first in the literature, describing the electron microscopic localization of Raf kinases in the brain (Mihály et al., 1991; Mihály & Rapp, 1994). Comparison of the Raf kinase localization in a rodent and a carnivore, did not reveal differences: the localization pattern of RPI was consistent and similar in the two species.

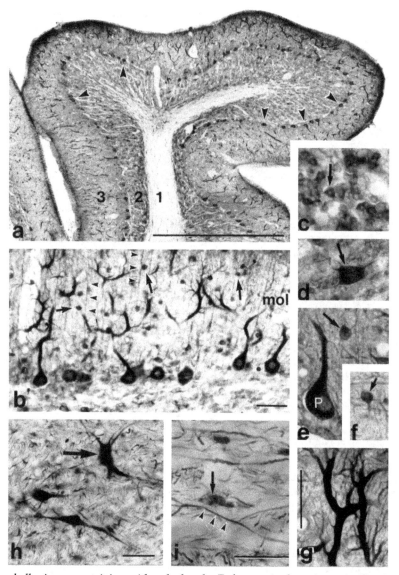

Fig. 4. Cerebellar immunostaining with polyclonal v-Raf serum in the guinea pig. Fig. 4a: Immunostaining depicts the layers of the cerebellar cortex: (1) white matter; (2) granular layer; (3) molecular layer; arrowheads: Purkinje cell layer; bar: 0.5 mm. Fig. 4b: the Purkinje cell RPI was detected not only in cell bodies, but also in dendrites (Fig. 4g). Arrowheads: radial glia processes; arrows: small neurons in the molecular layer (mol). Bar: 50 μm in Fig. 4b; 5 μm in Fig. 4g. The RPI in granule cells (Fig. 4c), Golgi cells (Fig. 4d), Purkinje cells (Fig. 4e) and outer stellate cells (Fig. 4f) are pointed by arrows. Bar as in Fig. 4g. Fig. 4h: the B-Raf staining of the neurons of the fastigial nucleus (arrow). Bar: 10 μm. Fig. 4i: white matter RPI apparently localized in glia cells (arrow). Arrowheads point to white matter axons containing RPI. Bar: 50 μm.

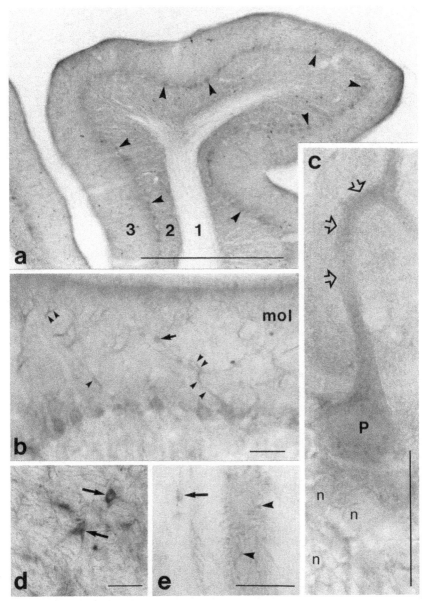

Fig. 5. Cerebellar immunostaining with anti-C-Raf serum in the guinea pig. Fig. 5a shows the overall weak staining (arrowheads: Purkinje cell layer; 1,2,3: as in Fig. 4a; bar: 0.5 mm). Fig. 5b: Purkinje cells contain C-Raf, which is localized in cell bodies and dendrites (arrowheads). Arrow points to outer stellate cell, which displays weak staining. Bar: 50 μm. Fig. 5c: Purkinje cell (P) staining with high magnification. Arrows point to dendrites (n: unstained granule cells; bar: 50 μm). Fig. 5d: neurons of the fastigial nucleus (arrows) containing C-Raf. Fig. 5e: white matter containing weakly stained glia cell (arrow) and weak axonal staining (arrowheads). Bars: 50 μm.

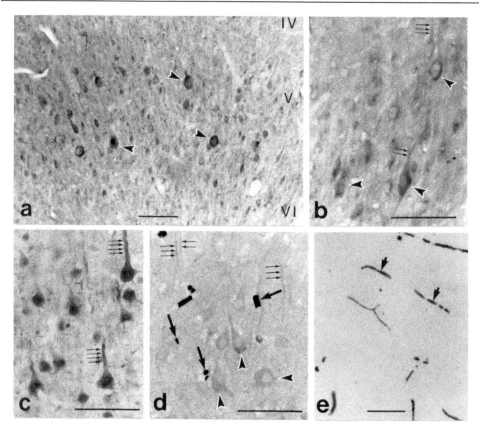

Fig. 6. B-Raf localization in the cat neocortex (a, b), rat neocortex (c) and human hippocampus (d, e). Bars: 50 μm. Roman numerals show neocortex layers. Arrowheads point to large nerve cells with immunostaining. Thin arrows in (b), (c) and (d) point to immunostained apical dendrites. Thick arrows in (d) and (e) point to red blood cells remaining in capillaries. Fig. 6e: control section was incubated in normal horse serum, the anti-B-Raf serum was omitted.

Neurons contain strong Raf kinase immunoreactivity in the cytoplasm of the cell body and dendrites (Figs 8-11). Not every neuron did stain: in the granule cell layer of the dentate fascia, cca. 20% of the granule cells remained unstained. This feature was visible already under the light microsocope, and was applicable to other cortical areas, as well (Mihály et al., 1993). As to the staining of the dendrites, very strong RPI was seen in dendritic spines (Figs 8-9; Fig. 11). The Raf-kinase-like staining was strong in postsynaptic densities, and in spine apparatuses (Fig. 9; Fig. 11; Mihály et al., 1991). Not only the spine apparatuses, but also dendritic subsynaptic cisternae contained RPI (Fig. 10d). In these membrane cisternal organelles the immunprecipitate was localized between the membrane cisternae; the cavity of the cisternae did not contain RPI (Fig. 11). We did not observe immunostained axon terminals in the tissues studied so far. None of the synapsing axons terminating on dendrites, dendritic spines or neuron somas did contain Raf-kinase-like immunostaining (see Figs 8-11). There are no comprehensive electron microscopic data about the ultrastructural localization of C-Raf, so this issue requires and deserves further, extensive experimentation.

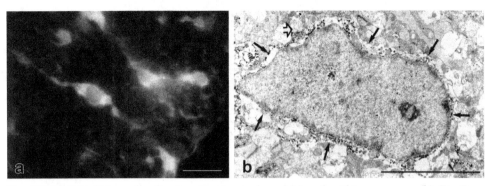

Fig. 7. Glial B-Raf kinase localization in rat spinal cord (a) and rat hippocampus (b). Fig. 7a displays RPI in astroglia-like cells of the white matter (bar: 10 µm). Fig. 7b shows the ultrastructural localization of B-Raf immunoreactivity in the cytoplasm of astrocyte-like cell of the rat hippocampus (solid arrows point to the contours of the astrocyte; empty arrow point to a synapse in the vicinity of the astrocyte; bar: 5 µm).

4.1 Electron microscopic localization of RPI in astrocytes

We observed several immunostained astrocytes in the cerebral cortex of rats (Fig. 7) and cats (Mihály & Rapp, 1994). The cytoplasmic staining of the cell body was conspicuous: very often in the vicinity of bundles of intermediate GFAP filaments (Mihály & Rapp, 1994). RPI was also observed in perivascular glial processes, and glial processes in the neuropil, around synapses (Mihály & Rapp, 1994). The RPI containing astrocytes were detected in the white matter, too – similarly to light microscopy observations. No RPI containing oligodendroglia cells were observed – in some sections we have seen RPI containing astrocytes together with unstained oligodendroglial cells (Mihály & Rapp, 1994). However, no systematic studies were performed to prove or disprove the lack of Raf proteins in oligodendroglia cells. The endothelial cells and pericytes were not labelled with any of the antibodies.

5. Comparison of B-Raf and C-Raf brain immunolocalization patterns

The localization of the two isozymes did not show substantial differences. The main observation in this respect was, that C-Raf kinase was expressed only in a few neurons and glia cells – compared to the expression of B-Raf. The staining was also weaker in most of the experiments. The C-Raf staining was specific, because the absorption controls did not show immunostaining (Mihály et al., 1993). The C-Raf was detected in the cerebellum and the cerebral cortex: these structures are compared in Table 1. B-Raf antibodies stained several neurons, irrespective of the size and type. However, staining was stronger in large cells (Figs 2-4). The B-Raf was also localized in astrocytes, as strong cytoplasmic staining. Under the electron microscope, the neuronal immunostaining was cytoplasmic, and exceptionally strong in dendritic spines, spine apparatuses and postsynaptic densities. Axon terminals did not stain. The C-Raf was localized in large neurons and their proximal dendritic branches. A few glia-like cells also displayed faint C-Raf-like staining (Fig. 5). We did not perform systematic electron microscopic studies for the localization of C-Raf kinase: only one series of experiments was done, in which we investigated the ultrastructural localization of C-Raf in guinea pig hippocampal slices (Mihály et al., 1991). In this electron microscopic study we observed neuronal and dendritic localization of C-Raf, thus corroborating the light microscopy findings (Mihály et al., 1991).

Isozymes	Neuronal localization	Glial localization	Other (e.g.: endothelium)
B-Raf	++++	++++	None
C-Raf	+	+	None

Table 1. Comparison of B-Raf and C-Raf localization and expression in the cerebellum and cerebral cortex of the rat and the guinea pig. The staining intensities are characterized by (+) symbols; (++++) means strong immunostaining; (+) means weak, but consistent staining.

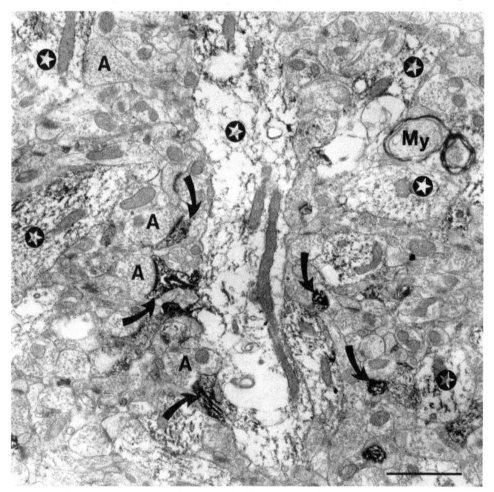

Fig. 8. B-Raf immunostaining in the rat cerebral cortex. Dendrites (stars) are labeled, axons (A) do not stain. The immunostaining of the dendritic spines (arrows) is extremely strong. My: myelinated axon. Bar: 1 μm.

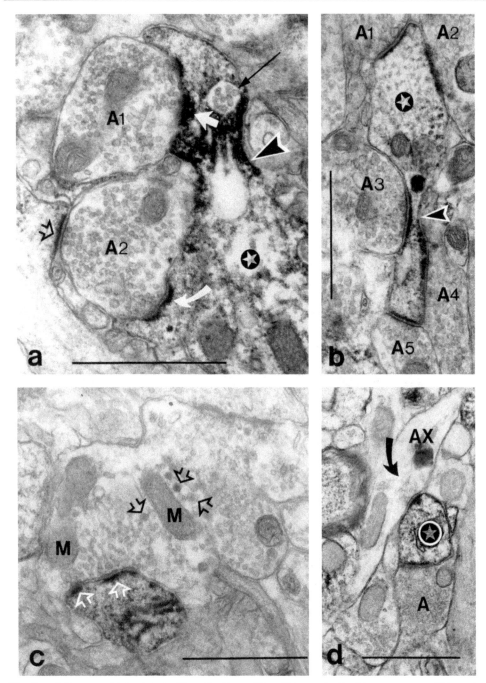

Fig. 9. a-d. Immunohistochemical localization of B-raf kinase in the hippocampus of the rat. In every case, postsynaptic dendrites and dendritic spines are immunoreactive. In (a) thick

arrows point to postsynaptic densities (thin arrow shows multivesicular body); arrowhead points to neck of spine, asterisk is in dendritic stem (A1, A2 are axon terminals); bar: 1 µm. In (b) dendrite (asterisk) is seen from which spine originates (arrowhead); A1-A5 are presynaptic axons; bar: 1 µm. In (c) mossy fiber terminal is unstained (M: mitochondria, black arrows point to dense core vesicles), postsynaptic dendrite (spine) is strongly immunoreactive (white arrows point to postsynaptic densities); bar: 1 µm. In (d) unstained axons (AX, A) synapsing with strongly stained small dendrite (asterisk); bar: 1 µm.

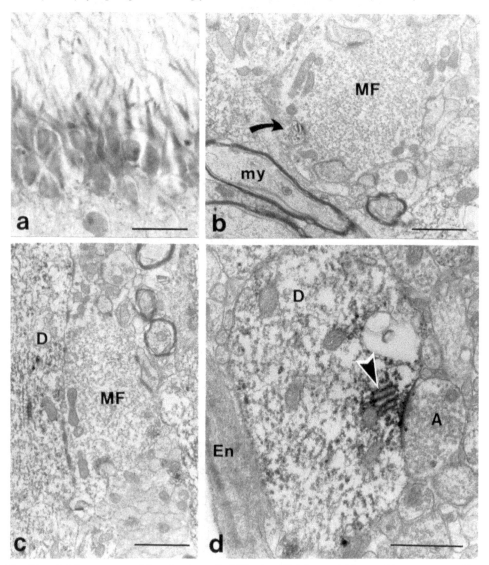

Fig. 10. a-d. Ultrastructural localization of RPI in the CA3 sector of the rat hippocampus. Fig. 9a displays the light microscopic appearance of the stratum pyramidale and stratum

lucidum after staining with polyclonal v-Raf antibodies. Cell bodies and apical dendrites are immunostained; mossy fibers do not stain (bar: 100 µm). Fig. 9b and c show two unstained mossy fiber ending (MF) synapsing with RPI containing dendrites (arrow and D). My: myelinated axon (unstained). Bars: 1 µm. Fig. 9d: large dendrite (D) containing RPI and immunostained subsynaptic cisternae (arrowhead). Synapsing axon (A) is unstained. En: unstained endothelial cell, bar: 1 µm.

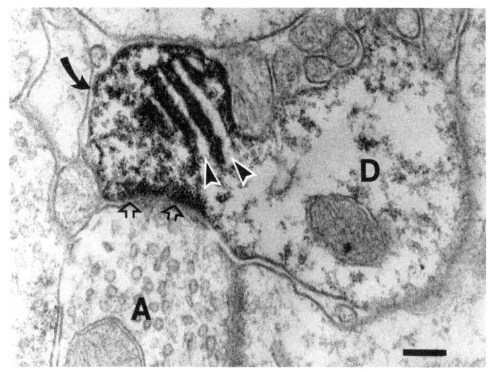

Fig. 11. Typical spine apparatus localization of B-Raf kinase in the rat cerebral cortex. The dendritic spine is pointed by the large solid arrow. The membranes of the spine apparatus are pointed by arrowheads. The postsynaptic density (empty arrows) is containing RPI, too. A: axon terminal; D: dendrite, containg RPI. Bar: 0.1 µm.

6. Discussion of the immunohistochemical findings

6.1 Neuronal B-Raf kinases

The immunolocalization of Raf kinases shows consistent patterns in different mammalian species (rat, guinea pig, cat and human). The main Raf molecule of the brain is certainly the B-Raf isozyme. The 95 kDa protein was detected in rat and guinea pig (Mihály et al., 1991; 1996; 2007; Mihály & Endrész, 2000). This observation conforms the literature data which give the same molecular weight in other experiments (Stephens et al., 1992; Dwivedi et al., 2006). Based on the specificity of the polyclonal v-Raf serum we refer to the immunostaining detected as to the B-Raf kinase immunostaining. This statement is strongly supported by the repeated Western blot experiments, which gave very consistent results: the polyclonal v-Raf serum and the B-Raf antibodies resulted in the same 95 kDa band in brain homogenates (Mihály et al., 1991; 2007; Mihály & Endrész, 2000). We did not perform immunoblotting with the C-Raf antibody because the polyclonal anti-SP 63 serum was thoroughly investigated (Schultz et al., 1985; 1988; Morice et al., 1999). The synthetic SP 63 peptide is the C-terminal part of the C-Raf protein (Schultz et al., 1985; 1988). This 12 amino acids peptide was used for immunization in rabbits (Schultz et al., 1985; 1988). The polyclonal serum was affinity purified, tested with immunoprecipitation and was found to be specific for the murine and human C-Raf terminal sequences in cell cultures and in tissue sections (Schultz et al., 1985; 1988; Mihály et al., 1993). This immunostaining is therefore referred to as the immunolocalization of C-Raf kinase. Neuronal B-Raf and C-Raf localization at the light microscopic level was found to be very similar in other experiments, years later (Morice et al., 1999). We think therefore, that our descriptions concerning the brain localization of Raf kinases are the first in the literature, and give a precise picture of neuronal B-Raf kinases (Mihály et al., 1991; 1993).

The main features of neuronal B-Raf kinase localization are as follows:

1. Cytoplasmic localization in the cell body was a general feature. This reflects the existence of a standard pool of the B-Raf kinase in the neuron. In case of growth factor signals, this cytoplasmic kinase pool can be mobilized quickly, in form of recruitment to the cell membranes or to the nucleus, if necessary.
2. Localization in dendrites was another characteristic feature of neuronal B-Raf. Dendritic staining was extremely strong in dendritic spines and in postsynaptic densities, meaning that a certain amount of B-Raf was already attached to the membrane, in proximity of the synaptic receptors. This localization suggested a very effective signal transduction process, mediated by the B-Raf.

6.2 Neuronal signal transduction with B-Raf: experimental proofs

The first observations pointing to the possible significance of Raf kinases in learning and memory were those of Mihály et al., (1990). In these experiments long term potentiation (LTP) was induced in rats, then the animals were subjected to immunohistochemical detection of Raf kinase with the polyclonal v-Raf antibodies (Mihály et al., 1990). The RPI was investigated in the granule cell layer of the dentate gyrus, by counting the immunoreactive neuronal cell bodies (Mihály et al., 1990). The number of RPI-containing cell bodies increased significantly in successfully potentiated animals (Mihály et al., 1990). This observation has been discussed either as the post-LTP increase of Raf kinase expression in dentate granule cells; or as the translocation of the Raf kinase from the distal dendrites to the cell body, following LTP (Mihály et al., 1990). We were not able to decide between these two possibilities in these in vivo experiments (Mihály et al., 1990). Years later, with help of in situ hybridization, it turned out that 24 h following LTP, the B-Raf expression in the dentate gyrus increased significantly (Thomas et al., 1994). Recently, experiments on B-Raf knockout mice proved, that the B-Raf is absolutely necessary for learning and memory consolidation (Chen et al., 2006; Valluet et al., 2010). However, the detailed mechanisms of the participation of B-Raf in LTP are unknown.

6.3 The possible intracellular translocation of B-Raf in neurons

Since the early immunohistochemical experiments on LTP (Mihály et al., 1990), the possibility of intracellular Raf translocation was an important issue, because the translocation could be one intracellular mechanism for signal transduction. The translocation was observed in vitro (Leevers et al., 1994) and in vivo (Oláh et al., 1991) in different experimental conditions. The recruitment to the cell membrane is considered to be a necessary step of Raf activation (Matallanas et al., 2011). However, the translocation to the cell nucleus is difficult to explain, although the experimental facts seem to be firm (Oláh et al., 1991). Another, translocation-like phenomenon was observed in sensory ganglion cells (Mihály et al., 1996). Primary sensory neurons contain B-Raf in their cell body, in form of homogeneous immunoreactivity (Mihály et al., 1996). Two-eight days after the transection of the peripheral nerve, this cytoplasmic staining pattern displays translocation towards the periphery, and at 8 days, most of the RPI was localized beneath the cell membrane of the sensory ganglion cells (Mihály et al., 1996). The significance of this phenomenon is not known, possibly it is connected to the process of chromatolysis (Mihály et al., 1996). The phenomenon of the translocation has been proved in cell cultures (see Mor & Philips, 2006), but not in vivo, in plasticity or pathological conditions. Further experiments are needed for the understanding of the possible in vivo translocation processes.

6.4 Possible functions of C-Raf in the neuron

Although C-Raf was found to be ubiquitous (Storm et al., 1990; Matallanas et al., 2011), not every neuron displayed immunoreactivity in our studies. These findings are in conform with other immunohistochemical studies (Morice et al., 1999). On the other hand, it seems, that C-Raf expression can be induced by some (yet unknown) conditions. Hippocampal slices maintained in vitro do not express C-Raf at the beginning of the in vitro incubation. Incubation of the slice for 2-4 h in oxygenized environment induced the appearance of the C-Raf staining in the nerve cells (Mihály et al., 1991). Upregulation of C-Raf was observed in Alzheimer-brains, too (Mei et al., 2006). In vitro grown neurons were protected by C-Raf inhibitors against the toxicity of amyloid peptides (Echeverria et al., 2008). It seems therefore, that the upregulation of C-Raf in the cell causes cell damage and death: it can happen through the activation of apoptosis signals (Echeverria et al., 2008). No data exist about the causes of C-Raf upregulation: excess excitatory amino acids (excitotoxic effects), hypoxic conditions, and ageing may operate through unknown signaling pathways in brain cells.

7. Conclusions

Raf protein kinases are widespread in the mammalian brain and spinal cord. Raf protein kinases are activated by several cytokines and growth factors through their membrane receptors. We investigated the immunohistochemical localization of B-Raf kinase and C-Raf kinase in neurons and glial cells. The B-Raf kinase is localized in the cell body, in dendrites, dendritic spines, spine apparatuses, subsynaptic cisternae and postsynaptic densities. The localization does not depend on neuronal type, but the intensity of the immunostaining is greater in large neurons. Neuronal B-Raf was not found in presynaptic axons, suggesting, that in these cells B-Raf kinase is coupled to postsynaptic signaling pathways. The similarities of the presence and localization of B-Raf in three mammalian orders (rodents, carnivores and hominids) suggest that this protein kinase has an important, and phylogenetically conserved function in normal, adult neurons. Data are available that learning and memory consolidation are processes in which B-Raf participates importantly. The C-Raf kinase has a similar CNS localization, but the amount of C-Raf in the cells is generally lower, than that of B-Raf. On the other hand, results from different experiments suggest, that the neuron adapts and alters the C-Raf expression, and thus regulates the amount of C-Raf present in the cytoplasm and in the membrane compartments. Therefore, we can state, that the regulation of the two isozymes (B-Raf and C-Raf) are different (although the downstream cellular events are similar):

1. CNS cells have a constant cytoplasmic pool of B-Raf, which translocates to different neighbouring organelles, in order to participate in signaling. The short-term regulation of B-Raf depends on membrane receptor stimulation.
2. CNS cells have another signaling molecule, the C-Raf, which displays low cytoplasmic concentration in normal neurons. However, the amount of cytoplasmic C-Raf is probably regulated by gene expression, the mechanisms of which are largely unknown. Therefore, C-Raf can be regulated twice: first by gene expression (upstream), second by membrane receptor activation (downstream). The significance and effects of this double regulation in the brain is not known: the effects probably manifest on the system level (e.g.: as alterations of the viability of neuronal networks).

8. Acknowledgement

This review was supported by the project „TÁMOP-4.2.1/B-09/1/KONV-2010-0005 – Creating the Center of Excellence at the University of Szeged". Financial support is from the European Union and the European Regional Fund. The author is thankful to Mrs. Katica Lakatos (Department of Anatomy) for technical help. We are thankful to Professor Dr. Ulf R. Rapp and Dr. Jakob Troppmair, Frederick Cancer Research and Development Center, Frederick, MD, USA, for providing the Raf antibodies for these experiments (see Acknowledgements in our referred articles).

9. References

App, H., Hazan, R., Zilberstein, A., Ullrich, A., Schlessinger, J., Rapp, U.R. (1991). Epidermal growth factor (EGF) stimulates association and kinase activity of Raf-1 with the EGF receptor. *Molecular Cell Biology*, Vol.11, pp. 913-919
Blackshear, P.J., Haupt, D.M.N., App, H., Rapp, U.R. (1990). Insulin activates the Raf-1 protein kinase. *Journal of Biological Chemistry*, Vol.265, pp. 12131-12134
Carrol, M.P., Clark-Lewis, I., Rapp, U.R., May, W.S. (1990). Interleukin-3 and granulocyte-macrophage colony-stimulating factor mediate rapid phosphorylation and activation of cytosolic c-Raf. *Journal of Biological Chemistry*, Vol.265, pp. 19812-19817
Chen, A.P., Ohno, M., Giese, K.P., Kühn, R., Chen, R.L., Silva, A.J. (2006). Forebrain-specific knockout of B-raf kinase leads to deficits in hippocampal long-term potentiation, learning and memory. *Journal of Neuroscience Research*, Vol.83, pp. 28-38

Dwivedi, Y., Rizavi, H.S., Conley, R.R., Pandey, G.N. (2006). ERK MAP kinase signaling in post-mortem brain of suicide subjects: differential regulation of upstream Raf kinases Raf-1 and B-Raf. *Molecular Psychiatry*, Vol.11, pp. 86-98

Echeverria, V., Burgess, S., Gamble-George, J., Arendash, G.W., Citron, B.A. (2008). Raf inhibition protects cortical cells against β-amyloid toxicity. *Neuroscience Letters*, Vol.444, pp. 92-96

Heidecker, G., Kolch, W., Morrison, D.K., Rapp, U.R. (1992). The role of Raf-1 phosphorylation in signal transduction. *Advances in Cancer Research*, Vol.58, pp. 53-73

Klysik, J., Theroux, S.J., Sedivy, J.M., Moffit, J.S., Boekelheide, K. (2008). Signaling crossroads: the function of Raf kinase inhibitory protein in cancer, the central nervous system and reproduction. *Cell Signal*, Vol.20, pp. 1-9

Kolch, W., Schultz, A.M., Oppermann, H., Rapp, U.R. (1988). Preparation of Raf-oncogene-specific antiserum with Raf protein produced in E. coli. *Biochimica Biophysica Acta*, Vol.949, pp. 233-239

Leevers, S.J., Paterson, H.F., Marshall, C.J. (1994). Requirement for *ras* in *raf* activation is overcome by targeting *raf* to the plasma membrane. *Nature*, Vol.369, pp. 411-414

Matallanas, D., Birtwistle, M., Romano, D., Zebisch, A., Rauch, J., von Kriegsheim, A., Kolch, W. (2011). Raf family kinases: old dogs have learned new tricks. *Genes & Cancer*, Vol.2, No.3, pp. 232-260

Mei, M., Su, B., Harrison, K., Chao, M., Siedlak, S.L., Previll, L.A., Jackson, L., Cai, D.X., Zhu, X. (2006). Distribution, levels and phosphorylation of Raf-1 in Alzheimer's disease. *Journal of Neurochemistry*, Vol.99, pp. 1377-1388

Mihály, A., Oláh, Z., Krug, M., Matthies, H., Rapp, U.R., Joó, F. (1990) Transient increase of raf protein kinase-like immunoreactivity in the rat dentate gyrus during long-term potentiation. *Neuroscience Letters*, Vol.116, pp. 45-50

Mihály, A., Kuhnt, U., Oláh, Z., Rapp, U.R. (1991) Induction of raf-1 protein immunoreactivity in guinea pig hippocampal slices during the *in vitro* maintenance. *Histochemistry* Vol.96, pp. 99-105

Mihály, A., Oravecz, T., Oláh, Z., Rapp, U.R. (1991) Immunohistochemical localization of raf protein kinase in dendritic spines and spine apparatuses of the rat cerebral cortex. *Brain Research*, Vol.547, pp. 309-314

Mihály, A., Endrész, V., Oravecz, T., Rapp, U.R., Kuhnt, U. (1993) Immunohistochemical detection of raf protein kinase in cerebral cortical areas of adult guinea pigs and rats. *Brain Research*, Vol.627, pp. 225-238

Mihály, A., Rapp, U.R. (1994) Expression of the raf protooncogene in glial cells of the adult rat cerebral cortex, brain stem and spinal cord. *Acta histochemica*, Vol.96, pp. 155-164

Mihály, A., Priestley, J.V., Molnár, E. (1996) Expression of raf serine/threonine protein kinases in the cell bodies of primary sensory neurons of the adult rat. *Cell & Tissue Research*, Vol.285, pp. 261-271

Mihály, A., Endrész, V. (2000) Neuronal expression of Raf protooncogene in the brain stem of adult guinea pig. *Acta histochemica*, Vol.102, pp. 203-217

Mihály, A., Karcsú-Kiss, G., Bakos, M., Bálint, E., (2007) Cellular distribution of B-raf protein kinase in the brainstem of the adult rat. A fluorescent immunohistochemical study. *Acta Biologica Szegediensis*, Vol.51, pp. 7-15

Mor, A., Philips, M.R. (2006). Compartmentalized Ras/MAPK signaling. *Annual Review of Immunology*, Vol.24, pp. 771-800

Morice, C., Nothias, F., König, S., Vernier, P., Baccarini, M., Vincent, J-D., Barnier, J.V. (1999). Raf-1 and B-Raf proteins have similar regional distributions but differential subcellular localization in adult rat brain. *European Journal of Neuroscience*, Vol.11, pp. 1995-2006

Morrison, D.K., Kaplan, D.R., Escobedo, J.A., Rapp, U.R., Roberts, T.M., Williams, L.T. (1989). Direct activation of the serine/threonine kinase activity of Raf-1 through tyrosine phosphorylation by the PDGF β-receptor. *Cell*, Vol.58, pp. 649-657

Niault, T.S., Baccarini, M. (2010). Targets of Raf in tumorigenesis. *Carcinogenesis*, Vol.31, No.7, pp. 1165-1174

Oláh, Z., Komoly, S., Nagashima, N., Joó, F., Rapp, U.R., Anderson, W.B. (1991). Cerebral ischemia induces transient intracellular redistribution and intranuclear translocation of the raf protoncogene product in hippocampal pyramidal cells. *Experimental Brain Research*, Vol.84, pp. 403-410

Oshima, M., Shitanandam, G., Rapp, U.R., Guroff, G. (1991). The phosphorylation and activation of B-Raf in PC 12 cells stimulated by nerve growth factor. *Journal of Biological Chemistry*, Vol.266, pp. 23753-23760

Rapp, U.R., Cleveland, J.L., Bonner, T.I., Storm, S.M. (1988). The raf oncogenes, In: *The Oncogene Handbook*, E.P. Reddy, A.M. Skalka, T. Curran, (Eds), pp. 213-251, Elsevier, Amsterdam, The Netherlands

Schultz, A.M., Copeland, T.D., Mark, G.E., Rapp, U.R., Oroszlan, S. (1985). Detection of the myristylated gag-raf transforming protein with raf-specific antipeptide sera. *Virology*, Vol.146, pp. 78-89

Schultz, A.M., Copeland, T.D., Oroszlan, S., Rapp, U.R. (1988). Identification and characterization of c-raf phosphoproteins in transformed murine cells. *Oncogene*, Vol.2, pp. 187-193

Stephens, R.M., Sithanandam, G., Copeland, T.D., Kaplan, D.R., Rapp, U.R., Morrison, D.K. (1992). 95-kilodalton B-Raf serine/threonine kinase: identification of the protein and its major autophosphorylation site. *Molecular Cell Biology*, Vol.12, pp. 3733-3742

Storm, S.M., Cleveland, J.L., Rapp, U.R. (1990). Expression of Raf family proto-oncogenes in normal mouse tissues. *Oncogene*, Vol.5, pp. 345-351

Swanson, L.W. (1992). *Brain Maps: Structure of the Rat Brain*, Elsevier, New York, USA

Thomas, K.L., Laroche, S., Errington, M.L., Bliss, T.V., Hunt, S.P. (1994). Spatial and temporal changes in signal transduction pathways during LTP. *Neuron*, Vol.13, pp. 737-745

Yang, H., Lu, D., Raizada, M.K. (1997). Angiotensin II-induced phosphorylation of the AT1 receptor from rat brain neurons. *Hypertension*, Vol.30, pp. 351-357

Valluet, A., Hmitou, I., Davis, S., Druillennec, S., Larcher, M., Laroche, S., Eychéne, A. (2010). B-raf alternative splicing is dispensable for development but required for learning and memory associated with the hippocampus int he adult mouse. *PLoS ONE*, Vol.5, pp. 1-9

Zebisch, A., Troppmair, J. (2006). Back to the roots: the remarkable RAF oncogene story. *Cellular and Molecular Life Sciences*, Vol.63, pp. 1314-1330

9

Protein Kinases in Spinal Plasticity: A Role for Metabotropic Glutamate Receptors

J. Russell Huie and Adam R. Ferguson
Brain and Spinal Injury Center (BASIC), University of California, San Francisco,
USA

1. Introduction

Neural plasticity is characterized by the lasting modulation of synaptic strength which alters the central nervous system's (CNS) capacity to encode and store information. Changes in synaptic plasticity have implications for brain-dependent learning and memory as well as a number of other forms of CNS information processing, including alterations in spinal cord function (reviewed in Patterson & Grau, 2001). As protein kinases have been shown to greatly affect neurotransmitter receptor dynamics, their role in synaptic plasticity is essential. A large body of work has highlighted the link between group I metabotropic glutamate receptor (mGluR) activation, subsequent protein kinase activation, and synaptic plasticity (Gereau & Heinemann, 1998; Sheng et al., 2002; Gallagher et al., 2004). Upon glutamate binding to mGluRs, a G-protein coupled to the receptor sets off an intracellular cascade, activating phospholipase C (PLC), diacyl glycerol (DAG), and ultimately protein kinases. Once activated, protein kinases can then exert a modulatory effect on both excitatory and inhibitory receptors that ultimately affects synaptic strength. In this way, mGluRs and protein kinases both play critical roles in a number of forms of neural plasticity, including those that are thought to underlie learning and memory (Bortolotto & Collingridge, 1993; Wang, et al. 2004). What many in neuroscience have overlooked is the fact that, just like in the brain, the *spinal cord* also exhibits the capacity for an amazing amount of plasticity, including simple forms of learning and memory that are relevant to pain, motor learning, and recovery of function after spinal cord damage (Woolf & Salter, 2000; Grau et al., 2006; Raineteau & Schwab, 2001). And, just as in the brain, spinal plasticity has been shown across a number of preparations to be mediated by mGluR/protein kinase signaling (Giles et al., 2007; Ferguson et al., 2008a).

This chapter will detail the specific role of protein kinase C (PKC) as an intermediary between initial mGluR activity and long-term changes in synaptic strength, and how this critical interaction affects a number of spinal processes. We will highlight how PKC and its isoforms provide a critical link between initial glutamatergic input and the alterations in receptor phosphorylation and trafficking that lead to spinal plasticity. We will first look at the role of mGluR/PKC signaling in the dorsal horn of the spinal cord, and the implications of this process on pain. We will then explore how the mGluR/PKC pathway also exerts modulatory control over the changes in plasticity (or *metaplasticity)* in the spinal cord, as evidenced in a spinal cord learning preparation. In addition, we will consider how the reorganization and reclamation of appropriate function after spinal cord injury is affected in large part by various forms of mGluR-mediated, protein kinase-dependent plasticity. Finally, we will investigate the neurobiological consequences and possible therapeutic potential to be found in altering protein kinase activity.

2. Nociceptive plasticity: mGluRs, protein kinase C, and pain

mGluR-mediated activation of protein kinases can modulate synaptic strength, by altering presynaptic and postsynaptic signaling. While these forms of synaptic plasticity have been suggested to underlie learning and memory in the hippocampus, similar mechanisms at work in the dorsal horn of the spinal cord produce a decidedly different form of learning that contributes to neuropathic pain. As the dorsal horn is the locus for the integration of incoming sensory information, sensitization of these neurons and their synaptic targets can have a dramatic behavioral effect (Mendell, 1966; Woolf, 1983). If a strong nociceptive signal is relayed from the periphery to the superficial dorsal horn, neurons in this area become sensitized, in a fashion that is mechanistically very similar to long-term potentiation (LTP), an electrophysiological mechanism believed to underlie learning and memory in the hippocampus (Sandkuhler & Liu, 1998; Bliss & Collingridge, 1993). This effect has been termed *central sensitization* (Woolf, 1983). As in any form of potentiation, subsequent input following sensitization can elicit a response even if it is much weaker than the initial input. That is, even those stimuli that would not normally be considered painful may now have the capacity to elicit a nociceptive response *(allodynia)*. Likewise, normally painful stimuli can now induce a much more robust nociceptive response *(hyperalgesia)*. These phenomena are conserved across species, and this makes sense from an evolutionary standpoint. To be rapidly sensitized to a painful stimulus is indeed adaptive, in that an organism will 'learn' to avoid this stimulus. Although this type of plasticity is essential for self-preservation, it can be problematic if dysregulated. Neural insult, whether it be an injury to the peripheral nervous system (e.g., nerve damage) or central nervous system (e.g., spinal cord injury), can produce an unregulated barrage of nociceptive input that may have no external initiator. This can produce a persistent pain response, and lasting nociceptive plasticity, that is generated wholly within the organism (neuropathic pain) (Kim & Chung, 1992; Christensen & Hulsebosch; Willis & Coggeshall, 1991; Willis & Westlund, 1997; Lindsey et al., 2000). A number of studies have outlined an important role for spinal protein kinase C (PKC) in mediating persistent pain states, including hyperalgesia, allodynia, and neuropathic pain (Coderre, 1992; Sun et al., 2004; Hua et al. 1999). As a major activator of PKC, group I mGluRs have been implicated as a driver of nociceptive plasticity (Yashpal et al., 2001; Fisher & Coderre, 1996; Adwanikar et al., 2004).

mGluRs activate PKC in two ways: both directly, through the G-coupled protein-PLC-DAG pathway, as well as indirectly by freeing intracellular Ca^{++} via activation of phosphoinositide-3 kinase (PI3K) pathway. *In vitro* work has shed light on how PKC works to sensitize spinal neurons and elicit persistent nociceptive behavior. PKC appears to have the capacity to induce sensitization through both presynaptic and postsynaptic effects. Presynaptically, PKC has been shown to phosphorylate voltage-gated calcium channels, thus increasing intracellular calcium and promoting neurotransmitter release (Yang et al., 2005). Postsynaptically, PKC can facilitate excitatory tone through actions on ionotropic glutamate receptors (Lu et al., 1999; Li et al., 1999). Following post-synaptic binding of glutamate to mGluRs, and subsequent PKC activation, PKC then phosphorylates NMDA receptors, increasing their open probability (Liao et al., 2001). Further, PKC has been shown to induce the rapid exocytosis of AMPA receptors (Li et al., 1999). Taken together, these excitatory effects have been suggested to play a critical role in sensitizing dorsal horn neurons (Ji et al., 2003).

In the brain, however, mGluR activation can have contradictory effects. For example, in hippocampal culture, mGluR activation has been shown to cause *internalization* of AMPA receptors leading to long-term *depression* (LTD) of postsynaptic potentials rather than sensitization (Huber et al., 2000; Oliet et al., 1997; Moult et al., 2002). This effect has been shown to depend on activation of the immediate early gene Arc/Arg 3.1 (Waung et al., 2008) as well as PKC (Camodeca et al., 1999; Oliet et al., 1997). On the other hand, mGluR-induced activation of PKC has also been implicated in LTP (Jia et al., 1998; Balschun et al., 1999; Aiba, et al., 1994; Conquet et al., 1994; Bortolotto, et al., 1994; Anwyl et al., 2009). This confusion about the bi-directional role of mGluRs and downstream PKC in hippocampal-dependent synaptic plasticity has been reviewed elsewhere (Bortolotto et al., 1999; Malenka & Bear, 2004) and the role of mGluRs in LTP and LTD remains a topic of ongoing research in the hippocampal plasticity literature (Mockett et al., 2011). It is clear that much more work is required to reconcile the hippocampal literature with the spinal plasticity literature regarding mGluR-PKC activation.

Tests of the necessity of mGluRs and PKC *in vivo* have confirmed the critical role of this pathway in spinally-mediated nociceptive plasticity. In order to study persistent inflammation, researchers will often give a subcutaneous injection of a noxious substance to the periphery, and then assess the effect this treatment has on the activity of dorsal horn neurons within nociceptive fields (Harris & Ryall, 1988; LaMotte et al., 1992). To investigate the specific role of mGluRs on this type of nociceptive plasticity, Young et al. (1995) gave an injection of the noxious substance algogen (mustard oil) to the hindpaws of rats, a substance known to evoke sustained activity in neurons of the dorsal horn. This treatment was followed by microinjections of the mGluR antagonist CHPG directly into the dorsal horn. They found that blocking mGluR activity with CHPG strongly inhibited mustard oil-evoked activity in these nociceptive areas. Beyond the necessity for mGluRs in sustained nociceptive activity, they also demonstrated the sufficiency for an mGluR agonist (ACPD) to *produce* sustained activity in these neurons (Young et al., 1995). Interestingly, Munro et al. found that treatment with a PKC inhibitor (chelerythrine or GF109203X) was able to inhibit both mustard oil- and ACPD-evoked activity (Munro et al. 1994). Together, these findings provide a strong case for mGluR/PKC signaling in mediating at least one form of long-term nociceptive plasticity. In 2001, Yashpal and colleagues demonstrated a further role for mGluR/PKC interaction in chronic neuropathic pain. Following a chronic constriction of the sciatic nerve (known to produce long-term nociceptive activity in the dorsal horn), they found that membrane localization of PKC was increased. Behaviorally, such an injury manifests as a chronic and robust hypersensitivity to both touch (mechanical allodynia) and temperature (thermal hyperalgesia). But, if an mGluR inhibitor was given prior to and after the injury, they found a decrease in the membrane-bound PKC expression, as well as a decrease in the subjects' injury-induced mechanical and thermal hypersensitivity.

Based on the mechanistic similarities, many researchers have drawn the distinct parallels between learning/memory and pain (Sandkuhler, 2000; Ji et al., 2003; Latremoliere and

Woolf, 2009). It is perhaps not surprising then that recent breakthroughs in our understanding of long-term memory storage in the brain continue to shed light on persistent pain syndromes that are mediated by the spinal cord. The protein kinase PKMζ (an isoform of PKC) is unique in that, unlike most enzymes, it can remain active for extremely long periods of time. This persistence is possibly due to the fact that activated PKMζ inhibits the protein PIN1, an inhibitor of PKMζ mRNA translation. Thus, by blocking the action of its regulator, PKMζ effectively creates a positive feedback loop, perpetuating PKMζ activity (Sacktor, 2011). Once PKMζ becomes active, it can promote trafficking of the major fast excitatory ionotropic glutamate receptor (AMPAR) to post-synaptic membranes, which in turn aids in potentiating the synapse (Sacktor, 2008; Migues et al., 2010; Malinow & Malenka, 2002). Further, PKMζ has been shown to also *inhibit* the internalization of these trafficked AMPARs (Yao et al., 2008). In this way, synaptic strength is not only induced, but also *maintained*. This finding has led many to investigate PKMζ has a critical component in long-term memory. Researchers have found that inhibiting PKMζ produces profound behavioral effects. A number of recent studies have shown that treatment with a PKMζ inhibitor effectively eliminates long-term memories from a number of learning paradigms (Pastalkova et al., 2006; Shema et al. 2007; Parsons & Davis, 2011; Madronal et al., 2010).

While these findings provide compelling evidence for the necessity for PKMζ in long-term memory, erasing long-term memories may not be a likely sought-after therapy. But consider the problem of chronic pain. As we have discussed, long-term neuropathic pain bears a striking mechanistic resemblance to memory, yet it is often regarded to be biologically dysfunctional. Therefore, inhibiting PKMζ may prove to be a very attractive therapeutic tool in overcoming persistent neuropathic pain. Asiedu and colleagues have recently shown this idea to be entirely possible. They initially primed rat subjects with a peripheral intraplantar injection of the inflammatory cytokine IL-6 or vehicle to the hindpaw. This treatment has previously been shown to induce allodynia for up to 3 days after injection (Asiedu et al., 2011). They then injected the mGluR agonist DHPG intrathecally to the spinal cord 6 days after initial peripheral injection, and found that DHPG injection produced a markedly enhanced nociceptive response in those subjects that had been primed days earlier with IL-6. This finding suggested that the maintenance of peripheral nociceptive sensitization was mediated centrally in the spinal cord and led them to investigate the possibility that PKMζ might mediate the storage of this nociceptive 'memory'. They found that intrathecal administration of the PKMζ inhibitor ZIP attenuated the capacity for DHPG to evoke the expression of nociceptive behavior. This lends support for the argument that, as in learning and memory in the brain, this persistent nociceptive sensitization reflects an LTP-like mechanism. Given that PKMζ has been shown to maintain LTP by inhibiting the internalization of AMPARs, and this effect can be disrupted by the peptide pep2m (Yao et al. 2008), Asiedu and colleagues hypothesized that this same mechanism is involved in the maintenance of nociceptive hypersensitivity. To test this, they gave an intrathecal injection of pep2m, and found that this treatment also blocked the DHPG evoked expression of sensitization (Asiedu et al., 2011). Together these findings suggest that the maintenance of nociception involves a PKMζ-dependent process within the spinal cord, and lends confirmatory evidence that the maintenance of LTP, memory, and nociception may be mediated by a common mechanism. In the future, PKMζ inhibition could hold very promising therapeutic potential for those suffering from chronic pain.

3. Learning in the spinal cord: Metaplasticity is PKC-dependent

The spinal cord supports a number of other forms of plasticity beyond just nociception. Throughout development, ventral motor neurons undergo a great amount of plasticity, as complex motor skills and locomotion are honed. Although the spinal cord was once believed to be fairly hard-wired after development, we now understand that the capacity for ongoing plasticity in spinal motor neurons persists throughout life (Edgerton et al. 2001; Courtine et al. 2008; De Leon et al. 2001; Wolpaw, 2007; Grau et al., 2006). This is evident in the spinal cord injury literature from the previous decade, where researchers have demonstrated the ability for spinal cord injured-subjects to regain locomotor function through the use of behavioral training, often in combination with pharmacological agents that facilitate plasticity (Wernig et al., 2000; Rossignol, 2007; Edgerton & Harkema, 2011). Promoting this kind of adaptive, use-dependent spinal plasticity is essential in order to realize functional recovery after injury.

Despite advances in our awareness of the spinal cord's capacity for plasticity, the underlying mechanisms dictating use-dependent spinal cord plasticity still require investigation. In order to better understand the unique role of the spinal cord in neural plasticity, outside of any supraspinal input, researchers have developed an *in vivo*, behavioral method for measuring plasticity in the isolated spinal cord. Building upon earlier work from Chopin and Buerger, Grau and colleagues demonstrated that following a complete spinal transection, spinal neurons below the lesion were able to support a simple form of instrumental (response-outcome) learning (Buerger and Fennesey, 1970; Chopin & Buerger, 1976; Grau et al., 1998). In this preparation, transected rats receive an electrical shock to the tibilias anterior muscle of their hindlimb whenever that limb is in an unflexed position (see Figure 1A). This stimulation causes a flexion of the hindlimb, at which point the stimulation is terminated. When the limb again falls to a resting, unflexed position, another shock is delivered.

Without input from the brain, spinalized subjects will learn to keep the hindlimb flexed in order to reduce exposure to the stimulation (Fig 1C). This form of spinal learning can also be inhibited: if subjects are given electrical stimulation that is not contingent upon limb position (*intermittent stimulation*), they will later fail to learn to keep their hindlimb flexed when tested with response-contingent stimulation (*controllable stimulation*; Figure 1B; Crown et al., 2002). Although these subjects are not learning the target response, they are still exhibiting a form of plasticity. Essentially, they have *learned* from the exposure to intermittent stimulation, that their limb position is not related to stimulation exposure, and thus fail even when later tested with controllable stimulation. This phenomenon has been considered analogous to the phenomenon of *learned helplessness* (Grau et al., 1998; Seligman & Maier, 1967). In contrast to the maladaptive effects of intermittent stimulation, training with controllable stimulation can *enhance* future learning (Grau et al., 1998). Subjects that have previously learned this instrumental task can be tested in the future with a more difficult response criterion (one that untrained subjects would not be able to exhibit), and this prior training *facilitates* learning. While both the learning deficit and the facilitation of learning are forms of plasticity, they are something more. Both of these phenomena affect lasting *change* in plasticity: a lasting alteration in the threshold at which learning occurs, either shifting the threshold up (in the case of intermittent stimulation inducing a future learning deficit) or down (in the instance of instrumental training facilitating future learning). In essence, these experience-dependent spinal changes represent a *plasticity* of plasticity. Abraham and Bear (1996) first described this type of plasticity of plasticity, characterizing it as "a higher-order form of synaptic plasticity" that they termed *metaplasticity* (Abraham and Bear, 1996). Uncontrollable stimulation induces a lasting alteration that undermines spinal learning, that can be described as a metaplastic inhibition of adaptive plasticity. Importantly, the same stimulation parameters that induce spinal learning deficits also undermine long-term recovery of locomotor function following a spinal contusion injury (Grau et al., 2004). Thus, a better understanding of the neurobiology underlying metaplastic inhibition of adaptive plasticity in the spinal cord will aid in the development of strategies to aid in functional recovery after spinal cord injury.

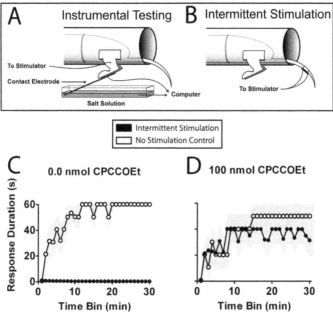

Fig. 1. Instrumental learning model of spinal plasticity, and the role of mGluRs in this phenomenon. A) Spinalized rat subjects are given an electrical shock each time their hindlimb is in an unflexed position (*controllable stimulation*). Over time, they learn to increase their response durations in order to reduce exposure to the stimulation, thus encoding an instrumental (response-outcome) relationship. B) If uncontrollable, intermittent stimulation is administered prior to instrumental testing, the subjects are unable to learn the relationship. C) In subjects that are given vehicle treatment (0.0 nmol of mGluR antagonist CPCCOEt), intermittent stimulation produces a significant spinal learning deficit. D) Treatment with 100 nmol CPCCOEt blocks the deficit induced by intermittent stimulation, suggesting a necessary role for mGluR activity in this effect. Adapted from Ferguson et al., 2008a.

Previous work has shown that metaplasticity in the hippocampus involves ionotropic glutamate receptor trafficking (Hellier et al., 2007). Given that mGluRs modulate ionotropic receptor function and trafficking through a PKC-mediated mechanism, we recently investigated the role of this mechanism in the metaplastic inhibition of spinal instrumental learning (Ferguson et al., 2008a). We first tested whether group I mGluRs are necessary for the metaplastic inhibition of spinal learning. We found that intermittent stimulation had no effect on future spinal learning if given after an intrathecal injection of an mGluR antagonist (CPCCOEt or MPEP, Fig. 1D). We next considered the contribution of PKC to this learning deficit. We found that in response to intermittent stimulation (which produces a lasting metaplastic inhibition of spinal learning) PKC activity in the spinal cord was significantly increased (Figure 2A). Similarly, if PKC inhibitors (bisindolylmaleimide or chelerythrine) were delivered intrathecally prior to intermittent stimulation, subjects exhibited no metaplastic inhibition of learning when tested 24 hours later. These data provided strong evidence that both mGluR and PKC activity are necessary in producing this form of spinal metaplasticity. To further examine the role of mGluR/PKC in this phenomenon, we tested whether pharmacological activation of mGluRs was sufficient to produce metaplastic inhibition of spinal learning. We found that a single bolus of the mGluR agonist DHPG was able to produce a spinal learning deficit that lasted at least 24 hours. We also found that PKC activity blockade (with chelerythrine or bisindolylmaleimide) prior to administration of DHPG, prevented metaplastic inhibition of spinal learning. These findings suggest an essential role for the mGluR/PKC pathway in mediating metaplasticity in the spinal cord.

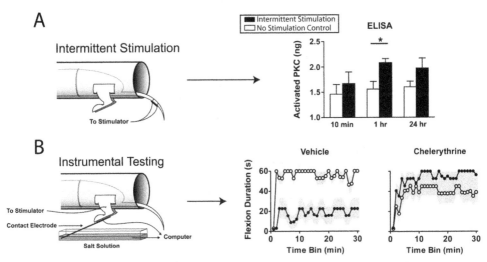

Fig. 2. Role for PKC in the metaplastic inhibition of spinal learning. A) Intermittent stimulation (which induces a lasting metaplastic inhibition of spinal learning) produces an increase in the expression of activated PKC that is significantly greater than unstimulated controls at 1 hour. B) When tested for spinal instrumental learning, vehicle-treated subjects that had received intermittent stimulation fail to learn; intrathecal injection of the PKC inhibitor chelerythrine blocks intermittent stimulation from producing a learning deficit. Adapted from Ferguson et al., 2008a.

As discussed above, PKC is known to alter the open-channel probability of NMDARs. Spinal learning, like many other forms of plasticity, has been characterized by its dependence on subtle, precise alterations in NMDAR function. Thus, it is likely that increased potentiation of NMDARs by the mGluR/PKC pathway upsets a delicate balance, pushing NMDARs (and subsequent intracellular calcium levels) beyond the range in which spinal learning can occur. Other work has also implicated other protein kinase activity in both the adaptive and maladaptive metaplastic changes in spinal learning. Inhibition of calcium/calmodulin-dependent protein kinase II (CaMKII) has been shown to block the development of the long-term inhibition of spinal learning if given after uncontrollable shock, and also blocks the facilitation effect of instrumental training if given prior to training (Baumbauer et al., 2007; Gomez-Pinilla et al., 2007). These findings suggest that protein kinase activity may engage a common mechanism in different forms of spinal metaplasticity. Future work will be necessary to elucidate how the various protein kinases interact and integrate to produce these lasting behavioral changes. Further, if these mechanisms are at work in the injured spinal cord, we can begin to develop therapeutic strategies that can reduce maladaptive metaplasticity, and promote the adaptive plasticity necessary for successful rehabilitation.

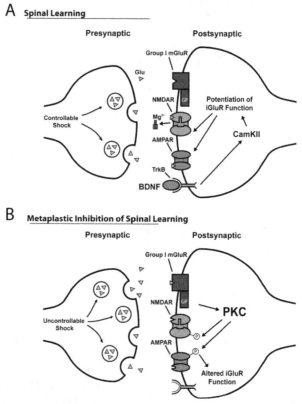

Fig. 3. Possible cellular mechanism for A) spinal learning and B) metaplastic inhibition of spinal learning. Both forms of spinal plasticity have been shown to involve protein kinase signaling. In contrast to controllable stimulation, uncontrollable stimulation is believed to engage group I metabotropic glutamate receptors, which leads to the downstream activation of PKC. PKC in turn alters ionotropic glutamate receptor function, which is believed to induce a lasting saturation of the synapse, inhibiting future learning. Adapted from Ferguson et al., 2008a.

4. Spinal cord injury: Protein kinase modulation as therapy

The previous sections have focused on spinal plasticity in isolated systems: how the mGluR/PKC pathway modulates nociceptive processing in the dorsal horn, and how it mediates the metaplastic inhibition of a ventral motor learning task. These models yield insight into the mechanisms by which plasticity in the spinal cord occur, and demonstrate that protein kinase activity is an essential step in mediating long-term neural modifications. In this final section, we will consider the therapeutic potential in altering protein kinase activity for spinal cord injury and related CNS disorders.

Within the wave of secondary processes following spinal cord injury, high levels of glutamate release can have a devastating effect on cell survival (Crowe et al., 1997; McAdoo et al., 1999; Ferguson et al., 2008b). As mGluR activation leads to PLC-mediated release of intracellular calcium stores, PKC activity increases, and in turn ionotropic glutamate receptors can be further potentiated. While this pathway can induce long-term spinal plasticity, the neural microenvironment around the spinal lesion is more vulnerable to excitotoxicity, and this cascade can ultimately lead to cellular degradation and excitotoxic cell death (Choi, 1992; Gereau & Heinemann, 1998; Mills et al., 2001). This has lead researchers to investigate the effect of PKC inhibition on cell survival after injury. Hara et al. showed that following ischemic injury, treatment with the PKC inhibitor staurosporine produced a neuroprotective effect (Hara et al. 1990). This group later showed that the broad protein kinase inhibitor fasudil was also effective in improving locomotor function and tissue sparing following a spinal cord injury (Hara et al., 2000).

As secondary injury processes develop, a glial scar formed by chondroitin sulfate proteoglycans (CSPGs) is created to protect the damaged tissue (Fawcett & Asher, 1999). This scar formation, along with myelin-associated proteins, exerts inhibitory effects on axonal regeneration (McKerracher et al. 1994; Chen et al. 2000). Interestingly, PKC has been shown to be a key signaling mediator that is activated by these inhibitory agents (Sivanskaran et al., 2004). Sivasankaran and colleagues used immobilized substrates coated in either inhibitory myelin proteins or CSPGs to assay neurite outgrowth. They tested a range of PKC inhibitors, and found that inhibiting PKC activity stimulated neurite outgrowth on both the inhibitory myelin protein and CSPG substrates. Further, they were able to confirm these *in vitro* findings in an *in vivo* model of spinal cord injury. Rat subjects were given a dorsal hemisection, followed by an osmotic infusion of the PKC inhibitor Go6976 over the next 14 days. Results showed axonal regeneration in the dorsal column, with fibers crossing the lesion gap and extending as far as 6 mm (Sivasankaran et al., 2004). While this treatment appears promising, they showed very little axonal regeneration of the the corticospinal tract (CST), which is thought to be necessary in order to maximize functional recovery of descending motor control in primates, including humans (Blesch & Tuszynski, 2009). Further work will be needed to determine whether the regeneration promoted by PKC inhibition results in functional connectivity and improved behavioral outcomes.

Recently, many spinal cord researchers have begun focusing on the signaling pathways that are activated by the inhibitory myelin proteins. Interestingly, many of these inhibitory proteins act through their receptors to activate a small GTPase called Rho (Niederost et al. 2002). Rho is known to be important for regulating cytoskeletal structure and guiding axons in the developing CNS, and thus has been become a target of interest for those that seek to promote the regeneration of axons across spinal cord lesions after injury (Hall, 1998; Dubreuil et al., 2003). Many researchers have focused on inhibiting or altering Rho function after spinal cord injury, and have shown this treatment to be effective in blocking the growth inhibitory factors that are rampant after spinal cord injury (McKerracher et al. 2006; Fehlings et al., 2011). Others have looked further downstream, to the protein kinase that is activated by Rho. Rho-associated protein kinase (ROCK) has been shown to mediate the retraction of neurites *in vitro*, and experimental activation of ROCK is known to regulate myelin phosphotase, an essential component of axonal sprouting (Hirose et al., 1998; Kimura et al. 1996). Thus, ROCK has become an attractive therapeutic target, as specific ROCK inhibition is believed to mitigate the inhibitory effects of those myelin-derived proteins that undermine axonal regeneration. In 2000, Bito and colleagues used cultured, immature cerebellar granule neurons to directly study the effects of ROCK inhibition on neurite outgrowth. By co-transfecting these cells with green fluorescent protein (GFP) and an active form of Rho (V14Rho), they observed a marked retardation of axonal growth. When they then introduced the ROCK inhibitor Y-27632, they found that inhibiting ROCK attenuated the stunted growth, and produced significant axon genesis.

Building on these findings, Dergham and colleagues (2002) tested this same ROCK inhibitor on a variety of substrates coated with myelin inhibitory proteins or CSPG. They too found that *in vitro* administration of Y-27632 promoted the growth of primary neurons across these substrates. Further, they extended these findings to an *in vivo* spinal cord injury model. They gave mouse subjects dorsal hemisections, followed by spinal injection of Y-27632. They found that this ROCK inhibitor not only attenuated axonal dieback, but promoted the regeneration of neurons within the corticospinal tract, generating sprouting that stretched 2-3 mms across the lesion site. The ROCK-inhibited subjects also exhibited a long-term improvement in locomotor function. In 2003, Fournier published a study that extended these findings to rats, showing similar results (Fournier et al., 2003). They found that ROCK inhibition with Y-27632 was not only sufficient to promote neurite outgrowth *in vitro*, but as with the Dergham study, they showed that ROCK inhibition after hemisection could promote CST axons to regenerate across the lesion, as well as produce significant behavioral improvement in comparision to vehicle-treated subjects.

Taken together as a group, these data indicate a strong role for PKC as well as Rho kinases in morphological regeneration of the CST that is thought to be necessary for recovery of function after spinal cord injury (Blesch & Tuszynski, 2009; Nielson et al., 2010). However, recent findings have revealed that, even in the absence of CST regeneration through a spinal cord lesion, there may be substantial sprouting of surviving CST fibers below the lesion site in primates, which has been linked to improved recovery of function in forelimb control (Rosenzweig et al., 2010). This suggests a role for local spinal plasticity in restoring recovery of function in CST-dependent function. The role of mGluRs and PKC flux in these effects is largely unknown, and may represent a fruitful area for further study.

5. Conclusions

This chapter has reviewed the role of mGluRs and down-stream PKC activity as a major factor in a number of forms of plasticity throughout the spinal cord. Its ubiquity indicates a potential common mechanism for a host of complex processes, in both the intact and injured spinal cord. As a critical link between mGluR activation and ionotropic GluR trafficking and phosphorylation, PKC activity can mediate either the potentiation or depression of synaptic strength, the promotion of neural regeneration or the exacerbation of excitotoxicity. In the future, targeting PKC activity within the appropriate circumstances, and at the right time, will be essential to tailoring effective treatments for both central and peripheral injuries, as well as in the promotion of use-dependent spinal plasticity.

6. Acknowledgements

The authors wish to thank Aiwen Liu for assistance in preparing this chapter. Supported by NIH Grants NS069537 and NS067092 to A.R.F. Correspondence should be addressed to J. Russell Huie (huier@neurosurg.ucsf.edu) or Adam Ferguson (adam.ferguson@ucsf.edu).

7. References

Abraham, W. C., & Bear, M. F. (1996). Metaplasticity: the plasticity of synaptic plasticity. *Trends Neurosci, 19*(4), 126-130.

Adwanikar, H., Karim, F., & Gereau, R. W. t. (2004). Inflammation persistently enhances nocifensive behaviors mediated by spinal group I mGluRs through sustained ERK activation. *Pain, 111*(1-2), 125-135.

Aiba, A., Chen, C., Herrup, K., Rosenmund, C., Stevens, C. F., & Tonegawa, S. (1994). Reduced hippocampal long-term potentiation and context-specific deficit in associative learning in mGluR1 mutant mice. *Cell, 79*(2), 365-375.

Anwyl, R. (2009). Metabotropic glutamate receptor-dependent long-term potentiation. *Neuropharmacology, 56*(4), 735-740.

Asiedu, M. N., Tillu, D. V., Melemedjian, O. K., Shy, A., Sanoja, R., Bodell, B., et al. Spinal protein kinase M zeta underlies the maintenance mechanism of persistent nociceptive sensitization. *J Neurosci, 31*(18), 6646-6653.

Balschun, D., Manahan-Vaughan, D., Wagner, T., Behnisch, T., Reymann, K. G., & Wetzel, W. (1999). A specific role for group I mGluRs in hippocampal LTP and hippocampus-dependent spatial learning. *Learn Mem, 6*(2), 138-152.

Baumbauer, K. M., Young, E. E., Hoy, K. C., Jr., Abood, A., & Joynes, R. L. (2007). Administration of a Ca-super(2+)/calmodulin-dependent protein kinase II (CaMKII) inhibitor prevents the learning deficit observed in spinal rats after noncontingent shock administration. *Behav Neurosci, 121*(3), 570-578.

Bito, H., Furuyashiki, T., Ishihara, H., Shibasaki, Y., Ohashi, K., Mizuno, K., et al. (2000). A critical role for a Rho-associated kinase, p160ROCK, in determining axon outgrowth in mammalian CNS neurons. *Neuron, 26*(2), 431-441.

Blesch, A., & Tuszynski, M. H. (2009). Spinal cord injury: plasticity, regeneration and the challenge of translational drug development. *Trends Neurosci, 32*(1), 41-47.

Bliss, T. V., & Collingridge, G. L. (1993). A synaptic model of memory: long-term potentiation in the hippocampus. *Nature, 361*(6407), 31-39.

Bortolotto, Z. A., Bashir, Z. I., Davies, C. H., & Collingridge, G. L. (1994). A molecular switch activated by metabotropic glutamate receptors regulates induction of long-term potentiation. *Nature, 368*(6473), 740-743.

Bortolotto, Z. A., & Collingridge, G. L. (1993). Characterisation of LTP induced by the activation of glutamate metabotropic receptors in area CA1 of the hippocampus. *Neuropharmacology, 32*(1), 1-9.

Bortolotto, Z. A., Fitzjohn, S. M., & Collingridge, G. L. (1999). Roles of metabotropic glutamate receptors in LTP and LTD in the hippocampus. *Curr Opin Neurobiol, 9*(3), 299-304.

Buerger, A. A., & Fennessy, A. (1970). Learning of leg position in chronic spinal rats. *Nature, 225*(5234), 751-752.

Camodeca, N., Breakwell, N. A., Rowan, M. J., & Anwyl, R. (1999). Induction of LTD by activation of group I mGluR in the dentate gyrus in vitro. *Neuropharmacology, 38*(10), 1597-1606.

Chen, M. S., Huber, A. B., van der Haar, M. E., Frank, M., Schnell, L., Spillmann, A. A., et al. (2000). Nogo-A is a myelin-associated neurite outgrowth inhibitor and an antigen for monoclonal antibody IN-1. *Nature, 403*(6768), 434-439.

Choi, D. W. (1992). Excitotoxic cell death. *J Neurobiol, 23*(9), 1261-1276.

Chopin, S. F., & Buerger, A. A. (1976). Instrumental avoidance conditioning in the spinal rat. *Brain Res Bull, 1*(2), 177-183.

Christensen, M. D., & Hulsebosch, C. E. (1997). Chronic central pain after spinal cord injury. *J Neurotrauma, 14*(8), 517-537.

Coderre, T. J. (1992). Contribution of protein kinase C to central sensitization and persistent pain following tissue injury. *Neurosci Lett, 140*(2), 181-184.

Conquet, F., Bashir, Z. I., Davies, C. H., Daniel, H., Ferraguti, F., Bordi, F., et al. (1994). Motor deficit and impairment of synaptic plasticity in mice lacking mGluR1. *Nature, 372*(6503), 237-243.

Courtine, G., Song, B., Roy, R. R., Zhong, H., Herrmann, J. E., Ao, Y., et al. (2008). Recovery of supraspinal control of stepping via indirect propriospinal relay connections after spinal cord injury. *Nat Med, 14*(1), 69-74.

Crowe, M. J., Bresnahan, J. C., Shuman, S. L., Masters, J. N., & Beattie, M. S. (1997). Apoptosis and delayed degeneration after spinal cord injury in rats and monkeys. *Nat Med, 3*(1), 73-76.

Crown, E. D., Ferguson, A. R., Joynes, R. L., & Grau, J. W. (2002). Instrumental learning within the spinal cord: IV. Induction and retention of the behavioral deficit observed after noncontingent shock. *Behav Neurosci, 116*(6), 1032-1051.

de Leon, R. D., Roy, R. R., & Edgerton, V. R. (2001). Is the recovery of stepping following spinal cord injury mediated by modifying existing neural pathways or by generating new pathways? A perspective. *Phys Ther, 81*(12), 1904-1911.

Dergham, P., Ellezam, B., Essagian, C., Avedissian, H., Lubell, W. D., & McKerracher, L. (2002). Rho signaling pathway targeted to promote spinal cord repair. *J Neurosci, 22*(15), 6570-6577.

Dubreuil, C. I., Winton, M. J., & McKerracher, L. (2003). Rho activation patterns after spinal cord injury and the role of activated Rho in apoptosis in the central nervous system. *J Cell Biol, 162*(2), 233-243.

Edgerton, V. R., & Harkema, S. (2011). Epidural stimulation of the spinal cord in spinal cord injury: current status and future challenges. *Expert Rev Neurother, 11*(10), 1351-1353.

Edgerton, V. R., Leon, R. D., Harkema, S. J., Hodgson, J. A., London, N., Reinkensmeyer, D. J., et al. (2001). Retraining the injured spinal cord. *J Physiol, 533*(Pt 1), 15-22.

Fawcett, J. W., & Asher, R. A. (1999). The glial scar and central nervous system repair. *Brain Res Bull, 49*(6), 377-391.

Fehlings, M. G., Theodore, N., Harrop, J., Maurais, G., Kuntz, C., Shaffrey, C. I., et al. A phase I/IIa clinical trial of a recombinant Rho protein antagonist in acute spinal cord injury. *J Neurotrauma, 28*(5), 787-796.

Ferguson, A. R., Bolding, K. A., Huie, J. R., Hook, M. A., Santillano, D. R., Miranda, R. C., et al. (2008). Group I metabotropic glutamate receptors control metaplasticity of spinal cord learning through a protein kinase C-dependent mechanism. *J Neurosci, 28*(46), 11939-11949.

Ferguson, A. R., Christensen, R. N., Gensel, J. C., Miller, B. A., Sun, F., Beattie, E. C., et al. (2008). Cell death after spinal cord injury is exacerbated by rapid TNF alpha-induced trafficking of GluR2-lacking AMPARs to the plasma membrane. *J Neurosci, 28*(44), 11391-11400.

Fisher, K., & Coderre, T. J. (1996). The contribution of metabotropic glutamate receptors (mGluRs) to formalin-induced nociception. *Pain, 68*(2-3), 255-263.

Fournier, A. E., Takizawa, B. T., & Strittmatter, S. M. (2003). Rho kinase inhibition enhances axonal regeneration in the injured CNS. *J Neurosci, 23*(4), 1416-1423.

Gallagher, S. M., Daly, C. A., Bear, M. F., & Huber, K. M. (2004). Extracellular signal-regulated protein kinase activation is required for metabotropic glutamate receptor-dependent long-term depression in hippocampal area CA1. *J Neurosci, 24*(20), 4859-4864.

Gereau, R. W. t., & Heinemann, S. F. (1998). Role of protein kinase C phosphorylation in rapid desensitization of metabotropic glutamate receptor 5. *Neuron, 20*(1), 143-151.

Giles, P. A., Trezise, D. J., & King, A. E. (2007). Differential activation of protein kinases in the dorsal horn in vitro of normal and inflamed rats by group I metabotropic glutamate receptor subtypes. *Neuropharmacology, 53*(1), 58-70.

Gomez-Pinilla, F., Huie, J. R., Ying, Z., Ferguson, A. R., Crown, E. D., Baumbauer, K. M., et al. (2007). BDNF and learning: Evidence that instrumental training promotes learning within the spinal cord by up-regulating BDNF expression. *Neuroscience, 148*(4), 893-906.

Grau, J. W., Barstow, D. G., & Joynes, R. L. (1998). Instrumental learning within the spinal cord: I. Behavioral properties. *Behav Neurosci, 112*(6), 1366-1386.

Grau, J. W., Crown, E. D., Ferguson, A. R., Washburn, S. N., Hook, M. A., & Miranda, R. C. (2006). Instrumental learning within the spinal cord: underlying mechanisms and implications for recovery after injury. *Behav Cogn Neurosci Rev, 5*(4), 191-239.

Grau, J. W., Washburn, S. N., Hook, M. A., Ferguson, A. R., Crown, E. D., Garcia, G., et al. (2004). Uncontrollable stimulation undermines recovery after spinal cord injury. *J Neurotrauma, 21*(12), 1795-1817.

Hall, A. (1998). Rho GTPases and the actin cytoskeleton. *Science, 279*(5350), 509-514.

Hara, H., Onodera, H., & Kogure, K. (1990). Protein kinase C activity in the gerbil hippocampus after transient forebrain ischemia: morphological and autoradiographic analysis using [3H]phorbol 12,13-dibutyrate. *Neurosci Lett, 120*(1), 120-123.

Hara, M., Takayasu, M., Watanabe, K., Noda, A., Takagi, T., Suzuki, Y., et al. (2000). Protein kinase inhibition by fasudil hydrochloride promotes neurological recovery after spinal cord injury in rats. *J Neurosurg, 93*(1 Suppl), 94-101.

Harris, N. C., & Ryall, R. W. (1988). Mustard oil excites but does not inhibit nociceptive dorsal horn neurones in the rat: a presumed effect on A-delta fibres. *Br J Pharmacol, 94*(1), 180-184.

Hellier, J. L., Grosshans, D. R., Coultrap, S. J., Jones, J. P., Dobelis, P., Browning, M. D., et al. (2007). NMDA receptor trafficking at recurrent synapses stabilizes the state of the CA3 network. *J Neurophysiol, 98*(5), 2818-2826.

Hirose, M., Ishizaki, T., Watanabe, N., Uehata, M., Kranenburg, O., Moolenaar, W. H., et al. (1998). Molecular dissection of the Rho-associated protein kinase (p160ROCK)-

regulated neurite remodeling in neuroblastoma N1E-115 cells. *J Cell Biol, 141*(7), 1625-1636.

Hua, X. Y., Chen, P., & Yaksh, T. L. (1999). Inhibition of spinal protein kinase C reduces nerve injury-induced tactile allodynia in neuropathic rats. *Neurosci Lett, 276*(2), 99-102.

Huber, K. M., Kayser, M. S., & Bear, M. F. (2000). Role for rapid dendritic protein synthesis in hippocampal mGluR-dependent long-term depression. *Science, 288*(5469), 1254-1257.

Ji, R. R., Kohno, T., Moore, K. A., & Woolf, C. J. (2003). Central sensitization and LTP: do pain and memory share similar mechanisms? *Trends Neurosci, 26*(12), 696-705.

Jia, Z., Lu, Y., Henderson, J., Taverna, F., Romano, C., Abramow-Newerly, W., et al. (1998). Selective abolition of the NMDA component of long-term potentiation in mice lacking mGluR5. *Learn Mem, 5*(4-5), 331-343.

Kim, S. H., & Chung, J. M. (1992). An experimental model for peripheral neuropathy produced by segmental spinal nerve ligation in the rat. *Pain, 50*(3), 355-363.

Kimura, K., Ito, M., Amano, M., Chihara, K., Fukata, Y., Nakafuku, M., et al. (1996). Regulation of myosin phosphatase by Rho and Rho-associated kinase (Rho-kinase). *Science, 273*(5272), 245-248.

LaMotte, R. H., Lundberg, L. E., & Torebjork, H. E. (1992). Pain, hyperalgesia and activity in nociceptive C units in humans after intradermal injection of capsaicin. *J Physiol, 448*, 749-764.

Latremoliere, A., & Woolf, C. J. (2009). Central sensitization: a generator of pain hypersensitivity by central neural plasticity. *J Pain, 10*(9), 895-926.

Li, P., Kerchner, G. A., Sala, C., Wei, F., Huettner, J. E., Sheng, M., et al. (1999). AMPA receptor-PDZ interactions in facilitation of spinal sensory synapses. *Nat Neurosci, 2*(11), 972-977.

Liao, G. Y., Wagner, D. A., Hsu, M. H., & Leonard, J. P. (2001). Evidence for direct protein kinase-C mediated modulation of N-methyl-D-aspartate receptor current. *Mol Pharmacol, 59*(5), 960-964.

Lindsey, A. E., LoVerso, R. L., Tovar, C. A., Hill, C. E., Beattie, M. S., & Bresnahan, J. C. (2000). An analysis of changes in sensory thresholds to mild tactile and cold stimuli after experimental spinal cord injury in the rat. *Neurorehabil Neural Repair, 14*(4), 287-300.

Lu, W. Y., Xiong, Z. G., Lei, S., Orser, B. A., Dudek, E., Browning, M. D., et al. (1999). G-protein-coupled receptors act via protein kinase C and Src to regulate NMDA receptors. *Nat Neurosci, 2*(4), 331-338.

Madronal, N., Gruart, A., Sacktor, T. C., & Delgado-Garcia, J. M. (2011). PKMzeta inhibition reverses learning-induced increases in hippocampal synaptic strength and memory during trace eyeblink conditioning. *PLoS One, 5*(4), e10400.

Malenka, R. C., & Bear, M. F. (2004). LTP and LTD: an embarrassment of riches. *Neuron, 44*(1), 5-21.

Malinow, R., & Malenka, R. C. (2002). AMPA receptor trafficking and synaptic plasticity. *Annu Rev Neurosci, 25*, 103-126.

McAdoo, D. J., Xu, G. Y., Robak, G., & Hughes, M. G. (1999). Changes in amino acid concentrations over time and space around an impact injury and their diffusion through the rat spinal cord. *Exp Neurol, 159*(2), 538-544.

McKerracher, L., David, S., Jackson, D. L., Kottis, V., Dunn, R. J., & Braun, P. E. (1994). Identification of myelin-associated glycoprotein as a major myelin-derived inhibitor of neurite growth. *Neuron, 13*(4), 805-811.

McKerracher, L., & Higuchi, H. (2006). Targeting Rho to stimulate repair after spinal cord injury. *J Neurotrauma, 23*(3-4), 309-317.

Mendell, L. M. (1966). Physiological properties of unmyelinated fiber projection to the spinal cord. *Exp Neurol, 16*(3), 316-332.

Migues, P. V., Hardt, O., Wu, D. C., Gamache, K., Sacktor, T. C., Wang, Y. T., et al. (2011). PKMzeta maintains memories by regulating GluR2-dependent AMPA receptor trafficking. *Nat Neurosci, 13*(5), 630-634.

Mills, C. D., Xu, G. Y., McAdoo, D. J., & Hulsebosch, C. E. (2001). Involvement of metabotropic glutamate receptors in excitatory amino acid and GABA release following spinal cord injury in rat. *J Neurochem, 79*(4), 835-848.

Mockett, B. G., Guevremont, D., Wutte, M., Hulme, S. R., Williams, J. M., & Abraham, W. C. (2011). Calcium/calmodulin-dependent protein kinase II mediates group I metabotropic glutamate receptor-dependent protein synthesis and long-term depression in rat hippocampus. *J Neurosci, 31*(20), 7380-7391.

Moult, P. R., Schnabel, R., Kilpatrick, I. C., Bashir, Z. I., & Collingridge, G. L. (2002). Tyrosine dephosphorylation underlies DHPG-induced LTD. *Neuropharmacology, 43*(2), 175-180.

Munro, F. E., Fleetwood-Walker, S. M., & Mitchell, R. (1994). Evidence for a role of protein kinase C in the sustained activation of rat dorsal horn neurons evoked by cutaneous mustard oil application. *Neurosci Lett, 170*(2), 199-202.

Niederost, B., Oertle, T., Fritsche, J., McKinney, R. A., & Bandtlow, C. E. (2002). Nogo-A and myelin-associated glycoprotein mediate neurite growth inhibition by antagonistic regulation of RhoA and Rac1. *J Neurosci, 22*(23), 10368-10376.

Nielson, J. L., Sears-Kraxberger, I., Strong, M. K., Wong, J. K., Willenberg, R., & Steward, O. Unexpected survival of neurons of origin of the pyramidal tract after spinal cord injury. *J Neurosci, 30*(34), 11516-11528.

Oliet, S. H., Malenka, R. C., & Nicoll, R. A. (1997). Two distinct forms of long-term depression coexist in CA1 hippocampal pyramidal cells. *Neuron, 18*(6), 969-982.

Parsons, R. G., & Davis, M. (2011). Temporary disruption of fear-potentiated startle following PKMzeta inhibition in the amygdala. *Nat Neurosci, 14*(3), 295-296.

Pastalkova, E., Serrano, P., Pinkhasova, D., Wallace, E., Fenton, A. A., & Sacktor, T. C. (2006). Storage of spatial information by the maintenance mechanism of LTP. *Science, 313*(5790), 1141-1144.

Raineteau, O., & Schwab, M. E. (2001). Plasticity of motor systems after incomplete spinal cord injury. *Nat Rev Neurosci, 2*(4), 263-273.

Rosenzweig, E. S., Courtine, G., Jindrich, D. L., Brock, J. H., Ferguson, A. R., Strand, S. C., et al. Extensive spontaneous plasticity of corticospinal projections after primate spinal cord injury. *Nat Neurosci, 13*(12), 1505-1510.

Rossignol, S., Schwab, M., Schwartz, M., & Fehlings, M. G. (2007). Spinal cord injury: time to move? *J Neurosci, 27*(44), 11782-11792.

Sacktor, T. C. (2011). How does PKMzeta maintain long-term memory? *Nat Rev Neurosci, 12*(1), 9-15.

Sacktor, T. C. (2008). PKMzeta, LTP maintenance, and the dynamic molecular biology of memory storage. *Prog Brain Res, 169*, 27-40.

Sandkuhler, J. (2000). Learning and memory in pain pathways. *Pain, 88*(2), 113-118.

Sandkuhler, J., & Liu, X. (1998). Induction of long-term potentiation at spinal synapses by noxious stimulation or nerve injury. *Eur J Neurosci, 10*(7), 2476-2480.

Seligman, M. E., & Maier, S. F. (1967). Failure to escape traumatic shock. *J Exp Psychol, 74*(1), 1-9.

Shema, R., Sacktor, T. C., & Dudai, Y. (2007). Rapid erasure of long-term memory associations in the cortex by an inhibitor of PKM zeta. *Science, 317*(5840), 951-953.

Sheng, M., & Kim, M. J. (2002). Postsynaptic signaling and plasticity mechanisms. *Science, 298*(5594), 776-780.

Sivasankaran, R., Pei, J., Wang, K. C., Zhang, Y. P., Shields, C. B., Xu, X. M., et al. (2004). PKC mediates inhibitory effects of myelin and chondroitin sulfate proteoglycans on axonal regeneration. *Nat Neurosci, 7*(3), 261-268.

Sun, R. Q., Tu, Y. J., Lawand, N. B., Yan, J. Y., Lin, Q., & Willis, W. D. (2004). Calcitonin gene-related peptide receptor activation produces PKA- and PKC-dependent mechanical hyperalgesia and central sensitization. *J Neurophysiol, 92*(5), 2859-2866.

Wang, Q., Walsh, D. M., Rowan, M. J., Selkoe, D. J., & Anwyl, R. (2004). Block of long-term potentiation by naturally secreted and synthetic amyloid beta-peptide in hippocampal slices is mediated via activation of the kinases c-Jun N-terminal kinase, cyclin-dependent kinase 5, and p38 mitogen-activated protein kinase as well as metabotropic glutamate receptor type 5. *J Neurosci, 24*(13), 3370-3378.

Waung, M. W., Pfeiffer, B. E., Nosyreva, E. D., Ronesi, J. A., & Huber, K. M. (2008). Rapid translation of Arc/Arg3.1 selectively mediates mGluR-dependent LTD through persistent increases in AMPAR endocytosis rate. *Neuron, 59*(1), 84-97.

Wernig, A., Nanassy, A., & Muller, S. (2000). Laufband (LB) therapy in spinal cord lesioned persons. *Prog Brain Res, 128,* 89-97.

Willis, W. D., & Westlund, K. N. (1997). Neuroanatomy of the pain system and of the pathways that modulate pain. *J Clin Neurophysiol, 14*(1), 2-31.

Wolpaw, J. R. (2007). Spinal cord plasticity in acquisition and maintenance of motor skills. *Acta Physiol (Oxf), 189*(2), 155-169.

Woolf, C. J. (1983). Evidence for a central component of post-injury pain hypersensitivity. *Nature, 306*(5944), 686-688.

Woolf, C. J., & Salter, M. W. (2000). Neuronal plasticity: increasing the gain in pain. *Science, 288*(5472), 1765-1769.

Xiao, M. Y., Zhou, Q., & Nicoll, R. A. (2001). Metabotropic glutamate receptor activation causes a rapid redistribution of AMPA receptors. *Neuropharmacology, 41*(6), 664-671.

Yang, L., Liu, G., Zakharov, S. I., Morrow, J. P., Rybin, V. O., Steinberg, S. F., et al. (2005). Ser1928 is a common site for Cav1.2 phosphorylation by protein kinase C isoforms. *J Biol Chem, 280*(1), 207-214.

Yao, Y., Kelly, M. T., Sajikumar, S., Serrano, P., Tian, D., Bergold, P. J., et al. (2008). PKM zeta maintains late long-term potentiation by N-ethylmaleimide-sensitive factor/GluR2-dependent trafficking of postsynaptic AMPA receptors. *J Neurosci, 28*(31), 7820-7827.

Yashpal, K., Fisher, K., Chabot, J. G., & Coderre, T. J. (2001). Differential effects of NMDA and group I mGluR antagonists on both nociception and spinal cord protein kinase C translocation in the formalin test and a model of neuropathic pain in rats. *Pain, 94*(1), 17-29.

Young, M. R., Fleetwood-Walker, S. M., Mitchell, R., & Dickinson, T. (1995). The involvement of metabotropic glutamate receptors and their intracellular signalling pathways in sustained nociceptive transmission in rat dorsal horn neurons. *Neuropharmacology, 34*(8), 1033-1041.

10

Protein Kinases and Pain

Mani Indiana Funez, Fabiane Hiratsuka Veiga de Souza,
José Eduardo Pandossio and Paulo Gustavo Barboni Dantas Nascimento
School of Ceilandia, Brasilia University,
Brazil

1. Introduction

There is abundant evidence that protein kinases are involved in the physiopathology of acute and chronic pain. In the first section, we discuss the role of protein kinases in pain and the signalling pathways involved in both the acute and chronic states. The second section will present evidence supporting the contribution of protein kinase inhibition to pain control by different classes of drugs. Both well-known drugs and new molecules can control pain in the peripheral and central nervous systems. The third section highlights the progress in pharmaceutical development and protein kinase research for new pain control drugs in the first decade of the 21st century.

2. Role of protein kinases in acute and chronic pain

In this section, we will discuss the differential activation of protein kinases by pain mediators and the modulation of the acute and chronic pain processes by several kinases.

2.1 PKC

Protein Kinase C (PKC) is a family of phospholipid-dependent serine/threonine phosphotransferases; it can be divided into the following groups of isoforms: a) conventional or classical (α, βI, βII, γ), b) novel (δ, ϵ, η, θ), and c) atypical (ζ, λ (mouse)/ι (human)) isoforms (Nishizuka, 1992). Five subspecies of PKC, PKC-βI, PKC-βII, PKC-δ, PKC-ϵ, and PKC-ζ, are expressed in the dorsal root ganglion (DRG) of rats (Cesare et al. 1999). The PKC isoforms that are expressed in the DRG of mice include PKC-α, PKC-βI, PKC-βII, PKC-δ, PKC-ϵ, PKC-η, PKC-θ, PKC-ζ, and PKC-λ (Khasar et al., 1999a). Thus, there are some differences in the expression of DRG PKC isoforms between species.

Signal transduction through the PKC pathway has been strongly linked to pain. Inflammatory stimuli and mediators can activate PKC to induce pain. Nociceptive response caused by formalin injection into the mouse paw is characterised by two phases; the neurogenic response, which is due to direct nociceptor activation, and the inflammatory response, which is caused by inflammatory mediators (Hunskaar and Hole, 1987). In this model, PKC blockade by local treatment with chelerythrine inhibited the second phase of nociceptive response (Souza et al., 2002), which is driven largely by tissue inflammation, indicating a relationship between PKC activation and the inflammatory process. In the same way, mechanical sensitisation induced by the inflammatory mediator bradykinin in rats is inhibited by a PKC inhibitor (Souza et al., 2002). *In vitro* experiments conducted in DRG neurons strongly suggest that bradykinin-induced heat sensitisation is dependent on PKC activation because it can be reversed by pharmacological inhibition with staurosporine or phosphatase inhibitors (Burguess et al., 1989; Cesare and McNaughton, 1996). *In vitro* experiments have shown noticeable PKC-activity in rat DRGs after 3 hours of prostaglandin E_2 (PGE_2) paw administration. This activity was accompanied by paw sensitisation to mechanical stimuli, as measured by behavioural experiments (Sachs et al., 2009).

Carrageenan injection in rat or mouse paws is another tool used to study inflammatory sensitisation. In the same way, pharmacological inhibition of PKCε reduces carrageenan-induced mechanical sensitisation in mice (Khasar et al., 1999a). Phosphorylation of PKCε in DRG neurons is increased after carrageenan-induced acute sensitisation (Zhou et al. 2003), and a PKCε agonist sensitises nociceptors to mechanical stimuli (Aley and Levine, 2003). Inflammatory sensitisation has a "sympathetic" component that involves the release of amines such as epinephrine and dopamine (Coderre et al. 1984; Nakamura and Ferreira 1987). Evidence suggests that mechanical sensitisation induced either by epinephrine in rats (Khasar et al. 1999b) or dopamine in mice (Villarreal et al., 2009b) is blocked by PKCε-selective inhibition. Cesare et al. (1999) found that bradykinin exposure induces PKCε translocation from the cytosol to a membrane-associated position in cultured DRG neurons, thus contributing to heat sensitisation.

The mechanical sensitisation induced by PGE_2 involves the peripheral activation of PKCε in rats and mice, as shown by specific pharmacological inhibition (Sachs et al., 2009; Villareal et al., 2009b). In PKCε-mutant mice, the nociceptive threshold is preserved, whereas the nociceptive response was significantly impaired, as evaluated in a model of visceral pain using peritoneal administration of acetic acid (Khasar et al. 1999a). Kassuya et al. (2007), found a noticeable increase in membrane-bound PKCα expression of mouse paw tissue after PGE_2 administration (Kassuya et al., 2007). Thus, the PGE_2–induced pain-related effects during inflammation may be mediated by PKCε and PKCα.

Multiple voltage-gated sodium channel (VGSC) isoforms are expressed in DRG neurons. For example, isoforms Na_v 1.8 and Na_v 1.9 are responsible for tetrodotoxin-resistant (TTX-resistant) currents due to Na^+ channel blocker insensitivity. These sodium currents can be modulated by PKC phosphorylation, which is induced by inflammatory mediators (Gold et al., 1998; Khasar et al., 1999a). Using whole-cell voltage-clamp recordings from DRG neurons, Gold et al. (1998) found that PKC inhibitors decreased the density of tetrodotoxin-resistant sodium current, whereas the PKC activator PMA produces changes that are opposite, suggesting that PKC modulates it. In addition, it was show a relationship between inflammatory mediators-induced changes in TTX-resistant sodium currents and PKC activity (Gold et al., 1998; Khasar et al., 1999a).

PKC peripheral activation contributes to central pain processing. During serotonin-induced rat paw sensitisation, another pain-sensitising mediator associated with inflammation, the response of animals to thermal stimulation and c-fos activation in the dorsal horn is attenuated by intraplantar application of the PKC inhibitor chelerythrine (Chen et al., 2006). During inflammation or in naïve animals, activation of glutamate receptors mGluRs in the spinal dorsal horn modulates acute nociception. These receptors are coupled to Gq/II protein phospholipase C (PLC)-phosphoinositide (PI) hydrolysis and PKC pre- and postsynaptic activation (Neugebauer, 2002; Giles et al., 2007), suggesting that PKC modulates the synaptic transmission at the spinal level.

PKC activation is associated with chronic pain conditions. Mao et al. (1992) found an increase in membrane-bound PKC in the spinal cord of rats in a model of post-injury neuropathic pain. The role of PKC was confirmed using an intracellular inhibitor of PKC translocation/activation and analysing membrane-bound PKC translocation and pain behaviour. The data suggest a role for PKC in neuropathic pain states. Ahlgren and Levine (1994) found a reduction in streptozotocin-induced diabetic rat pain sensitisation after treatment with PKC inhibitors.

Using the partial sciatic nerve section model, Malmberg et al. (1997) verified that mice lacking PKCγ completely fail to develop neuropathic-associated sensitisation even though they respond normally to acute pain stimuli. In addition, PKCγ expression is restricted to a subset of dorsal horn neurons. Malmberg and co-workers suggest that targeting PKCγ is a promising tool for treating chronic pain. This isoform inhibition also attenuates opioid tolerance in the spinal cord (section 3).

The physiopathology of alcoholic neuropathy in rats seems to depend on PKCε activation and up-regulation in DRG neurons, as shown by selective pharmacological inhibition and western blot analysis performed after 70 days of ethanol administration (Dina et al., 2000). In addition, the role of PKCε in pain sensitisation is associated with neuropathy induced by the antineoplastic agent paclitaxel in rats (Dina et al., 2001).

The role of PKCε is well demonstrated during chronic inflammatory pain conditions. Aley et al. (2000) developed a model to study chronic inflammatory sensitisation that can be induced by a single episode of acute inflammation; after the induction, in a time-lapse of 5 days there is inflammatory-mediator prolonged-response. During this state, PKCε seems to be responsible for the maintenance of this "primed state" and the prolonged response to inflammatory mediators (Aley et al. 2000). Accordingly, the phosphorylation of PKCε in DRG neurons correlated with pain-associated prolonged inflammation after 3 days of the administration of Complete Freund's Adjuvant (CFA) to rat paws (Zhou et al., 2003).

Mechanical persistent inflammatory sensitisation can also be induced by intraplantar administration of inflammatory mediators like prostaglandins and sympathetic amines in rats and mice (Ferreira et al., 1990; Villarreal et al., 2009b). Studies suggest that PKC activity in the DRG is up-regulated by and is at least partially responsible for the persistent condition, as shown by analyses of PKC activity in rat DRGs (Villarreal et al. 2009a). Moreover, the local administration of a selective PKCε inhibitor abolished the persistent state induced by PGE_2 in rats and mice (Villarreal et al. 2009a; Villarreal et al., 2009b). Evaluation of the mechanisms downstream of PKCε activation found that $Na_v1.8$ mRNA levels in the DRG from rats was up-regulated and inhibition of PKCε activity reduced these levels (Villarreal et al., 2009a).

2.2 PKA

Cyclic adenosine-monophosphate (cAMP)-dependent protein kinase (PKA) is a serine/threonine phosphotransferase; in its inactive form, it is a tetrameric holoenzyme composed of two regulatory and two catalytic subunits (Taylor et al., 1990). When the second messenger cAMP is generated the PKA-regulatory subunits bind cAMP, and the holoenzyme separates into the regulatory subunits and the catalytic subunits (Taylor et al., 1990). The catalytic subunits can phosphorylate their biological targets and regulate many cellular functions. There are different regulatory (RIα, RIβ, RIIα, RIIβ) and catalytic (Cα, Cβ) subunits; α subunits are expressed in non-neuronal and neuronal tissue, whereas β subunits are expressed predominantly in neuronal cells (Cadd and McKnight, 1989).

cAMP/PKA signalling is involved in nociceptor sensitisation by inflammatory mediators. Ferreira and Nakamura (1979) provided evidence that sensitisation of rat hind-paws by prostaglandins is dependent on cAMP generation. Since this original study, many subsequent studies have shown that cAMP generation is induced by a plethora of inflammatory stimuli. The mechanical nociceptor sensitisation that occurs during inflammation or induced by either inflammatory mediators (PGE$_2$, dopamine, serotonin) is blocked by treatment with PKA inhibitors in rats and mice (Taiwo and Levine 1991; Taiwo et al., 1992; Aley and Levine, 1999; Aley et al., 2000; Sachs et al., 2009; Villarreal et al., 2009b).

Adenylyl cyclase (AC)/cAMP/PKA activation may be necessary to induce and maintain mechanical nociceptor sensitisation (Aley and Levine, 1999). Moreover, PGE$_2$–induced inflammatory sensitisation increased PKA activity in mouse paws (Kassuya et al., 2007); and in rat DRGs (Sachs et al., 2009), which correlates with the behavioural data. Accordingly, the intraplantar administration of the catalytic subunit of PKA (PKACS) induces mechanical nociceptor sensitisation (Aley and Levine, 1999; Aley and Levine, 2003). Supporting the animal model data, in vitro studies using sensory neurons that were cultured and bathed in classic inflammatory mediators showed that prostaglandins can sensitise these cells to bradykinin and that this effect is dependent on PKA activation (Cui and Nicol, 1995; Smith et al., 2000). The role of PKA in formalin-induced nociceptive pain and inflammatory sensitisation was demonstrated in experiments in mice with a null mutation in the type I regulatory subunit (RIβ) of PKA. This mutation dampens the response during nociceptive pain and thermal stimulation (Malmberg et al., 1997).

Once activated, the PKA substrate in the nociceptive pathways can be voltage-gated sodium channels. In fact, in vitro studies have shown that TTX-resistant sodium current is modulated via PKA activation during inflammation (England et al., 1996; Gold et al., 1998). Additionally, during inflammation, PKA enhances the gating of transient receptor potential vanilloid channel-1 (TRPV-1) via direct phosphorylation (Lopshire and Nicol, 1998; Rathee et al., 2002). Therefore, PKA can directly phosphorylate ion channels, thus increasing the excitability of sensory neurons and contributing to some pain conditions. Studies using a model of persistent inflammatory sensitisation in rats and mice show that PKA could exert a role in the maintenance of the chronic state. The persistent sensitisation is abolished by injection of PKA inhibitors, and PKA expression and activity were up-regulated in DRG (Villarreal et al., 2009a, 2009b). The contribution of PKA to sensitisation maintenance seems to be due to the regulation of the Na$_v$1.8 sodium channel expression (Villarreal et al., 2009a).

In the neuropathic pain model of sciatic nerve ligature, PKARIβ-null animals present nociceptive responses that are similar to control animals (Malmberg et al., 1997). However, in a model of paclitaxel-induced pain neuropathy, pharmacological inhibition of PKA attenuates the response to thermal stimulation (Dina et al., 2001). Other subunits of PKA, different from PKARIβ, may be activated during neuropathy because PKA inhibitors do not present selectivity.

2.3 MAPKs

Mitogen-Activated Protein Kinases (MAPKs) are protein-serine/threonine kinases. There are many subfamily isoforms known, and are currently 14 mammalian members. They are important for pain regulation and control and are divided in extracellular-signal-regulated kinases (ERKs, 7 isoforms), stress-activated protein kinases or c-Jun N-terminal kinases (JNKs, 3 isoforms) and p38 mitogen-activated protein kinases (p38 MAPK, 4 isoforms). These enzymes are activated by direct phosphorylation of two sites in the kinase activation loop, at a tyrosine and a threonine residue; separated by a single, variable residue (Pearson et al. 2001).

Classically, upon receptor-dependent tyrosine kinase activation on the cellular surface, a cascade of biochemical reactions culminates in small GTPase (Ras) activation. This molecular event initiates a series of catalytic phosphorylation-based signalling, involving kinases such as the proto-oncogene serine/threonine-protein kinase (C-Raf), mitogen-activated protein kinase kinase 1 (MEK1), MEK2 and MAPKs. The dual phosphorylation of these proteins leads to conformational changes, allowing their respective catalytic domains to be accessible to their substrates, which are mainly transcription factors that regulate diverse genes, and others proteins that are regulated by phosphorylation. In addition, the MAPKs interact with inactivating phosphatases, which finely tunes their cellular activity. The same hierarchical cascade exists for JNK and p38 MAPK activation, consisting of three consecutive steps of phosphorylation and activation of different kinases (MAPKKK → MAPKK → MAPK).

Extracellular mitogens such as growth factors (cytokines and hormones) and phorbol esters (e.g., 12-O-tetradecanoylphorbol-13-acetate, TPA) activate ERK1 and ERK2, which regulates cell proliferation and promotes effects such as induction or inhibition of differentiation, stimulation of secretory responses in a variety of cell types such as neutrophils, modulating membrane activity, and generating active oxygen species (Blumberg, 1988). The stress-activated protein kinases, or JNKs, and p38 MAPK signalling pathways are responsive to stress stimuli, such as cytokines, ultraviolet irradiation, heat shock, osmotic shock and cellular redox state, and are involved in cell differentiation and apoptosis. There are 10 isoforms of the three JNKs due to alternative splicing of JNK-1, JNK-2, and JNK-3, and there are four p38 MAPK isoforms.

These MAPKs are involved in processing cellular pain. Dai et al. (2002) demonstrated that ERK is activated in DRG neurons by electrical, thermal and chemical stimuli using electrophysiological recordings and western blot analysis. The peripheral stimulation of ERK1/2 and p38 MAPKs is involved in the nociceptor sensitisation produced by epinephrine, nerve grow factor (NGF) and capsaicin (Aley & Levine 2003, Zhu & Oxford, 2007). Activation of nuclear factor-kappaB (NF-κB), a transcription factor linked to inflammation, and p38 MAPK leads to the formation of various pro-inflammatory cytokines, such as TNF-α, IL-1β, and IL-6 (Doyle et al. 2011). TNF-α may induce acute peripheral mechanical sensitisation by acting directly on its receptor TNFR1, which is localised in primary afferent neurons, resulting in the p38-dependent modulation of TTX-resistant Na^+ channel currents (Jin & Gereau 2006; Zhang et al. 2011).

In neurons, synaptic activity-induced increases in the intracellular Ca^{2+} concentration activate MAPKs. Ca^{2+}/calmodulin-activated protein kinase (CaMKII) is essential for synaptic plasticity because it regulates transcriptional and translational modifications in gene expression and regulation. MAPKs are downstream effectors of multiple kinases, including CaMKII. Membrane depolarisation and calcium influx activate MAPK/ERK kinases. ERK and p38 MAPKs are up-regulated both in primary afferent nerves and the spinal cord in response to noxious stimulation, nerve injury and tissue injury. Inhibition of ERK or p38 MAPK phosphorylation or activity induces an antinociceptive effect in many of the animal pain models described throughout this section. Thus, in addition to the PKA and PKC signalling pathways, some cross-talk may exist with MAPK cascades upon inflammation or injury.

As an example, IL-6 exerts an important role in the development and maintenance of muscular sensitisation to nociception. The IL-6-mediated muscular pain response involves resident cell activation, polymorphonuclear cell infiltration, cytokine production, prostanoids and sympathomimetic amines release (Manjavachi et al. 2010). This response to IL-6 triggers the activation of intracellular pathways, especially MAPKs. Upon IL-6 stimulation, ERK, p38 MAPK and JNK phosphorylation is measurable by flux cytometry, and selective inhibitors of ERK and p38 MAPK partially reduced mechanical nociceptive behaviour (Manjavachi et al. 2010). Inflamed tissues release NGF that act upon nociceptors, activating the p38 MAPK cascade and leading to an increase of TRPV-1 translation and transport to nerve terminals, which contributes to the maintenance of nociceptive behaviour in animal models (Ji et al. 2002). Additionally, two separate p38 MAPK pharmacological inhibitors were effective at inhibiting the development of burn-induced sensitisation when administered as intrathecal pre-treatments (Sorkin et al. 2009).

A screen of MAPK activation in the dorsal horn in both phases of the formalin test demonstrated that p38 MAPK is activated in spinal microglia. Thus, a reduction in the level of spinal p38β, but not p38α, prevented the development of sensitisation following peripheral inflammation (Li et al., 2010). Any study of MAPK signalling must also consider the effect of nervous system cells other than neurons in the pain process.

The same kind of consideration is needed for chronic pain. Synaptic and nerve plasticity is a key element in pain chronification. Changes in structure and function as a result of input from the environment, lesions and pathologies may lead to neuropathic pain. These changes depend upon transcriptional and translational modifications in cell function that are mediated by MAPK signalling. Thus, MAPK modulation became a natural choice for research and the development of new drugs and pharmacological tools.

Pfizer Global Research and Development published a research paper in 2003 showing that the development of neuropathic pain is associated with an increase in the activity of the MAPK/ERK-kinase cascade within the spinal cord. They explored the chronic constriction injury model and the streptozocin-induced diabetic model to mimic neuropathic pain states. Global changes in gene expression and the effect of MAPK/ERK-kinase (MEK) inhibitor were analysed (Ciruela et al., 2003). These efforts lead to the selection of these kinases as targets of drug design for pain, with a focus on neuropathic pain.

The MAPK intracellular signalling cascades are also associated with synaptic long-term potentiation and memory and are associated with nociceptive behaviour in spinal cord injury (Crown et al., 2006). ERK 1/2 and p38 MAPK phosphorylation levels are upregulated in rat-spinal cords during mechanical sensitisation after spinal cord injury. Neurons are not the only cells involved in this process; microglial but not astrocytic p38α contributes to the maintenance of neuronal hyperexcitability in caudal areas after spinal cord injury (Gwak et al. 2009).

The IL-6/p38 MAPK/CX3C Receptor 1 signalling cascade is involved in neural–glial communication and plays an important role in triggering spinal glial activation and facilitating pain processing following peripheral nerve injury. Up-regulation of CX3CR1 expression by IL-6-p38 MAPK signalling enhances the responsiveness of microglia to chemokine CXCL1, or fractalkine, after nerve injury (Lee et al., 2010). TNF-α is important during the development of neuropathic pain by spinal nerve ligature (SNL) (Schäfers et al., 2003). The inhibition of spinal p38 MAPK activation prevents this event. However, the activation of ERK but not p38 MAPK is critically involved in the TNFα-induced increase in TRPV1 expression in cultured DRG neurons (Hensellek et al., 2007).

Injury to peripheral nerves may result in the formation of neuromas. Elevated levels of phosphorylated ERK1/2 can be identified in individual neuroma axons that also possess the voltage-gated sodium channel $Na_V 1.7$. Painful human neuromas show accumulation of this sodium channel, and its function is modulated by ERK1/2 phosphorylation (Persson et al., 2011).

MAPK expression analysed in the spinal cord after SNL showed differential activation in injured and uninjured DRG neurons. Uninjured neurons had only p38 MAPK detectable induction. In contrast ERK, p38 MAPK and JNK were activated in several populations of injured DRG neurons (Obata et al., 2004). Differential activation of MAPK in lesioned and sound primary nerve afferents may be linked to the pathogenesis of neuropathic pain after partial nerve injury (Svensson et al., 2003).

2.4 Interplay between pathways

The specificity of activation for each signalling pathway may be determined by the stimuli (Juntilla et al., 2008), and the crosstalk between them could be induced during pathological states (Noselli, 2000). Pimienta and Pascual (2007) described MAPK intracellular signalling as "different signalling cascades crosstalk with each other in a way that their functional compensation makes possible the simultaneous integration of multiple inputs".

Considering only one inflammatory mediator, PGE_2, in three models performed in the same species (mice) with analyses of not the same tissue, differences between the signalling pathways involved can be detected:

a. Acute nociception induced by high-dose PGE_2 administration is dependent on ERK signalling mechanisms because its overexpression was detected in hind paw by western immunoblotting analyses (Kassuya et al., 2007). This effect was reversed by EP receptor antagonists (Kassuya et al., 2007).

b. In the same way, PGE_2 is a final mediator of nociceptor sensitisation that acts on the peripheral nerve endings through the prostanoid receptors, leading to sensitisation of sensory nerves. PGE_2-induced acute mechanical sensitisation, which is also associated with kinase activation, was completely prevented by PKA and PKCε, but not by ERK, pharmacological inhibition (Villarreal et al. 2009b). The persistent pain state induced by chronic PGE_2 administration is completely abolished by PKA or PKCε inhibitors, but not by ERK inhibitors (Villarreal et al. 2009b).

Thus, we conclude that PKA, PKC and ERK are involved in the effects of PGE_2 (including nociceptor sensitisation and nociception). The inflammatory processes include several others mediators in addition to PGE_2. In the same work, Villarreal et al. (2009b) showed that dopamine-induced acute sensitisation involves PKA, PKC and ERK activation, whereas the dopamine-induced persistent sensitisation state is abolished by ERK inhibition and temporarily inhibited by PKA or PKCε inhibitors, suggesting that ERK plays the major role. So, what is the real meaning of these results?

The study of MAPK and other kinases must keep its momentum. The interplay between different signalling pathways is challenging to understand. The available experimental models allow individual probing of each mediator and the kinase transduction of its signalling. Biological systems and pathological states have multiple variables in a complex regulated environment that hinder our understanding of each molecule and their combined role. Nevertheless, the continuous efforts have already achieved interesting findings. In the next sections, additional mechanisms and protein kinases will be described in discussing the pharmacological mechanisms of different drug classes in pain control.

3. Role of protein kinases in pain control

Both acute and chronic pain are usually controlled by administration of pharmacological agents (analgesics and adjuvants) that attempt to tackle pain in both the central and peripheral divisions of the nociceptive pathway. Although a classical mechanism of action is well described for most drugs, additional mechanisms link their analgesic effects with some kinases-dependent pathways, mainly pathways related to PKA, PKC, MAPKs and cyclic guanosine monophosphate (cGMP)-dependent protein kinase (PKG). In this section, the involvement of protein kinases in the mechanisms of action of drugs used for pain control, such as opioids, dipyrone, general and local anaesthetics and antidepressants will be analysed.

3.1 Opioids

Among opioids, morphine is widely used as a classical opioid analgesic for the clinical management of acute and chronic pain. Despite its wide use, tolerance to the analgesic actions of morphine is an important side effect of prolonged exposure. Individuals who are tolerant to the effects of morphine require larger doses to elicit the same amount of analgesia. Thus, antinociceptive tolerance and the high doses required to achieve effects have limited the use of morphine.

Many factors have been related to morphine tolerance, such as a change of the descending pain modulatory pathway, receptor desensitisation, down-regulation of opioid functional receptors, release of excitatory neurotransmission and other adaptive changes in cell signalling pathways. Interestingly, PKC, especially PKCγ, plays a major role in the changes associated with morphine tolerance. Song et al. (2010) demonstrated that an isoform-specific inhibitor could successfully down-regulate PKCγ in the spinal cord and reverse the development of morphine tolerance in rats. This result not only implicates this PKC isoform in the opioid tolerance mechanism but also has potential applications in pain management.

Beyond the involvement of PKC in opioid tolerance, PKC is involved in inflammatory and neuropathic pain. The capacity of opioids to alleviate inflammatory pain is negatively regulated by the glutamate-binding N-methyl-D-aspartate receptor (NMDAR). And increased activity of this receptor complicates the clinical use of opioids for treating neuropathic pain. Rodríguez-Muñoz et al. (2011) indicated that morphine disrupts the glutamate-binding NMDAR complex by PKC-mediated phosphorylation and potentiates the NMDAR-CaMKII pathway, which is implicated in morphine tolerance. Inhibition of PKC restored the antinociceptive effect of morphine on the μ-opioid receptor (MOR). Thus, the opposing activities of the MOR and NMDAR in pain control affect their relation within neurons of structures such as the periaqueductal grey (PAG), a region that is implicated in the opioid control of nociception. This finding could be exploited in developing bifunctional drugs that would act exclusively on NMDARs associated with MORs.

MORs are not the only opioid receptors that influence PKC. Berg et al. (2011), who were investigating the regulation of the κ-opioid receptor (KOR) in rat primary sensory neurons *in vitro* and in a rat model of thermal sensitisation, showed that the application of a KOR agonist (U50488) did not inhibit AC activity or release of calcitonin gene-related peptide (CGRP) *in vitro* and did not inhibit thermal sensitisation *in vivo*. It is important to note that AC activity, CGRP release, and thermal sensitisation process are related to PKC activation (see section 2). However, after a 15-min pretreatment with bradykinin, the agonist became capable of inhibiting AC activity, CGRP release, and thermal sensitisation. The *in vitro* effects of bradykinin on the KOR agonist were abolished by a PKC inhibitor; thus, Berg and co-workers suggest that PKC activation mediates BK-induced regulation of the KOR system. More studies are necessary to understand the mechanisms by which peripheral KOR agonist efficacy is regulated and the relationship of the KOR agonist effects with PKC activation.

In this regard, formalin-induced inflammatory nociception may inhibit morphine tolerance in mice. In this model, conventional PKC (cPKC) is up-regulated and treatment with an antisense oligonucleotide (AS-ODN) directed against cPKC abolished the development of morphine tolerance, suggesting that cPKC is involved in morphine tolerance development (Fujita-Hamabe et al., 2010). Additionally, formalin-induced inflammatory nociception inhibit morphine tolerance by a mechanism involving KOR activation, down-regulation of cPKC, and up-regulation of MOR activity (Fujita-Hamabe et al., 2010). The data suggest a key role to cPKC in opioid-induced tolerance and that nociception-activated mechanisms may modulate opioid-response, improving it. In addition, studying the effects of chronic ethanol–induced neuropathy in rats, Narita et al. (2007) showed that chronic ethanol exposure dysregulated MOR but not DOR and KOR, and was related to PKC up-regulation in the spinal cord, which may explain the reduced sensitivity to the morphine antinociceptive effect. Taken together, these findings suggest the PKC activation disrupts MOR function, which could be counteracted by the KOR system. How the DOR participates remains unclear.

Like PKC, PKA may also play a role in morphine antinociceptive tolerance. Previous studies have shown that chronic exposure to morphine results in intracellular adaptations within neurons that cause an increase in PKA activity. Unexpectedly, sustained morphine treatment produces paradoxical pain sensitisation (opioid-induced hyperalgesia) and causes an increase in spinal pain-related neurotransmitter concentrations, such as CGRP, in experimental animals. Studies have also shown that PKA plays a major role in the regulation of presynaptic neurotransmitter (such as CGRP and substance P) synthesis and release. Tumati et al. (2011) previously showed that in cultured DRG neurons, sustained *in vitro* opioid agonist treatment up-regulates cAMP levels (AC superactivation) and augments CGRP release in a PKA-dependent manner. The authors also showed that selective knock-down of spinal PKA activity by intrathecal pretreatment of rats with a PKA-selective small interference RNA (siRNA) mixture significantly attenuates sustained morphine-mediated augmentation of spinal CGRP immunoreactivity, thermal and mechanical sensitisation and antinociceptive tolerance. These findings indicate that sustained morphine-mediated activation of spinal cAMP/PKA-dependent signalling may play an important role in opioid-induced pain sensitisation. More specifically, morphine acts acutely on MORs, which couple with G-proteins to inhibit AC and reduce PKA activity. However, during tolerance, MORs become uncoupled from G-proteins, AC inhibition is reduced, and PKA activity is increased. These findings also provide potential molecular targets for pharmacological intervention to prevent the development of such paradoxical pain sensitisation.

The majority of studies that have demonstrated an increase in PKA activity during opioid tolerance have been conducted in rats using brain regions associated with the reinforcing properties of opioids, such as the *locus coeruleus* and *nucleus accumbens*. Studying the expression of morphine antinociceptive tolerance at the behavioural level (tail-flick test) and the alterations in PKA activity at the cellular level in mouse brain (PAG, medulla, thalamus) and lumbar spinal cord, Dalton et al. (2005) support the hypothesis that an increase in PKA activity contributes to the tolerance to morphine-induced antinociception. However, the effect of chronic morphine treatment for 15 days on PKA activity was region-specific because increases in cytosolic PKA activity were observed in the lumbar spinal cord. In contrast, PKA activity/kinetics was not altered in the PAG, medulla or thalamus. These results demonstrate that spinal and supraspinal PKA activity are differentially altered during morphine tolerance. Thus, the neurons in mouse brain and lumbar spinal cord that make up the pain pathway from the brainstem to the spinal cord respond differently to chronic morphine treatment. To confirm these findings, future studies need to elucidate the differential responses to chronic morphine treatment using *in vivo* models of morphine antinociceptive tolerance concerning the PKA involvement.

Using a behavioural paw pressure test in rats, Yamdeu et al. (2011) demonstrated that up-regulation of NGF, through activation of the p38 MAPK pathway, lead to adaptive changes in sensory neuron opioid receptors that enhance susceptibility to local opioids. After intraplantar NGF treatment, this effect occurs in three consecutive steps: MOR expression is increased in DRG at 24 h, increased axonal MOR transport at 48 h, and increased MOR density at 96 h. Consequently, the dose-dependent peripheral antinociceptive effects of locally applied full opioid agonists such as fentanyl are potentiated, and the effects of partial opioid agonists such as buprenorphine are more efficacious, which is reversed by the intrathecal administration of p38 MAPK inhibitor SB203580. Thus, in rats, peripheral inflammation increases MOR expression in nociceptors by NGF activation of p38 MAPK. This mechanism may act as a counter-regulatory response to painful p38 MAPK–induced conditions, such as inflammatory pain, to facilitate exogenously or endogenously mediated opioid antinociception.

Recently, the roles of several MAPKs, including p38 MAPK and ERK, have been investigated in animal models of morphine tolerance and postoperative nociceptive sensitisation. It is unknown, however, whether prior morphine-induced MAPK activation affects the resolution of postoperative nociceptive sensitisation. Horvath et al. (2010) investigated the effect of morphine-induced antinociceptive tolerance on the resolution of postoperative nociceptive sensitisation. They hypothesised that prior chronic morphine administration would inhibit or delay the resolution of postoperative nociceptive sensitisation via enhanced spinal glial proteins expression and MAPK signalling. Chronic morphine treatment attenuated the resolution of postoperative nociceptive sensitisation, as determined by thermal and mechanical behavioural tests, and enhanced microglial p38 MAPK and ERK phosphorylation. To better understand these results, prior chronic morphine exposure could prime microglia, causing exacerbated MAPK signalling pathway activation following subsequent paw incision injury. This would cause more robust microglial responses in rats with a history of morphine tolerance versus naïve rats, and this response is manifested by further neuronal sensitisation, behavioural hypersensitivity and inhibition of the resolution of the postoperative-associated nociceptive condition. The Horvath and co-workers study indicates that microglial MAPKs play a role in the mechanisms by which morphine attenuates the resolution of postoperative pain and suggests that patients who abuse opioids or are on chronic opioid therapy may be more susceptible to developing chronic pain syndromes following acute injury.

In conclusion, protein kinases (PKs) exert a crucial role in pain control responses mediated by opioids, mainly in tolerance-induced mechanisms. Thus, PKs could be the key to better understanding opioid pharmacodynamics.

3.2 General and local anaesthetics

3.2.1 General anaesthetics

Ketamine is an NMDAR antagonist that is available for clinical use as a general anaesthetic. Ketamine presents analgesic effect in acute and chronic pain models in both animals and humans (Mathisen et al. 1995, Rabben et al., 1999; Visser & Schug, 2006; Pascual et al., 2010).

The involvement of kinases in the analgesic effect of ketamine has been investigated. Using a model of neuropathic pain induced by SNL in rats, Mei et al. (2011) showed that SNL induced ipsilateral JNK phosphorylation up-regulation in astrocytes, but not microglia or neurons, within the spinal dorsal horn. Intrathecal ketamine relieved SNL-induced mechanical sensitisation and produced a dose-dependent effect on the suppression of SNL-induced spinal astrocytic JNK phosphorylation but had no effect on JNK protein expression, suggesting that the inhibition of spinal JNK activation may be involved in the analgesic effects of ketamine in this model.

The inhibition of MAPK phosphorylation by ketamine has also been related to a reduction in cytokine gene expressions in lipopolysaccharide (LPS)-activated macrophages (Wu et al., 2008). A therapeutic concentration of ketamine can decrease LPS-induced JNK phosphorylation, thus inhibiting TNF-α and IL-6 gene expression, which leads to the suppression of LPS-induced macrophage activation (Wu et al., 2008). In addition, ketamine reduced IL-1β biosynthesis in LPS-stimulated macrophages through the suppression of Ras, Raf, MEK1/2, and ERK1/2/IKK phosphorylation and the subsequent translocation and transactivation of the transcription factor NF-κB (Chen et al., 2009). The involvement of TNF-α, IL-6 and IL-1β in inflammatory nociceptive sensitisation is well known. Thus, the inhibition of cytokine production by ketamine in different cells may be an additional mechanism that contributes toward its analgesic effect.

Beyond its own specific effects, ketamine also has analgesic effects when given in combination with opioids. As mentioned earlier, several studies have demonstrated that ERK1/2 is involved in nociception. However, activation of MOR by opioids leads to ERK1/2 phosphorylation (Fukuda et al. 1996; Gutstein et al. 1997; Gupta et al., 2011), and this can be potentiated by ketamine. Gupta et al. (2011) investigated whether the ability of ketamine to increase the duration of opioid-induced effects could be related to the modulation of opioid-induced signalling. The authors found that, in a cell culture model, ketamine increases the effectiveness of opioid-induced signalling by enhancing the level of opioid-induced ERK1/2 phosphorylation. Ketamine also delays the desensitisation and improves the resensitisation of ERK1/2 signalling. These effects were observed in heterologous cells expressing MOR, suggesting a non-NMDA receptor-mediated action of ketamine (Gupta et al., 2011). The authors concluded that the overall effect of ketamine appears to be keeping opioid-induced ERK1/2 signalling active for a longer time period, and this could account for the observed effects of ketamine on the duration of opioid-induced analgesia. However, these data were obtained from *in vitro* experiments, and the link with analgesia is not clearly understood. Data provided from *in vivo* studies could contribute to improve the understanding of opioid-induced analgesia and its potentiation by ketamine.

3.2.2 Local anaesthetics

Systemic or topical administration of lidocaine and other local anaesthetics reduce hypersensitivity states induced by both acute inflammation and peripheral nerve injury in animals and brings significant relief in some patients with neuropathic pain syndromes (Mao & Chen, 2000; Ma et al., 2003; Gu et al., 2008; Fleming & O'Connor, 2009; Suter et al., 2009; Buchanan& MacIvor, 2010; Suzuki et al., 2011).

The analgesic effect of lidocaine in neuropathic pain can be partially explained by its ability to attenuate MAPK activation. Intrathecal injection of lidocaine in rats with chronic constriction injury suppressed the phosphorylation of p38 MAPK in the activated microglia in the spinal cord (Gu et al., 2008). In ATP-activated cultured rat microglia, lidocaine inhibited p38 MAPK activation and attenuated the production of proinflammatory cytokines, including TNF-α, IL-1β and IL-6 (Su et al., 2010). Furthermore, lidocaine significantly inhibited LPS-induced Toll-like receptor 4, NF-κB, ERK and p38 MAPK activation, but not JNK activation in LPS-stimulated murine macrophages (Lee et al., 2008).

Spared nerve injury (SNI) induces mechanical sensitisation and p38 MAPK activation in spinal microglia. Bupivacaine microspheres induced a complete sensory and motor blockade and significantly inhibited p38 MAPK activation and microglial proliferation in the spinal cords of rats (Suter et al., 2009). Carrageenan-induced hind paw inflammation and sensitisation triggers phosphorylation of spinal p38 MAPK and enhances TNF and IL-1 production in the bilateral DRGs and spinal cord. Although bupivacaine inhibits oedema, hyperalgesia and the carrageenan-induced production of systemic cytokines (Beloeil et al., 2006a; Combettes et al., 2010), the inhibitory effects of bupivacaine on the expression of cytokines or phosphorylated p38 MAPK in spinal cord or DRGs have not been verified (Beloeil et al., 2006b).

ERK activation as a potential target for bupivacaine antinociception was also investigated. The activation of both ionotropic (AMPA, NMDA, TRPV1) and metabotropic (NK-1, bradykinin 2 receptor, mGluR) receptors results in ERK phosphorylation in superficial dorsal horn neurons in rats. Bupivacaine blocked ionotropic but not metabotropic, receptor–induced ERK activation by apparently blocking Ca^{2+} influx through the plasma membrane in the spinal cord (Yanagidate & Strichartz, 2006).

Taken together, the inhibition of MAPK activation by general and local anaesthetics seems to represent a common and important pathway to at least partially explain the mechanism of analgesic action exerted by these drugs through ion influx inhibition.

3.3 Antidepressants

Selected antidepressants suppress pain through diverse mechanisms and are now considered as an essential component of the therapeutic strategy for treatment of many types of persistent pain. Their main mechanism of action involves reinforcement of the descending inhibitory pathways by increasing the amount of norepinephrine and serotonin in the synaptic cleft at both the supraspinal and spinal levels. Based on this, tricyclic antidepressants (TCAs) are widely used for treating chronic pain, such as neuropathic and inflammatory pain. Intrathecal (i.t.) co-infusion of amitriptyline with morphine not only attenuates the development of morphine tolerance but also preserves its antinociceptive efficacy (Tai et al., 2006). Tai et al. (2007) showed that amitriptyline pretreatment reverses the spinal cord PKA and PKC upregulation and preserves morphine's antinociceptive effect in morphine-tolerant rats submitted to thermal behaviour test; this reversal may occur via preventing the up-regulation of PKA and PKC protein expression. It results in the trafficking of glutamate transporters from the cytosol to the plasma membrane of glial cells, thus reducing the excitatory amino acid (EAA) concentration in the cerebrospinal fluid (CSF) spinal cord by the morphine challenge. This study suggested that amitriptyline is a useful analgesic adjuvant in the treatment of patients who need long-term opioid administration for pain relief.

In addition to the traditionally used TCAs, such as amitriptyline, selective serotonin reuptake inhibitors (SSRIs) and mixed monoamine uptake inhibitors are also used as a first-line treatment for managing pain syndromes. As mentioned above, voltage-gated sodium channels (VGSCs) are subject to modulation by G protein-coupled receptor signalling cascades involving PKA- and PKC-mediated phosphorylation. Depending on the neuron type and its anatomical location, phosphorylation of the VGSCs by PKC may facilitate slow inactivation (Cantrell and Catterall, 2001). Activating the 5-HT2C subtype of serotonin receptors in prefrontal cortex neurons results in a negative shift in the voltage-dependence of fast inactivation accompanied with a reduction of the peak current due to a PKC-mediated phosphorylation process (Carr et al., 2002). Concurrent phosphorylation by PKA seems necessary for the maximal current reduction (Cantrell et al., 2002). These mechanisms can be activated by various neurotransmitters including serotonin (Cantrell and Catterall, 2001). Because SSRIs increase the extracellular concentration of serotonin it is logical that they would indirectly modulate sodium channels in the central nervous system. This action mediated by increased serotonin and, PKA and PKC activity, could account for the analgesic effect of SSRIs.

Thán et al. (2007) studied the pharmacological interaction between SSRIs and sodium channel blocking agents such as lamotrigine. They examined the interaction of VGSCs blockers and SSRIs at the level of spinal segmental neurotransmission in the rat hemisected spinal cord model. The reflex inhibitory action of VGSCs blocker was markedly enhanced when SSRI compounds were co-applied; and it was found serotonin receptors and PKC involvement in the modulation of sodium channel function (Thán et al., 2007).

In conclusion, it seems that antidepressants exert analgesic effects by a mechanism involving serotonin, PKA and PKC activation, and modulation of VGSCs. Understanding the PK dynamics in these processes would be key to improve pain management.

3.4 PKG signalling and pain control

As described in section 1, the activation of signalling pathways that are dependent on PKA, PKC and MAPKs is important for the sensitisation of nociceptors and pain processing. The PKG pathway, in turn, is related to the nitric oxide (NO)/cGMP/PKG/ATP-sensitive K^+ channel pathway, which plays an important role in peripheral antinociception (Rodrigues & Duarte, 2000; Sachs et al., 2004).

The relationship between the NO/cGMP pathway and peripheral antinociception was first demonstrated by Ferreira and co-workers (Durate et al., 1990; Ferreira et al., 1991). They showed that the antinociceptive effect of acetylcholine and morphine was blocked by a guanylyl cyclase inhibitor and an NO synthase inhibitor, and was potentiated by a specific cGMP phosphodiesterase inhibitor. Moreover, the antinociception achieved with these drugs was mimicked by NO donors such as sodium nitroprusside. The involvement of this pathway has also been demonstrated for other analgesics, such as dipyrone (Duarte et al., 1992), diclofenac (Tonussi et al., 1994), and some antinociceptive agents, such as *Crotalus durissus terrificus* snake-venom (Picolo et al., 2000), the potent κ-opioid receptor agonist bremazocine (Amarante & Duarte, 2002), xylazine (Romero & Duarte, 2009), the cannabinoid receptor agonist anandamida (Reis et al., 2009) and ketamine (Romero et al., 2011). In agreement with *in vivo* studies, data from electrophysiological experiments studying inflammatory sensitisation showed that capsaicin-induced elevations in intracellular Ca^{2+} levels of rat sensory neurons lead to an enhanced production of cGMP via the NO pathway. The elevated cGMP levels and the subsequent activation of PKG appear to inactivate the sensitisation, confirming the important regulatory role of this kinase in reversing the neuronal sensitisation (Lopshire & Nicol, 1997).

In addition to studies on the mechanism of antinociceptive action of analgesics, Duarte and co-workers showed that the ability of morphine and dipyrone to induce peripheral antinociception is dependent on the activation of ATP-sensitive K^+ channels (Rodrigues & Duarte, 2000). cGMP can directly or indirectly (via PKG stimulation) modulate the activity of ion channels. PKG is a protein kinase that is stimulated selectively but not exclusively by cGMP. Once stimulated, PKG inhibits phospholipase C activity, stimulates Ca^{2+}-ATPase activity, inhibits inositol 1,4,5-triphosphate, inhibits Ca^{2+} channels, and/or stimulates K^+ channels activity (Cury et al., 2011). Furthermore, Sachs et al. (2004) demonstrated that the antinociceptive effect of dipyrone on persistent inflammatory sensitisation is dependent on the PKG activation and its modulation of ATP-sensitive K^+ channels.

Taken together, these findings suggest the relevant role of PKG as an intermediate between cGMP generation and the opening of ATP-sensitive K⁺ channels. The activation of this modulatory pathway may be an interesting target for new drug development.

4. Conclusions: A perspective of promising drug targets

In this section, protein kinases will be viewed as targets for pain control drug development. Several pre-clinical and clinical trials will be reviewed, focusing on the effectiveness and adverse effects of such drugs.

The genomic analysis of the eukaryotic protein kinase superfamily together with drug design approaches such as the bioisosteric replacement of pharmacophoric groups of lead compounds and 3D-quantitative structure-activity relationship analysis provide several new chemical entities to be tested and developed as drug candidates.

The continuous progress in protein structure determination and improved resolution allows the identification of pharmacological targets. The experimental results from genetically modified animals support new hypotheses and help to validate new concepts to better understand the pathological genesis and natural processes of our body.

Such progress in medicinal chemistry, biochemistry and pharmacology paradoxically leads to poor results in terms of new pharmaceutical entities and therapeutics. The pharmaceutical innovation decrease in recent decades is due to many aspects that are beyond the scope of this chapter. As targets of pharmaceutics, protein kinases play an important role in this history, providing several new therapeutic cancer targets. Drug discovery companies have targeted protein kinase inhibitors, which have led to billion dollar merges and a new branch of research and development that spread beyond the boundaries of cancer therapeutics (Garber, 2003).

At the beginning of the second decade of the 21st century, there are synthetic and medicinal chemistry service companies with strong backgrounds in kinase targets and kinase inhibitor drug discovery; these companies can develop new compounds on demand. There are sixteen pharmaceuticals actually licensed as protein kinase inhibitors, mainly to treat different cancers. The first drug, Trastuzumab, was licensed in 1998; this drug is a monoclonal antibody targeting membrane receptors that activates the MAPK pathway as well as the PI3 Kinase/AKT pathways. After this initial drug, many small molecules followed, targeting kinases as mechanism of action, mainly as ATP competitors.

The International Federation of Pharmaceutical Manufacturers & Associations has listed in its Clinical Trials Portal three entries for clinical trials focusing on pain and protein kinases. Two of these trials involve p38 MAPK inhibitors from a large pharmaceutical company and are testing for neuropathic pain following nerve trauma and from lumbosacral radiculopathy. The third trial involves tyrosine kinase (TrkA) receptor expression in children with retrosternal pain.

Experimental evidence suggests that p38 MAPK is activated in spinal microglia after nerve injury and contributes to neuropathic pain development and maintenance (Ji & Suter, 2007). p38 MAPK phosphorylates targets that transduce cellular signals to molecules and transcription factors that are involved in regulating the biosynthesis of inflammatory cytokines such as IL-1 and TNF-α. The inhibitor dilmapimod was associated with a significant reduction in pain intensity in patients with neuropathic pain following nerve injury (Anand et al., 2011). The clinical efficacy of p38 MAPK inhibitors in acute pain was also demonstrated in an assay of acute postsurgical dental pain; these inhibitors increased the time to rescue medication and decreased pain intensity when compared with the placebo group (Tong et al. 2011).

Despite these clinical assays that are directly associated with pain, many other clinical and pre-clinical studies have some degree of relevance when pain management is the goal. The action of different protein kinases inhibitors in cancer, rheumatoid arthritis, postsurgical conditions, diabetes and so forth has significant impact on decreasing pain in subjects suffering from these pathologies.

In addition to these efforts, one long-sought goal is the development of inhibitors of PKC isoforms because this family of protein kinases is involved in the cellular signalling of nociception, anxiety and cognition (Van Kolen et al., 2008). Non-isoform-specific PKC inhibitors have proven to be too toxic for *in vivo* use. PKCε is the primary target for drug design. This isoform is activated during nerve sensitisation and phosphorylates ion channels in the peripheral nervous system such as TRPV-1, and N-type voltage-dependent calcium channels (VDCCs) in isolectin B4-positive nociceptors; in addition, it mediates interplay between other kinases that are important to nociceptor function, such as PKA and MAPK (Hucho et al., 2005). There are no specific ATP-binding competitors for PKCε. There are other compounds that target alternative domains, such as the pseudosubstrate sequence, which is responsible for keeping the kinase in an inactivate state. The lipid-binding, cellular localisation and actin-binding domains are also valid targets. The main goal is to develop isoform-specific inhibitors among the ten known isozymes and to provide tissue specificity because cardiac-specific PKCε inhibition blocks norepinephrine-mediated regulation of heart contraction (Johnson et al., 1996).

The pharmaceutical paradigm of "new targets for old drugs", where known medications are employed in new pathologies as an innovation strategy to keep new products flowing to the market also applies to protein kinases and pain control because many drugs utilised for chronic and neuropathic pain management, such as antidepressants and anaesthetics, depend on protein kinases for their mechanisms of action.

New lead drugs are also being proposed; these drugs utilise molecular hybridisation and bioisosteric replacement of pharmacophoric groups, where different bioactive molecular moieties of mechanistically diverse drugs are fused, giving birth to new chemical entities with dual activity profiles (Brando Lima et al., 2011) that incorporate protein kinase inhibition with another type of biological activity. The challenge of developing new molecular approaches to create drugs that manage pain is great, and as seen throughout this chapter, protein kinases are an important aspect of this problem.

5. References

Ahlgren, S.C. & Levine, J.D. (1994). Protein Kinase C Inhibitors Decrease Hyperalgesia and C-Fiber Hyperexcitability in the Streptozotocin-Diabetic Rat. *J Neurophysiol*, Vol.72, No.2, (August 1994), pp.684-692, ISSN 0022-3077.

Aley, K.O. & Levine, J.D. (1999). Role of Protein Kinase a in the Maintenance of Inflammatory Pain. *J Neurosci*, Vol.19, No.6, (March 1999), pp.2181-2186, ISSN 0270-6474.

Aley, K.O., Messing, R.O., Mochly-Rosen, D. & Levine, J.D. (2000). Chronic Hypersensitivity for Inflammatory Nociceptor Sensitization Mediated by the Epsilon Isozyme of Protein Kinase C. *J Neurosci*, Vol.20, No.12, (June 2000), pp.4680-4685, ISSN 0270-6474.

Aley, O. & Levine, J.D. (2003). Contribution of 5- and 12-Lipoxygenase Products to Mechanical Hyperalgesia Induced by Prostaglandin E(2) and Epinephrine in the Rat. *Exp Brain Res*, Vol.148, No.4, (February 2003), pp.482-487, ISSN 0014-4819.

Amarante, L.H. & Duarte, I.D. (2002). The Kappa-Opioid Agonist (+/-)-Bremazocine Elicits Peripheral Antinociception by Activation of the L-Arginine/Nitric Oxide/Cyclic Gmp Pathway. *Eur J Pharmacol*, Vol.454, No.1, (November 2002), pp.19-23, ISSN 0014-2999.

Anand, P., Shenoy, R., Palmer, J.E., Baines, A.J., Lai, R.Y., Robertson, J., Bird, N., Ostenfeld, T. & Chizh, B.A. (2011). Clinical Trial of the P38 Map Kinase Inhibitor Dilmapimod in Neuropathic Pain Following Nerve Injury. *Eur J Pain*, (May 2011), ISSN 1532-2149.

Beloeil, H., Ababneh, Z., Chung, R., Zurakowski, D., Mulkern, R.V. & Berde, C.B. (2006a). Effects of Bupivacaine and Tetrodotoxin on Carrageenan-Induced Hind Paw Inflammation in Rats (Part 1): Hyperalgesia, Edema, and Systemic Cytokines. *Anesthesiology*, Vol.105, No.1, (July 2006), pp.128-138, ISSN 0003-3022.

Beloeil, H., Ji, R.R. & Berde, C.B. (2006b). Effects of Bupivacaine and Tetrodotoxin on Carrageenan-Induced Hind Paw Inflammation in Rats (Part 2): Cytokines and P38 Mitogen-Activated Protein Kinases in Dorsal Root Ganglia and Spinal Cord. *Anesthesiology*, Vol.105, No.1, (July 2006), pp.139-145, ISSN 0003-3022.

Berg, K.A., Rowan, M.P., Sanchez, T.A., Silva, M., Patwardhan, A.M., Milam, S.B., Hargreaves, K.M. & Clarke, W.P. (2011). Regulation of Kappa-Opioid Receptor Signaling in Peripheral Sensory Neurons in Vitro and in Vivo. *J Pharmacol Exp Ther*, Vol.338, No.1, (July 2011), pp.92-99, ISSN 1521-0103.

Blumberg, P.M. (1988). Protein Kinase C as the Receptor for the Phorbol Ester Tumor Promoters: Sixth Rhoads Memorial Award Lecture. *Cancer Res*, Vol.48, No.1, (January 1988), pp.1-8, ISSN 0008-5472.

Brando Lima, A.C., Machado, A.L., Simon, P., Cavalcante, M.M., Rezende, D.C., Sperandio da Silva, G.M., Nascimento, P.G., Quintas, L.E., Cunha, F.Q., Barreiro, E.J., Lima, L.M. & Koatz, V.L. (2011). Anti-Inflammatory Effects of Lassbio-998, a New Drug Candidate Designed to Be a P38 Mapk Inhibitor, in an Experimental Model of Acute Lung Inflammation. *Pharmacol Rep*, Vol.63, No.4, (July 2011), pp.1029-1039, ISSN 1734-1140.

Buchanan, D.D. & F, J.M. (2010). A Role for Intravenous Lidocaine in Severe Cancer-Related Neuropathic Pain at the End-of-Life. *Support Care Cancer*, Vol.18, No.7, (July 2010), pp.899-901, ISSN 1433-7339.

Burgess, G.M., Mullaney, I., McNeill, M., Dunn, P.M. & Rang, H.P. (1989). Second Messengers Involved in the Mechanism of Action of Bradykinin in Sensory Neurons in Culture. *J Neurosci*, Vol.9, No.9, (September 1989), pp.3314-3325, ISSN 0270-6474.

Cadd, G. & McKnight, G.S. (1989). Distinct Patterns of Camp-Dependent Protein Kinase Gene Expression in Mouse Brain. *Neuron*, Vol.3, No.1, (July 1989), pp.71-79, ISSN 0896-6273.

Cantrell, A.R. & Catterall, W.A. (2001). Neuromodulation of Na+ Channels: An Unexpected Form of Cellular Plasticity. *Nat Rev Neurosci*, Vol.2, No.6, (June 2001), pp.397-407, ISSN 1471-003X.

Cantrell, A.R., Tibbs, V.C., Yu, F.H., Murphy, B.J., Sharp, E.M., Qu, Y., Catterall, W.A. & Scheuer, T. (2002). Molecular Mechanism of Convergent Regulation of Brain Na(+) Channels by Protein Kinase C and Protein Kinase a Anchored to Akap-15. *Mol Cell Neurosci*, Vol.21, No.1, (September 2002), pp.63-80, ISSN 1044-7431.

Cesare, P., Dekker, L.V., Sardini, A., Parker, P.J. & McNaughton, P.A. (1999). Specific Involvement of Pkc-Epsilon in Sensitization of the Neuronal Response to Painful Heat. *Neuron*, Vol.23, No.3, (July 1999), pp.617-624, ISSN 0896-6273.

Cesare, P. & McNaughton, P. (1996). A Novel Heat-Activated Current in Nociceptive Neurons and Its Sensitization by Bradykinin. *Proc Natl Acad Sci U S A*, Vol.93, No.26, (December 1996), pp.15435-15439, ISSN 0027-8424.

Chen, T.L., Chang, C.C., Lin, Y.L., Ueng, Y.F. & Chen, R.M. (2009). Signal-Transducing Mechanisms of Ketamine-Caused Inhibition of Interleukin-1 Beta Gene Expression in Lipopolysaccharide-Stimulated Murine Macrophage-Like Raw 264.7 Cells. *Toxicol Appl Pharmacol*, Vol.240, No.1, (October 2009), pp.15-25, ISSN 1096-0333.

Chen, X., Bing, F., Dai, P. & Hong, Y. (2006). Involvement of Protein Kinase C in 5-Ht-Evoked Thermal Hyperalgesia and Spinal Fos Protein Expression in the Rat. *Pharmacol Biochem Behav*, Vol.84, No.1, (May 2006), pp.8-16, ISSN 0091-3057.

Ciruela, A., Dixon, A.K., Bramwell, S., Gonzalez, M.I., Pinnock, R.D. & Lee, K. (2003). Identification of Mek1 as a Novel Target for the Treatment of Neuropathic Pain. *Br J Pharmacol*, Vol.138, No.5, (March 2003), pp.751-756, ISSN 0007-1188.

Coderre, T.J., Abbott, F.V. & Melzack, R. (1984). Effects of Peripheral Antisympathetic Treatments in the Tail-Flick, Formalin and Autotomy Tests. *Pain*, Vol.18, No.1, (Jananuary 1984), pp.13-23, ISSN 0304-3959.

Combettes, E., Benhamou, D., Mazoit, J.X. & Beloeil, H. (2010). Comparison of a Bupivacaine Peripheral Nerve Block and Systemic Ketoprofen on Peripheral Inflammation and Hyperalgesia in Rats. *Eur J Anaesthesiol*, Vol.27, No.7, (July 2010), pp.642-647, ISSN 1365-2346.

Crown, E.D., Ye, Z., Johnson, K.M., Xu, G.Y., McAdoo, D.J. & Hulsebosch, C.E. (2006). Increases in the Activated Forms of Erk 1/2, P38 Mapk, and Creb Are Correlated with the Expression of at-Level Mechanical Allodynia Following Spinal Cord Injury. *Exp Neurol*, Vol.199, No.2, (June 2006), pp.397-407, ISSN 0014-4886.

Cui, M. & Nicol, G.D. (1995). Cyclic Amp Mediates the Prostaglandin E2-Induced Potentiation of Bradykinin Excitation in Rat Sensory Neurons. *Neuroscience*, Vol.66, No.2, (May 1995), pp.459-466, ISSN 0306-4522.

Cury, Y., Picolo, G., Gutierrez, V.P. & Ferreira, S.H. (2011). Pain and Analgesia: The Dual Effect of Nitric Oxide in the Nociceptive System. *Nitric Oxide*, Vol.25, No.3, (October 2011), pp.243-254, ISSN 1089-8611.

Dai, Y., Iwata, K., Fukuoka, T., Kondo, E., Tokunaga, A., Yamanaka, H., Tachibana, T., Liu, Y. & Noguchi, K. (2002). Phosphorylation of Extracellular Signal-Regulated Kinase in Primary Afferent Neurons by Noxious Stimuli and Its Involvement in Peripheral

Sensitization. *J Neurosci*, Vol.22, No.17, (September 2002), pp.7737-7745, ISSN 1529-2401.

Dalton, G.D., Smith, F.L., Smith, P.A. & Dewey, W.L. (2005). Protein Kinase a Activity Is Increased in Mouse Lumbar Spinal Cord but Not Brain Following Morphine Antinociceptive Tolerance for 15 Days. *Pharmacol Res*, Vol.52, No.3, (September 2005), pp.204-210, ISSN 1043-6618.

Dina, O.A., Barletta, J., Chen, X., Mutero, A., Martin, A., Messing, R.O. & Levine, J.D. (2000). Key Role for the Epsilon Isoform of Protein Kinase C in Painful Alcoholic Neuropathy in the Rat. *J Neurosci*, Vol.20, No.22, (November 2000), pp.8614-8619, ISSN 1529-2401.

Dina, O.A., Chen, X., Reichling, D. & Levine, J.D. (2001). Role of Protein Kinase Cepsilon and Protein Kinase a in a Model of Paclitaxel-Induced Painful Peripheral Neuropathy in the Rat. *Neuroscience*, Vol.108, No.3, (December 2001), pp.507-515, ISSN 0306-4522.

Doyle, T., Chen, Z., Muscoli, C., Obeid, L.M. & Salvemini, D. (2011). Intraplantar-Injected Ceramide in Rats Induces Hyperalgesia through an Nf-Kappab- and P38 Kinase-Dependent Cyclooxygenase 2/Prostaglandin E2 Pathway. *FASEB J*, Vol.25, No.8, (Aug), pp.2782-2791, ISSN 1530-6860.

Duarte, I.D., dos Santos, I.R., Lorenzetti, B.B. & Ferreira, S.H. (1992). Analgesia by Direct Antagonism of Nociceptor Sensitization Involves the Arginine-Nitric Oxide-Cgmp Pathway. *Eur J Pharmacol*, Vol.217, No.2-3, (July 1992), pp.225-227, ISSN 0014-2999.

Durate, I.D., Lorenzetti, B.B. & Ferreira, S.H. (1990). Peripheral Analgesia and Activation of the Nitric Oxide-Cyclic Gmp Pathway. *Eur J Pharmacol*, Vol.186, No.2-3, (September 1990), pp.289-293, ISSN 0014-2999.

England, S., Bevan, S. & Docherty, R.J. (1996). Pge2 Modulates the Tetrodotoxin-Resistant Sodium Current in Neonatal Rat Dorsal Root Ganglion Neurones Via the Cyclic Amp-Protein Kinase a Cascade. *J Physiol*, Vol.495 (Pt 2), (September 1996), pp.429-440, ISSN 0022-3751.

Ferreira, S.H., Duarte, I.D. & Lorenzetti, B.B. (1991). Molecular Base of Acetylcholine and Morphine Analgesia. *Agents Actions Suppl*, Vol.32, (January 1991), pp.101-106, ISSN 0379-0363.

Ferreira, S.H., Lorenzetti, B.B. & De Campos, D.I. (1990). Induction, Blockade and Restoration of a Persistent Hypersensitive State. *Pain*, Vol.42, No.3, (September 1990), pp.365-371, ISSN 0304-3959.

Ferreira, S.H. & Nakamura, M. (1979). I - Prostaglandin Hyperalgesia, a Camp/Ca2+ Dependent Process. *Prostaglandins*, Vol.18, No.2, (August 1979), pp.179-190, ISSN 0090-6980.

Fleming, J.A. & O'Connor, B.D. (2009). Use of Lidocaine Patches for Neuropathic Pain in a Comprehensive Cancer Centre. *Pain Res Manag*, Vol.14, No.5, (September 2009), pp.381-388, ISSN 1203-6765.

Fujita-Hamabe, W., Nagae, R., Nawa, A., Harada, S., Nakamoto, K. & Tokuyama, S. (2010). Involvement of Kappa Opioid Receptors in the Formalin-Induced Inhibition of Analgesic Tolerance to Morphine Via Suppression of Conventional Protein Kinase C Activation. *J Pharm Pharmacol*, Vol.62, No.8, (August 2010), pp.995-1002, ISSN 2042-7158.

Fukuda, K., Kato, S., Morikawa, H., Shoda, T. & Mori, K. (1996). Functional Coupling of the Delta-, Mu-, and Kappa-Opioid Receptors to Mitogen-Activated Protein Kinase and

Arachidonate Release in Chinese Hamster Ovary Cells. *J Neurochem*, Vol.67, No.3, (September 1996), pp.1309-1316, ISSN 0022-3042.

Garber, K. (2003). Research Retreat: Pfizer Eliminates Sugen, Shrinks Cancer Infrastructure. *J Natl Cancer Inst*, Vol.95, No.14, (July 2003), pp.1036-1038, ISSN 1460-2105.

Giles, P.A., Trezise, D.J. & King, A.E. (2007). Differential Activation of Protein Kinases in the Dorsal Horn in Vitro of Normal and Inflamed Rats by Group I Metabotropic Glutamate Receptor Subtypes. *Neuropharmacology*, Vol.53, No.1, (July 2007), pp.58-70, ISSN 0028-3908

Gold, M.S., Levine, J.D. & Correa, A.M. (1998). Modulation of Ttx-R Ina by Pkc and Pka and Their Role in Pge2-Induced Sensitization of Rat Sensory Neurons in Vitro. *J Neurosci*, Vol.18, No.24, (December 1998), pp.10345-10355, ISSN 0270-6474.

Gu, Y.W., Su, D.S., Tian, J. & Wang, X.R. (2008). Attenuating Phosphorylation of P38 Mapk in the Activated Microglia: A New Mechanism for Intrathecal Lidocaine Reversing Tactile Allodynia Following Chronic Constriction Injury in Rats. *Neurosci Lett*, Vol.431, No.2, (January 2008), pp.129-134, ISSN 0304-3940.

Gupta, A., Devi, L.A. & Gomes, I. (2011). Potentiation of Mu-Opioid Receptor-Mediated Signaling by Ketamine. *J Neurochem*, Vol.119, No.2, (October 2011), pp.294-302, ISSN 1471-4159.

Gutstein, H.B., Rubie, E.A., Mansour, A., Akil, H. & Woodgett, J.R. (1997). Opioid Effects on Mitogen-Activated Protein Kinase Signaling Cascades. *Anesthesiology*, Vol.87, No.5, (November 1997), pp.1118-1126, ISSN 0003-3022.

Gwak, Y.S., Unabia, G.C. & Hulsebosch, C.E. (2009). Activation of P-38alpha Mapk Contributes to Neuronal Hyperexcitability in Caudal Regions Remote from Spinal Cord Injury. *Exp Neurol*, Vol.220, No.1, (November 2009), pp.154-161, ISSN 1090-2430.

Hensellek, S., Brell, P., Schaible, H.G., Brauer, R. & Segond von Banchet, G. (2007). The Cytokine Tnfalpha Increases the Proportion of Drg Neurones Expressing the Trpv1 Receptor Via the Tnfr1 Receptor and Erk Activation. *Mol Cell Neurosci*, Vol.36, No.3, (November 2007), pp.381-391, ISSN 1044-7431.

Horvath, R.J., Landry, R.P., Romero-Sandoval, E.A. & DeLeo, J.A. (2010). Morphine Tolerance Attenuates the Resolution of Postoperative Pain and Enhances Spinal Microglial P38 and Extracellular Receptor Kinase Phosphorylation. *Neuroscience*, Vol.169, No.2, (August 2010), pp.843-854, ISSN 1873-7544.

Hucho, T.B., Dina, O.A. & Levine, J.D. (2005). Epac Mediates a Camp-to-Pkc Signaling in Inflammatory Pain: An Isolectin B4(+) Neuron-Specific Mechanism. *J Neurosci*, Vol.25, No.26, (June 2005), pp.6119-6126, ISSN 1529-2401.

Hunskaar, S. & Hole, K. (1987). The Formalin Test in Mice: Dissociation between Inflammatory and Non-Inflammatory Pain. *Pain*, Vol.30, No.1, (July 1987), pp.103-114, ISSN 0304-3959.

Ji, R.R., Samad, T.A., Jin, S.X., Schmoll, R. & Woolf, C.J. (2002). P38 Mapk Activation by Ngf in Primary Sensory Neurons after Inflammation Increases Trpv1 Levels and Maintains Heat Hyperalgesia. *Neuron*, Vol.36, No.1, (September 2002), pp.57-68, ISSN 0896-6273.

Ji, R.R. & Suter, M.R. (2007). P38 Mapk, Microglial Signaling, and Neuropathic Pain. *Mol Pain*, Vol.3, (November 2007), pp.33, ISSN 1744-8069.

Jin, X. & Gereau, R.W.t. (2006). Acute P38-Mediated Modulation of Tetrodotoxin-Resistant Sodium Channels in Mouse Sensory Neurons by Tumor Necrosis Factor-Alpha. *J Neurosci*, Vol.26, No.1, (January 2006), pp.246-255, ISSN 1529-2401.

Johnson, J.A., Gray, M.O., Chen, C.H. & Mochly-Rosen, D. (1996). A Protein Kinase C Translocation Inhibitor as an Isozyme-Selective Antagonist of Cardiac Function. *J Biol Chem*, Vol.271, No.40, (October 1996), pp.24962-24966, ISSN 0021-9258.

Junttila, M.R., Li, S.P. & Westermarck, J. (2008). Phosphatase-Mediated Crosstalk between Mapk Signaling Pathways in the Regulation of Cell Survival. *FASEB J*, Vol.22, No.4, (April 2007), pp.954-965, ISSN 1530-6860.

Kassuya, C.A., Ferreira, J., Claudino, R.F. & Calixto, J.B. (2007). Intraplantar Pge2 Causes Nociceptive Behaviour and Mechanical Allodynia: The Role of Prostanoid E Receptors and Protein Kinases. *Br J Pharmacol*, Vol.150, No.6, (March 2007), pp.727-737, ISSN 0007-1188.

Khasar, S.G., Lin, Y.H., Martin, A., Dadgar, J., McMahon, T., Wang, D., Hundle, B., Aley, K.O., Isenberg, W., McCarter, G., Green, P.G., Hodge, C.W., Levine, J.D. & Messing, R.O. (1999a). A Novel Nociceptor Signaling Pathway Revealed in Protein Kinase C Epsilon Mutant Mice. *Neuron*, Vol.24, No.1, (September 2000), pp.253-260, ISSN 0896-6273.

Khasar, S.G., McCarter, G. & Levine, J.D. (1999b). Epinephrine Produces a Beta-Adrenergic Receptor-Mediated Mechanical Hyperalgesia and in Vitro Sensitization of Rat Nociceptors. *J Neurophysiol*, Vol.81, No.3, (March, 1999), pp.1104-1112, ISSN 0022-3077.

Lee, K.M., Jeon, S.M. & Cho, H.J. (2010). Interleukin-6 Induces Microglial Cx3cr1 Expression in the Spinal Cord after Peripheral Nerve Injury through the Activation of P38 Mapk. *Eur J Pain*, Vol.14, No.7, (August 2009), pp.682 e681-612, ISSN 1532-2149.

Lee, P.Y., Tsai, P.S., Huang, Y.H. & Huang, C.J. (2008). Inhibition of Toll-Like Receptor-4, Nuclear Factor-Kappab and Mitogen-Activated Protein Kinase by Lignocaine May Involve Voltage-Sensitive Sodium Channels. *Clin Exp Pharmacol Physiol*, Vol.35, No.9, (September 2008), pp.1052-1058, ISSN 1440-1681.

Lopshire, J.C. & Nicol, G.D. (1997). Activation and Recovery of the Pge2-Mediated Sensitization of the Capsaicin Response in Rat Sensory Neurons. *J Neurophysiol*, Vol.78, No.6, (December 1998), pp.3154-3164, ISSN 0022-3077.

Lopshire, J.C. & Nicol, G.D. (1998). The Camp Transduction Cascade Mediates the Prostaglandin E2 Enhancement of the Capsaicin-Elicited Current in Rat Sensory Neurons: Whole-Cell and Single-Channel Studies. *J Neurosci*, Vol.18, No.16, (August 1998), pp.6081-6092, ISSN 0270-6474.

Ma, W., Du, W. & Eisenach, J.C. (2003). Intrathecal Lidocaine Reverses Tactile Allodynia Caused by Nerve Injuries and Potentiates the Antiallodynic Effect of the Cox Inhibitor Ketorolac. *Anesthesiology*, Vol.98, No.1, (January 2002), pp.203-208, ISSN 0003-3022.

Malmberg, A.B., Chen, C., Tonegawa, S. & Basbaum, A.I. (1997). Preserved Acute Pain and Reduced Neuropathic Pain in Mice Lacking Pkcgamma. *Science*, Vol.278, No.5336, (October 1997), pp.279-283, ISSN 0036-8075.

Manjavachi, M.N., Motta, E.M., Marotta, D.M., Leite, D.F. & Calixto, J.B. (2010). Mechanisms Involved in Il-6-Induced Muscular Mechanical Hyperalgesia in Mice. *Pain*, Vol.151, No.2, (November 2010), pp.345-355, ISSN 1872-6623.

Mao, J. & Chen, L.L. (2000). Systemic Lidocaine for Neuropathic Pain Relief. *Pain*, Vol.87, No.1, (July 2000), pp.7-17, ISSN 0304-3959.

Mao, J., Price, D.D., Mayer, D.J. & Hayes, R.L. (1992). Pain-Related Increases in Spinal Cord Membrane-Bound Protein Kinase C Following Peripheral Nerve Injury. *Brain Res*, Vol.588, No.1, (August 1992), pp.144-149, ISSN 0006-8993.

Mathisen, L.C., Skjelbred, P., Skoglund, L.A. & Oye, I. (1995). Effect of Ketamine, an Nmda Receptor Inhibitor, in Acute and Chronic Orofacial Pain. *Pain*, Vol.61, No.2, (May 1995), pp.215-220, ISSN 0304-3959.

Mei, X.P., Zhang, H., Wang, W., Wei, Y.Y., Zhai, M.Z., Xu, L.X. & Li, Y.Q. (2011). Inhibition of Spinal Astrocytic C-Jun N-Terminal Kinase (Jnk) Activation Correlates with the Analgesic Effects of Ketamine in Neuropathic Pain. *J Neuroinflammation*, Vol.8, No.1, (January 2011), pp.6, ISSN 1742-2094.

Nakamura, M. & Ferreira, S.H. (1987). A Peripheral Sympathetic Component in Inflammatory Hyperalgesia. *Eur J Pharmacol*, Vol.135, No.2, (March 1987), pp.145-153, ISSN 0014-2999.

Narita, M., Miyoshi, K. & Suzuki, T. (2007). Functional Reduction in Mu-Opioidergic System in the Spinal Cord under a Neuropathic Pain-Like State Following Chronic Ethanol Consumption in the Rat. *Neuroscience*, Vol.144, No.3, (February 2006), pp.777-782, ISSN 0306-4522.

Neugebauer, V. (2002). Metabotropic Glutamate Receptors--Important Modulators of Nociception and Pain Behavior. *Pain*, Vol.98, No.1-2, (July 2002), pp.1-8, ISSN 0304-3959.

Nishizuka, Y. (1992). Intracellular Signaling by Hydrolysis of Phospholipids and Activation of Protein Kinase C. *Science*, Vol.258, No.5082, (October 1992), pp.607-614, ISSN 0036-8075.

Noselli, S. (2000). Signal Transduction: Are There Close Encounters between Signaling Pathways? *Science*, Vol.290, No.5489, (October 2000), pp.68-69, ISSN 00368075.

Obata, K., Yamanaka, H., Dai, Y., Mizushima, T., Fukuoka, T., Tokunaga, A. & Noguchi, K. (2004). Differential Activation of Mapk in Injured and Uninjured Drg Neurons Following Chronic Constriction Injury of the Sciatic Nerve in Rats. *Eur J Neurosci*, Vol.20, No.11, (December 2004), pp.2881-2895, ISSN 0953-816X.

Pascual, D., Goicoechea, C., Burgos, E. & Martin, M.I. (2010). Antinociceptive Effect of Three Common Analgesic Drugs on Peripheral Neuropathy Induced by Paclitaxel in Rats. *Pharmacol Biochem Behav*, Vol.95, No.3, (May 2010), pp.331-337, ISSN 1873-5177.

Pearson, G., Robinson, F., Beers Gibson, T., Xu, B.E., Karandikar, M., Berman, K. & Cobb, M.H. (2001). Mitogen-Activated Protein (Map) Kinase Pathways: Regulation and Physiological Functions. *Endocr Rev*, Vol.22, No.2, (April 2001), pp.153-183, ISSN 0163-769X.

Persson, A.K., Gasser, A., Black, J.A. & Waxman, S.G. (2011). Nav1.7 Accumulates and Co-Localizes with Phosphorylated Erk1/2 within Transected Axons in Early Experimental Neuromas. *Exp Neurol*, Vol.230, No.2, (August 2011), pp.273-279, ISSN 1090-2430.

Picolo, G., Giorgi, R. & Cury, Y. (2000). Delta-Opioid Receptors and Nitric Oxide Mediate the Analgesic Effect of Crotalus Durissus Terrificus Snake Venom. *Eur J Pharmacol*, Vol.391, No.1-2, (March 2000), pp.55-62, ISSN 0014-2999.

Pimienta, G. & Pascual, J. (2007). Canonical and Alternative Mapk Signaling. *Cell Cycle*, Vol.6, No.21, (November 2007), pp.2628-2632, ISSN 1551-4005.

Rabben, T., Skjelbred, P. & Oye, I. (1999). Prolonged Analgesic Effect of Ketamine, an N-Methyl-D-Aspartate Receptor Inhibitor, in Patients with Chronic Pain. *J Pharmacol Exp Ther*, Vol.289, No.2, (May 1999), pp.1060-1066, ISSN 0022-3565.

Rathee, P.K., Distler, C., Obreja, O., Neuhuber, W., Wang, G.K., Wang, S.Y., Nau, C. & Kress, M. (2002). Pka/Akap/Vr-1 Module: A Common Link of Gs-Mediated Signaling to Thermal Hyperalgesia. *J Neurosci*, Vol.22, No.11, (June 2002), pp.4740-4745, ISSN 1529-2401.

Reis, G.M., Pacheco, D., Perez, A.C., Klein, A., Ramos, M.A. & Duarte, I.D. (2009). Opioid Receptor and No/Cgmp Pathway as a Mechanism of Peripheral Antinociceptive Action of the Cannabinoid Receptor Agonist Anandamide. *Life Sci*, Vol.85, No.9-10, (August 2009), pp.351-356, ISSN 1879-0631.

Rodrigues, A.R. & Duarte, I.D. (2000). The Peripheral Antinociceptive Effect Induced by Morphine Is Associated with Atp-Sensitive K(+) Channels. *Br J Pharmacol*, Vol.129, No.1, (January 2000), pp.110-114, ISSN 0007-1188.

Rodriguez-Munoz, M., Sanchez-Blazquez, P., Vicente-Sanchez, A., Berrocoso, E. & Garzon, J. (2011). The Mu-Opioid Receptor and the Nmda Receptor Associate in Pag Neurons: Implications in Pain Control. *Neuropsychopharmacology*, (August 2011), ISSN 1740-634X.

Romero, T.R. & Duarte, I.D. (2009). Alpha(2)-Adrenoceptor Agonist Xylazine Induces Peripheral Antinociceptive Effect by Activation of the L-Arginine/Nitric Oxide/Cyclic Gmp Pathway in Rat. *Eur J Pharmacol*, Vol.613, No.1-3, (June 2009), pp.64-67, ISSN 1879-0712.

Romero, T.R., Galdino, G.S., Silva, G.C., Resende, L.C., Perez, A.C., Cortes, S.F. & Duarte, I.D. (2011). Ketamine Activates the L-Arginine/Nitric Oxide/Cyclic Guanosine Monophosphate Pathway to Induce Peripheral Antinociception in Rats. *Anesth Analg*, Vol.113, No.5, (November 2011), pp.1254-1259, ISSN 1526-7598.

Sachs, D., Cunha, F.Q. & Ferreira, S.H. (2004). Peripheral Analgesic Blockade of Hypernociception: Activation of Arginine/No/Cgmp/Protein Kinase G/Atp-Sensitive K+ Channel Pathway. *Proc Natl Acad Sci U S A*, Vol.101, No.10, (March 2004), pp.3680-3685, ISSN 0027-8424.

Sachs, D., Villarreal, C., Cunha, F., Parada, C. & Ferreira, S. (2009). The Role of Pka and Pkcepsilon Pathways in Prostaglandin E2-Mediated Hypernociception. *Br J Pharmacol*, Vol.156, No.5, (March 2009), pp.826-834, ISSN 1476-5381.

Smith, J.A., Davis, C.L. & Burgess, G.M. (2000). Prostaglandin E2-Induced Sensitization of Bradykinin-Evoked Responses in Rat Dorsal Root Ganglion Neurons Is Mediated by Camp-Dependent Protein Kinase A. *Eur J Neurosci*, Vol.12, No.9, (September 2000), pp.3250-3258, ISSN 0953-816X.

Song, Z., Zou, W., Liu, C. & Guo, Q. (2010). Gene Knockdown with Lentiviral Vector-Mediated Intrathecal Rna Interference of Protein Kinase C Gamma Reverses Chronic Morphine Tolerance in Rats. *J Gene Med*, Vol.12, No.11, (November 2010), pp.873-880, ISSN 1521-2254.

Sorkin, L., Svensson, C.I., Jones-Cordero, T.L., Hefferan, M.P. & Campana, W.M. (2009). Spinal P38 Mitogen-Activated Protein Kinase Mediates Allodynia Induced by First-

Degree Burn in the Rat. *J Neurosci Res*, Vol.87, No.4, (March 2008), pp.948-955, ISSN 1097-4547.

Souza, A.L., Moreira, F.A., Almeida, K.R., Bertollo, C.M., Costa, K.A. & Coelho, M.M. (2002). In Vivo Evidence for a Role of Protein Kinase C in Peripheral Nociceptive Processing. *Br J Pharmacol*, Vol.135, No.1, (January 2002), pp.239-247, ISSN 0007-1188.

Su, D., Gu, Y., Wang, Z. & Wang, X. (2010). Lidocaine Attenuates Proinflammatory Cytokine Production Induced by Extracellular Adenosine Triphosphate in Cultured Rat Microglia. *Anesth Analg*, Vol.111, No.3, (September 2010), pp.768-774, ISSN 1526-7598.

Suter, M.R., Berta, T., Gao, Y.J., Decosterd, I. & Ji, R.R. (2009). Large a-Fiber Activity Is Required for Microglial Proliferation and P38 Mapk Activation in the Spinal Cord: Different Effects of Resiniferatoxin and Bupivacaine on Spinal Microglial Changes after Spared Nerve Injury. *Mol Pain*, Vol.5, (September 2009), pp.53, ISSN 1744-8069.

Suzuki, N., Hasegawa-Moriyama, M., Takahashi, Y., Kamikubo, Y., Sakurai, T. & Inada, E. (2011). Lidocaine Attenuates the Development of Diabetic-Induced Tactile Allodynia by Inhibiting Microglial Activation. *Anesth Analg*, Vol.113, No.4, (October 2011), pp.941-946, ISSN 1526-7598.

Svensson, C.I., Marsala, M., Westerlund, A., Calcutt, N.A., Campana, W.M., Freshwater, J.D., Catalano, R., Feng, Y., Protter, A.A., Scott, B. & Yaksh, T.L. (2003). Activation of P38 Mitogen-Activated Protein Kinase in Spinal Microglia Is a Critical Link in Inflammation-Induced Spinal Pain Processing. *J Neurochem*, Vol.86, No.6, (September 2003), pp.1534-1544, ISSN 0022-3042.

Tai, Y.H., Wang, Y.H., Tsai, R.Y., Wang, J.J., Tao, P.L., Liu, T.M., Wang, Y.C. & Wong, C.S. (2007). Amitriptyline Preserves Morphine's Antinociceptive Effect by Regulating the Glutamate Transporter Glast and Glt-1 Trafficking and Excitatory Amino Acids Concentration in Morphine-Tolerant Rats. *Pain*, Vol.129, No.3, (June 2007), pp.343-354, ISSN 1872-6623.

Tai, Y.H., Wang, Y.H., Wang, J.J., Tao, P.L., Tung, C.S. & Wong, C.S. (2006). Amitriptyline Suppresses Neuroinflammation and up-Regulates Glutamate Transporters in Morphine-Tolerant Rats. *Pain*, Vol.124, No.1-2, (September 2006), pp.77-86, ISSN 1872-6623.

Taiwo, Y.O., Heller, P.H. & Levine, J.D. (1992). Mediation of Serotonin Hyperalgesia by the Camp Second Messenger System. *Neuroscience*, Vol.48, No.2, (January 1992), pp.479-483, ISSN 0306-4522.

Taiwo, Y.O. & Levine, J.D. (1991). Further Confirmation of the Role of Adenyl Cyclase and of Camp-Dependent Protein Kinase in Primary Afferent Hyperalgesia. *Neuroscience*, Vol.44, No.1, (January 1991), pp.131-135, ISSN 0306-4522.

Taylor, S.S., Buechler, J.A. & Yonemoto, W. (1990). Camp-Dependent Protein Kinase: Framework for a Diverse Family of Regulatory Enzymes. *Annu Rev Biochem*, Vol.59, (January 1990), pp.971-1005, ISSN 0066-4154.

Than, M., Kocsis, P., Tihanyi, K., Fodor, L., Farkas, B., Kovacs, G., Kis-Varga, A., Szombathelyi, Z. & Tarnawa, I. (2007). Concerted Action of Antiepileptic and Antidepressant Agents to Depress Spinal Neurotransmission: Possible Use in the

Therapy of Spasticity and Chronic Pain. *Neurochem Int*, Vol.50, No.4, (March 2007), pp.642-652, ISSN 0197-0186.

Tong, S.E., Daniels, S.E., Black, P., Chang, S., Protter, A. & Desjardins, P.J. (2011). Novel P38{Alpha} Mitogen-Activated Protein Kinase Inhibitor Shows Analgesic Efficacy in Acute Postsurgical Dental Pain. *J Clin Pharmacol*, (June 2011), ISSN 1552-4604.

Tonussi, C.R. & Ferreira, S.H. (1994). Mechanism of Diclofenac Analgesia: Direct Blockade of Inflammatory Sensitization. *Eur J Pharmacol*, Vol.251, No.2-3, (January 1994), pp.173-179, ISSN 0014-2999.

Tumati, S., Roeske, W.R., Largent-Milnes, T.M., Vanderah, T.W. & Varga, E.V. (2011). Intrathecal Pka-Selective Sirna Treatment Blocks Sustained Morphine-Mediated Pain Sensitization and Antinociceptive Tolerance in Rats. *J Neurosci Methods*, Vol.199, No.1, (July 2011), pp.62-68, ISSN 1872-678X.

Van Kolen, K., Pullan, S., Neefs, J.M. & Dautzenberg, F.M. (2008). Nociceptive and Behavioural Sensitisation by Protein Kinase Cepsilon Signalling in the Cns. *J Neurochem*, Vol.104, No.1, (January 2007), pp.1-13, ISSN 1471-4159.

Villarreal, C.F., Sachs, D., Funez, M.I., Parada, C.A., de Queiroz Cunha, F. & Ferreira, S.H. (2009a). The Peripheral Pro-Nociceptive State Induced by Repetitive Inflammatory Stimuli Involves Continuous Activation of Protein Kinase a and Protein Kinase C Epsilon and Its Na(V)1.8 Sodium Channel Functional Regulation in the Primary Sensory Neuron. *Biochem Pharmacol*, Vol.77, No.5, (March 2008), pp.867-877, ISSN 1873-2968.

Villarreal, C.F., Funez, M.I., Figueiredo, F., Cunha, F.Q., Parada, C.A. & Ferreira, S.H. (2009b). Acute and Persistent Nociceptive Paw Sensitisation in Mice: The Involvement of Distinct Signalling Pathways. *Life Sci*, Vol.85, No.23-26, (December 2009), pp.822-829, ISSN 1879-0631.

Visser, E. & Schug, S.A. (2006). The Role of Ketamine in Pain Management. *Biomed Pharmacother*, Vol.60, No.7, (August 2006), pp.341-348, ISSN 0753-3322.

Wu, G.J., Chen, T.L., Ueng, Y.F. & Chen, R.M. (2008). Ketamine Inhibits Tumor Necrosis Factor-Alpha and Interleukin-6 Gene Expressions in Lipopolysaccharide-Stimulated Macrophages through Suppression of Toll-Like Receptor 4-Mediated C-Jun N-Terminal Kinase Phosphorylation and Activator Protein-1 Activation. *Toxicol Appl Pharmacol*, Vol.228, No.1, (April 2008), pp.105-113, ISSN 0041-008X.

Yamdeu, R.S., Shaqura, M., Mousa, S.A., Schafer, M. & Droese, J. (2011). P38 Mitogen-Activated Protein Kinase Activation by Nerve Growth Factor in Primary Sensory Neurons Upregulates Mu-Opioid Receptors to Enhance Opioid Responsiveness toward Better Pain Control. *Anesthesiology*, Vol.114, No.1, (January 2010), pp.150-161, ISSN 1528-1175.

Yanagidate, F. & Strichartz, G.R. (2006). Bupivacaine Inhibits Activation of Neuronal Spinal Extracellular Receptor-Activated Kinase through Selective Effects on Ionotropic Receptors. *Anesthesiology*, Vol.104, No.4, (April 20006), pp.805-814, ISSN 0003-3022.

Zhang, X.C., Kainz, V., Burstein, R. & Levy, D. (2011). Tumor Necrosis Factor-Alpha Induces Sensitization of Meningeal Nociceptors Mediated Via Local Cox and P38 Map Kinase Actions. *Pain*, Vol.152, No.1, (January 2010), pp.140-149, ISSN 1872-6623.

Zhou, Y., Li, G.D. & Zhao, Z.Q. (2003). State-Dependent Phosphorylation of Epsilon-Isozyme of Protein Kinase C in Adult Rat Dorsal Root Ganglia after Inflammation and Nerve Injury. *J Neurochem*, Vol.85, No.3, (May 2003), pp.571-580, ISSN 0022-3042.

Zhu, W. & Oxford, G.S. (2007). Phosphoinositide-3-Kinase and Mitogen Activated Protein Kinase Signaling Pathways Mediate Acute Ngf Sensitization of Trpv1. *Mol Cell Neurosci*, Vol.34, No.4, (April 2007), pp.689-700, ISSN 1044-7431.

Permissions

The contributors of this book come from diverse backgrounds, making this book a truly international effort. This book will bring forth new frontiers with its revolutionizing research information and detailed analysis of the nascent developments around the world.

We would like to thank Gabriela Da Silva Xavier, for lending her expertise to make the book truly unique. She has played a crucial role in the development of this book. Without her invaluable contribution this book wouldn't have been possible. She has made vital efforts to compile up to date information on the varied aspects of this subject to make this book a valuable addition to the collection of many professionals and students.

This book was conceptualized with the vision of imparting up-to-date information and advanced data in this field. To ensure the same, a matchless editorial board was set up. Every individual on the board went through rigorous rounds of assessment to prove their worth. After which they invested a large part of their time researching and compiling the most relevant data for our readers. Conferences and sessions were held from time to time between the editorial board and the contributing authors to present the data in the most comprehensible form. The editorial team has worked tirelessly to provide valuable and valid information to help people across the globe.

Every chapter published in this book has been scrutinized by our experts. Their significance has been extensively debated. The topics covered herein carry significant findings which will fuel the growth of the discipline. They may even be implemented as practical applications or may be referred to as a beginning point for another development. Chapters in this book were first published by InTech; hereby published with permission under the Creative Commons Attribution License or equivalent.

The editorial board has been involved in producing this book since its inception. They have spent rigorous hours researching and exploring the diverse topics which have resulted in the successful publishing of this book. They have passed on their knowledge of decades through this book. To expedite this challenging task, the publisher supported the team at every step. A small team of assistant editors was also appointed to further simplify the editing procedure and attain best results for the readers.

Our editorial team has been hand-picked from every corner of the world. Their multi-ethnicity adds dynamic inputs to the discussions which result in innovative outcomes. These outcomes are then further discussed with the researchers and contributors who give their valuable feedback and opinion regarding the same. The feedback is then collaborated with the researches and they are edited in a comprehensive manner to aid the understanding of the subject.

Apart from the editorial board, the designing team has also invested a significant amount of their time in understanding the subject and creating the most relevant covers. They scrutinized every image to scout for the most suitable representation of the subject and create an appropriate cover for the book.

The publishing team has been involved in this book since its early stages. They were actively engaged in every process, be it collecting the data, connecting with the contributors or procuring relevant information. The team has been an ardent support to the editorial, designing and production team. Their endless efforts to recruit the best for this project, has resulted in the accomplishment of this book. They are a veteran in the field of academics and their pool of knowledge is as vast as their experience in printing. Their expertise and guidance has proved useful at every step. Their uncompromising quality standards have made this book an exceptional effort. Their encouragement from time to time has been an inspiration for everyone.

The publisher and the editorial board hope that this book will prove to be a valuable piece of knowledge for researchers, students, practitioners and scholars across the globe.

List of Contributors

Barbara Peruzzi
Regenerative Medicine Unit, Ospedale Pediatrico Bambino Gesù, Rome, Italy

Nadia Rucci and Anna Teti
Department of Experimental Medicine, University of L'Aquila, Italy

Ana Maneva
Medical University, Pharmaceutical Faculty, Department of Chemistry and Biochemistry, Plovdiv, Bulgaria

Lilia Maneva-Radicheva
Medical University, Medical Faculty, Department of Chemistry and Biochemistry, Sofia, Bulgaria

Etienne Roux
Univ. de Bordeaux, Adaptation Cardiovasculaire à l'ischémie, Pessac, France
INSERM, Adaptation Cardiovasculaire à l'ischémie, Pessac, France

Prisca Mbikou
Institute of Biomedical Technologies, Auckland University of Technology, Auckland, New Zealand

Ales Fajmut
University of Maribor, Medical Faculty, Faculty of Natural Sciences and Mathematics and Faculty of Health Sciences, Slovenia

Yuansheng Gao, Dou Dou, Xue Qin, Hui Qi and Lei Ying
Department of Physiology and Pathophysiology, Peking University Health Science Center, China
Key Laboratory of Molecular Cardiovascular Science (Peking University), Ministry of Education, China

Adriana Castello Costa Girardi
University of São Paulo Medical School, Brazil

Luciene Regina Carraro-Lacroix
Hospital for Sick Children, Canada

Gabriela Da Silva Xavier
Imperial College London, Section of Cell Biology, Division of Diabetes, Endocrinology and Metabolism, London, UK

Takehiko Matsushita and Shunichi Murakami
Department of Orthopaedics, Case Western Reserve University, USA

András Mihály
University of Szeged, Faculty of Medicine, Department of Anatomy, Histology and Embryology, Hungary

J. Russell Huie and Adam R. Ferguson
Brain and Spinal Injury Center (BASIC), University of California, San Francisco, USA

Mani Indiana Funez, Fabiane Hiratsuka Veiga de Souza, José Eduardo Pandossio and Paulo Gustavo Barboni Dantas Nascimento
School of Ceilandia, Brasilia University, Brazil